SPATIAL DEVELOPMENT PLANNING
A Dynamic Convex Programming Approach

Studies in
Regional Science
and Urban Economics

Editors

ÅKE ANDERSSON
WALTER ISARD

Volume 2

NORTH-HOLLAND PUBLISHING COMPANY – AMSTERDAM · NEW YORK · OXFORD

Spatial
Development Planning
A Dynamic Convex Programming Approach

MASAHISA FUJITA
University of Pennsylvania

1978

NORTH-HOLLAND PUBLISHING COMPANY – AMSTERDAM · NEW YORK · OXFORD

Library of Congress Catalog Card Number 78-19135

ISBN North-Holland for this volume 0-444-85157-7

Publishers

NORTH-HOLLAND PUBLISHING COMPANY
AMSTERDAM · NEW YORK · OXFORD

Distributors for the U.S.A. and Canada

ELSEVIER NORTH-HOLLAND, INC.
52 VANDERBILT AVENUE
NEW YORK, N.Y. 10017

Library of Congress Cataloging in Publication Data

Fujita, Masahisa.
  Spatial development planning.
  (Studies in regional science and urban economics; v. 2)
  Bibliography: p. 327
  Includes index.
  1.   Regional economics – Mathematical models.
2. Dynamic programming.   I. Title.   II. Series.
HT391.F77   309.2'2'0151   78-19135
ISBN 0-444-85157-7

PRINTED IN THE NETHERLANDS

INTRODUCTION TO THE SERIES

Regional Science and Urban Economics are two interrelated fields of research which have developed very rapidly in the last three decades. The main theoretical foundation of these fields comes from economics but in recent years the interdisciplinary character has become more pronounced. The editors desire to have the interdisciplinary character of regional science as well as the development of spatial aspects of theoretical economics fully reflected in this book series. Material presented in this book series will fall in three different groups:

- interdisciplinary textbooks at the advanced level,
- monographs reflecting theoretical or applied work in spatial analysis,
- proceedings reflecting the advancement of the frontiers of regional science and urban economics.

In order to ensure homogeneity in this interdisciplinary field, books published in this series will be:

- theoretically oriented, i.e., analyse problems with a large degree of generality,
- employ formal methods from mathematics, econometrics, operations research and related fields, and
- focus on immediate or potential uses for regional and urban forecasting, planning and policy.

The Editors

PREFACE

It is a pleasure to publish this book on Spatial Development Planning by Dr. Masahisa Fujita as the second volume in the series Studies in Regional Science and Urban Economics. Dr. Fujita's analysis ploughs new ground in the field of Regional Science – in particular through the introduction of investment and other dynamic processes into multi-region, multi-commodity systems analysis. In a most rigorous fashion he neatly unifies and further develops in his dynamic convex programming approach the analytical advances made by Leontief, von Neumann and neoclassical growth theorists in order to identify efficient sets of prices, trading patterns and investment plans for regions. More important, he probes the very difficult area of non-convex structures. While his analysis of non-convex structures, explicitly treating scale economies and diseconomies, pertains to a multi-region system of one aggregate commodity only, it represents an important advance. For many decades, a number of regional scientists have worked on this area with little success. Fujita's advances in treating scale economies and diseconomies should therefore be studied carefully by theorists concerned with this development factor.

The book as a whole is a "must" for those studying advanced regional science theory.

Walter Isard

TO

Beethoven

Symphony No. 1

ACKNOWLEDGMENTS

This book is an outgrowth of my doctoral thesis in the Department of Regional Science at the University of Pennsylvania. In connection with this work, my indebtedness to others is indeed great.

First of all, I am grateful to Professor Walter Isard for his continuous inspiration and encouragement throughout my years as a graduate student and the completion of this study. I am also enormously indebted to my thesis supervisor, Professor Tony E. Smith, for his comprehensive help, sincere interest and constructive criticism. In addition, I acknowledge the contribution of Professor Benjamin H. Stevens and Colin Gannon who were most helpful to me from the beginning of the study as members of my dissertation committee. Moreover, discussions with and continuous encouragement from Professor Ronald E. Miller are appreciated.

Drs. Tatsuhiko Kawashima, Kunio Kudo, Sung W. Hong and Hisayoshi Morisugi and Professor Keisuke Suzuki have provided valuable discussions throughout; Professors Noboru Sakashita, Hiroyuki Yamada and Yoshitaka Ohtsuki offered valuable suggestions and comments. To all of them, I express my sincere thanks.

I wish to acknowledge a special debt of gratitude to Professor R. Tyrrell Rockafellar for allowing me to summarize, in Appendix B, a part of his monograph, *Monotone Processes of Convex and Concave Type*, which provided a basis for the mathematical development in this book. I am also indebted to the *Journal of Regional Science*, *Journal of Mathematical Economics* and *Regional Science and Urban Economics* for permission to include in this book some of my articles which were originally published by them.

I would further like to express my deepest appreciation to Professors Eiji Kometani and Kozo Amano for their help and moral support throughout the study. I am also grateful to Janet Kohlhase for her patient work in editing this book and to Hideaki Ogawa for his excellent drafting of figures. Finally, I would like to express my warmest thanks to my host family, Donald and Suzanne Rudalevige, for their encouragement during my stay in the United States.

Philadelphia, October 1977   Masahisa Fujita

CONTENTS

Part II. DEVELOPMENT IN SPACE SYSTEMS WITH CONVEX STRUCTURES

PART III. DEVELOPMENT IN SPACE SYSTEMS WITH NON-CONVEX STRUCTURES

Chapter 6. Regional allocation of investment under conditions of variable returns to scale

LIST OF BASIC SYMBOLS

$x \in A$ x belongs to A (x is an element of A)

$x \notin A$ x does not belong to A (x is not an element of A)

$A \subset B$ A is contained in B (set-theoretic inclusion)

$A \not\subset B$ A is not contained in B

$A \subsetneq B$ $A \subset B$ but $A \neq B$

$A = B$ A is equal to B (that is, $A \subset B$ and $B \subset A$)

$\{x|P\}$ set of all elements x having a property P

$\emptyset$ empty set

$\{a\}$ set consisting of a single element a

$A \cup B$ set-theoretic union of A and B

$\bigcup_{\lambda \in \Lambda} A_\lambda$ set-theoretic union of A_λ indexed by $\lambda \in \Lambda$, often denoted by $\cup\{A_\lambda|\lambda \in \Lambda\}$

$A \cap B$ set theoretic intersection of A and B

$\bigcap_{\lambda \in \Lambda} A_\lambda$ set-theoretic intersection of A_λ indexed by $\lambda \in \Lambda$

$A \times B \times C \times \cdots$ Cartesian product of sets $A, B, C, \ldots$

$A + B = \{a + b|a \in A, b \in B\}$ vectorial sum of subsets A and B of $\mathbf{R}^n$

λA $\{\lambda a|a \in A\}$

$\mathbf{R}^n$ n-dimensional Euclidean space

$\mathbf{R}^n_+$ the nonnegative orthant of $\mathbf{R}^n$

$x \cdot y = \sum_{i=1}^{n} x_i y_i$ inner product of $x = (x_i)$ and $y = (y_i)$

$\|x\|$ norm of vector x

$x \geqq y$ $x_i \geqq y_i$ for all i where $x = (x_i)$ and $y = (y_i)$

$x \geq y$ $x_i \geqq y_i$ for all i and $x_i > y_i$ for some i where $x = (x_i)$ and $y = (y_i)$

$x > y$ $x_i > y_i$ for all i where $x = (x_i)$ and $y = (y_i)$

$[a, b]$	a closed interval with end points a and b
$[a, b)$	an interval with end point a but without end point b
$(a, b]$	an interval with end point b but without end point a
(a, b)	an open interval without end points a and b
m.t.cc.	monotone transformation of concave type
m.t.cv.	monotone transformation of convex type
generalized m.t.cc.	generalized monotone transformation of concave type
graph T	graph of mapping (transformation) T, that is, graph $T = \{(x, y) \mid y \in T(x), \ x \in X, \ y \in Y\}$ for mapping T from X into Y
$\lim_{i \to \infty} x^i$	limit of sequence $\{x^i\}$
$\lim_{x \to a} f(a)$	limit of a function (or mapping) $f(x)$ as x tends to a
$\sup A$	supremum of A, that is, the least upper bound of A
$\inf A$	infimum of A, that is, the greatest lower bound of A
$\Rightarrow$	"implies"
$\equiv$	"is by the definition equal to" or "is identically equal to"
B-property α	"property α in Appendix B"
B-definition α	"definition α in Appendix B"

Chapter 1

INTRODUCTION

1.1. Nature of the study

This book is concerned with the study of economic development planning in spatial systems. The study stems from three streams of thought in normative economic planning: the non-spatial optimal growth theory by economists, the spatial programming theory by regional scientists, and the theory of regional investment allocation originated by A. Rahman.

During the 1960's, great progress was made concerning the theory of economic growth. And when we limit ourselves to optimum growth theory, we find that "Ramsey-type" problems were extensively studied by the neoclassical school and "von Neumann-type" problems were extensively developed by activity analysts and the von Neumann school under the title of the turnpike theorem.[1] But, unfortunately, no spatial aspects were explicitly considered in these studies.

The importance of spatial aspects in actual economic planning has been well recognized, and many regional scientists have extensively investigated optimal problems in space systems. Earlier examples of

[1] For a general view of recent economic growth theory see, for example, Burmeister and Dobell (1970), and articles in Łos and Łos (1974) and Cass and Shell (1976).

Ramsey-type problems are concerned with finding the optimal allocation of outputs between consumption and investment at each time so as to maximize the utility function of consumption over some time period. On the other hand, von Neumann-type problems are mainly concerned with determining the optimal path for an economic system in which the objective is to maximize the quantity of some bundle of commodities which are in fixed proportions, or to maximize the value of outputs at fixed prices. In the first type of problems, models are generally highly aggregated and continuous time is used. In the latter type of problems, models are generally disaggregated and discrete time is used. In recent works, both approaches are being unified.

such works are Isard (1958), Lefeber (1958), Moses (1960) and Stevens (1958).[2] These studies effectively demonstrate the usefulness of mathematical programming frameworks for the planning of efficient resource utilization over space. These frameworks are quite general; yet most of these studies involve static or single-period programming.[3] Static programming analyses impose a serious limitation on the study of spatial development planning since locational investment decisions cannot be explained satisfactorily within static models.

There are several studies addressed to dynamic programming problems in space systems, and they constitute the third stream of thought guiding this study. An aggregate investment allocation model was first analyzed in Rahman (1963). Later, this problem was reformulated and generalized by many authors.[4] These works pioneered the study of the spatial allocation of investment, and provided valuable contributions in this direction. But all of them are highly aggregated planning models, that is, only a single homogeneous good, "national income," is considered. Thus, commodity trade cannot be incorporated in these models since at least two different goods are needed to generate a trading situation.[5]

This study is an attempt to synthesize the above three streams of thought. But, of course, this is a formidable task, and it is hoped that this book can take a small first step in that direction.

In moving in that direction, we first observe that most of the studies mentioned above can be handled within the framework of *monotone programming* (or its extensions). The precise problem formulation for monotone programming is given in Chapter 2. However, in short, it is the maximization of a linearly homogeneous objective function

---

[2]Here, we exclude the programming models for analyzing competitive spatial equilibrium, for example, Samuelson (1952) and Takayama and Judge (1970, 1971). Of course, the two types of studies are closely related theoretically.

[3]The terms, "static" and "single period," are used synonymously here.

[4]For some of them, see the references in section 6.1.

[5]These investment allocation models contrast with the static programming models. In the latter group of problems, the efficient locational patterns of production and transportation of intermediate and consumption goods are studied under given spatial distributions of natural resources and capital, and the allocation of investment goods cannot be explored. On the contrary, only investment allocation was considered in the first group of problems. In short, static programming models deal mainly with the complementary aspects of regions under given economic conditions, while investment allocation models emphasize the competitive aspects of regions for attaining the maximum rate of growth of the system.

subject to technological restrictions having the property of constant returns to scale. Examples are programming problems involving input-output models, the von Neumann model of production and Cobb–Douglas production functions.

We call those space systems which can be incorporated within the framework of monotone programming *monotone space systems* (of which precise definition is given in section 1.3). And the first objective of this book is to investigate the general character of optimal growth paths in this class of space systems.

Then, since mathematical properties of monotone space systems are implied by those of monotone programming, we must initially explore the mathematical theory of monotone programming. Thus, the second objective of this book is to develop the mathematical theory of monotone programming.

On the other hand, since monotone programming is characterized by the convexity of functions and constraint-sets describing it, the planning problems involving non-convex properties (e.g., indivisibility, increasing returns to scale and the non-convexity due to external diseconomies) cannot be studied within the framework of monotone space systems. Thus the third objective of this book is to attempt to extend the study of spatial development planning beyond monotone space systems. That is, by using a specific model, we will study the effects of scale economies and diseconomies on growth processes of spatial systems.[6]

1.2. Plan of the study

As noted before, the mathematical properties of monotone space systems are implied by those of monotone programming. Hence we first exploit the mathematical theory of monotone programming, and this is done in Part I (Chapter 2). Starting from works in Rockafellar (1967), we first obtain a duality theorem, existence theorem, reciprocity principle, max-min principle and maximum principle in monotone programming for both final state problems and consumption stream problems. Then, we extend these theorems to "generalized monotone

---

[6]It is desirable to extend the analysis of monotone space systems to general dynamic space systems with non-convexity properties. But this will clearly be quite difficult. Hence, in this book we take a more limited approach.

programming." Finally, these theorems are extended to monotone
programming with infinite horizons. For simplicity of representation
and since these theorems have a wide variety of applications besides
programs in space systems, analyses in this chapter are conducted
without reference to space systems.

In Part II, consisting of Chapters 3, 4 and 5, we investigate the
character of optimal growth paths in monotone space systems.

In Chapter 3 we examine in detail how to construct meaningful
monotone space systems. It is shown that the construction of monotone space systems can be accomplished both efficiently and in full
generality by using the combinatorial properties of monotone transformations. In addition, we show that at least three different types of
monotone space systems can be constructed, depending on how
production activities and transportation activities are combined.

In Chapter 4, the properties of the optimum growth paths for the
monotone space systems constructed in Chapter 3 are clarified by
applying the mathematical theorems obtained in Chapter 2. Special
emphasis is given to the duality between the optimal commodity path
and the optimal price path. Stepwise procedures for optimal paths are
explained by using the maximum principle, and interpretations of dual
variables are given in detail. It is shown that the "singular control
problem" is important theoretically as well as computationally in the
analysis of space systems.

Only quite general characteristics of optimal growth paths are studied in Chapter 4. Then, by describing the space system in more detail
in Chapter 5, more specific properties of optimal growth paths are
investigated. The optimum growth of two-region, two-commodity
monotone space systems is studied for the case of final state problems,
both with and without the requirement of transport inputs for commodity movements. It is shown that when the planning period is
sufficiently long, the optimal commodity path exhibits the classical
turnpike behaviour around the maximum balanced growth path in the
space system. Moreover, it is shown that investment goods should be
allocated between regions in a "balanced manner" throughout most of
the planning period and that the price variables which sustain this
"balanced allocation" of investment should satisfy certain well-defined
spatial relations.

In Part III (Chapter 6), we study the effects of scale economies and
diseconomies on the growth processes of space systems, and explore
some of the limitations of the convexity assumption inherent in mono-

tone space systems. More specifically, we develop a generalization of the Rahman model and, within the context of this model, examine the optimum growth implications of the following alternative assumptions of production: increasing, constant, decreasing and variable returns to scale.

Finally, in Chapter 7, some remarks are given for further development of the study.

Throughout this book, an "intuitive understanding" of the problem is emphasized and, for this purpose, graphical representation of the problem is given wherever possible. For convenience of readers, some basic mathematics necessary for the understanding of the analyses in this book is summarized in Appendix A.

Chapters 2 and 6 can be read independently of the rest of the book.

1.3. Space systems and spatial programming problems

In this section, the conventions used in describing space systems and spatial programming problems are explained (for a more detailed discussion of this topic, see Chapter 3). This will help to unify the analyses in the subsequent chapters.

Let us assume there are m physically distinct commodities in our system and we are given n predetermined locations (regions) in the system. Thus we are given two index sets:

commodity index set $J_i = (1, 2, \ldots, i, \ldots, m)$,

location index set $J_l = (1, 2, \ldots, l, \ldots, r, \ldots, s, \ldots, n)$.

For simplicity, both index sets are assumed to be time invariant, and a discrete representation of the time is adopted. Thus time is broken into successive periods of equal length, and this is shown by

time index set $J_t = (0, 1, 2, \ldots, t, \ldots, N)$.

For simplicity, N is considered to be a finite number (except in Section 2.6). Moreover, it may be convenient to distinguish time t and period t. As shown in Figure 1.1, the whole of the time lying between

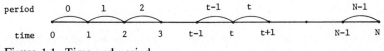

Figure 1.1. Time and period.

time t and $t + 1$ is called *period t*. Note that the phrases "the end of period $t - 1$" or "the beginning of period t" are used more often than "time t."

The *state of the space system in location l* at the beginning of period t is denoted by a vector x_l^t in $\mathbf{R}_+^m$. In detail,

$$x_l^t = (x_{1l}^t, x_{2l}^t, \ldots, x_{il}^t, \ldots, x_{ml}^t), \quad l = 1, 2, \ldots, n,$$

where x_{il}^t represents the amount of commodity i in location l at the beginning of period t. Thus, the *state of the space system* at the beginning of period t is denoted by a vector X_t in $\mathbf{R}_+^{mn}$ defined as follows:[7]

$$X_t = (x_l^t)_1^n.$$

Vectors X_t may be interpreted as the *input vector* for activities in period t. And, given input vector X_t, the set of all feasible states of the space system, that is, the set of all feasible *output vectors* at the end of period t is represented by

$$T_t(X_t),$$

which is a subset in $\mathbf{R}_+^{mn}$. This set is restricted by, among other things, technology available to this space system in period t.[8] Now, T_t is a multivalued function from $\mathbf{R}_+^{mn}$ into $\mathbf{R}_+^{mn}$ termed the *space transformation*. The space transformation T_t defines how a state of the space system at the beginning of period t is transformed into a state at the end of that period. Each vector in the set $T_t(X_t)$ can be denoted by

$$Z_t = (z_l^t)_1^n \in \mathbf{R}_+^{mn} \quad \text{with } z_l^t = (z_{1l}^t, z_{2l}^t, \ldots, z_{il}^t, \ldots, z_{ml}^t) \in \mathbf{R}_+^m,$$

where z_{il}^t represents the amount of commodity i in location l at the end of period t.

A *space system* is a collection of three index sets, J_i, J_l and J_t, and a set of space transformations, $T_t (t = 0, 1, \ldots, N - 1)$. And a monotone space system is defined as follows:

---

[7]Here we introduce a conventional representation of a vector which relates to the space system. When a vector component for location l is denoted by x_l, whatever it represents, the corresponding vector for the space system is denoted by $(x_l)_1^n$. Thus, $(x_l)_1^n = (x_1, x_2, \ldots, x_l, \ldots, x_n)$. $\mathbf{R}_+^{mn}$ represents the non-negative orthant of the mn-dimensional Euclidian space.

[8]We can consider this set is also restricted by institutional and/or political constraints, depending on the problem.

Definition 1.1. A *monotone space system* is a space system such that each of its space transformations T_t, $t = 0, 1, \ldots, N-1$, is a monotone transformation of concave type from $\mathbf{R}_+^{mn}$ into $\mathbf{R}_+^{mn}$.

According to this definition, the mathematical properties of monotone space systems are implied by the character of *monotone transformations of concave type* (for the definition and a detailed explanation of these transformations, see Section 2.2.1).

Next, to define a programming problem in a space system, we divide each vector z_l^t as follows:

$$z_l^t = x_l^{t+1} + (c_l^{t+1}, 0), \qquad c_l^{t+1} \in \mathbf{R}_+^{m_c}, \quad l = 1, 2, \ldots, n,$$

where vector c_l^{t+1} represents the part of z_l^t used as the consumption in the next period and x_l^{t+1} is used as the input vector for activities in location l in period $t + 1$. Each component of vector 0 in the above equation denotes a commodity which is prohibited to be consumed, for example, man. Hence, $m \geqq m_c \geqq 0$.

Let us denote the consumption vector and the input vector of the space system in period $t + 1$, respectively, by C_{t+1} and X_{t+1}. Then, in detail,

$$X_{t+1} = (x_l^{t+1})_1^n \in \mathbf{R}_+^{mn},$$
$$C_{t+1} = (c_l^{t+1})_1^n \in \mathbf{R}_+^{m_c n}.$$

A pair of vectors (X_{t+1}, C_{t+1}) represents the result of all the activities, including the consumption decision, during period t, and each vector (X_{t+1}, C_{t+1}) is evaluated by a numerical function, $f_{t+1}(X_{t+1}, C_{t+1})$, on $\mathbf{R}_+^{(m+m_c)n}$. This function f_{t+1}, $t = 0, 1, \ldots, N-1$, may be called the *welfare function* for period $t + 1$.

Thus, given a space system and a family of welfare functions f_t, $t = 1, 2, \ldots, N$, a *programming problem* is defined as the maximization of the total welfare achievable over the planning horizon subject to the initial state of the space system and the transformation condition in each period. The process of a space system in each period may be depicted as in Figure 1.2.

$$f_{t+1}(X_{t+1}, C_{t+1})$$

$$\longrightarrow X_t \longrightarrow T_t(X_t) \longrightarrow (X_{t+1}, C_{t+1}) \longrightarrow X_{t+1} \longrightarrow$$

Figure 1.2. Process in a spatial system.

PART I

DYNAMIC CONVEX PROGRAMMING

Chapter 2

DUALITY AND MAXIMUM PRINCIPLE IN DYNAMIC CONVEX PROGRAMMING

2.1. Introduction

It is well known in economics that under certain assumptions on technologies every efficient program has an associated set of "efficient" price vectors. This fact was first studied extensively by Koopmans (1951) and later by many authors [among others, Malinvaud (1953), Gale (1967a) and Radner (1967)]. But it appears that these economic studies are not much concerned with how these efficient price vectors are obtained.

Rockafellar demonstrates in his article (1967) that efficient price vectors can be obtained as the solution to a dual program in the case of "monotone programming." The duality theorem in monotone programming is different in its representation from the ordinary duality theorem in nonlinear programming since the dual problem in monotone programming does not involve any variables explicitly in the primal problem, and thus, the dual program can be solved independently of the primal problem. This property was previously known to hold only in linear programming.

The purpose of this chapter is to extend the duality and related theorems of monotone programming, which were originally studied in Rockafellar (1967) for the case of single-period programs, into multi-period programming and "generalized monotone programming."[1]

---

[1] It is appropriate to give here a brief historical note on studies of monotone programming. Starting from von Neumann (1945), monotone transformations (i.e., monotone processes or superlinear point-set maps) were extensively studied in economics under such topics as the von Neumann model of production, the closed linear model of production or activity analysis [see among others, Koopmans (1951),

These theorems are applied in later chapters to the analyses of spatial development planning.

In Section 2.2.1, definitions of terms and notation in monotone programming and basic results in monotone programming are summarized from Rockafellar (1967); economic examples are also included. These definitions and basic results are also summarized in Appendix B and numbered sequentially for reference purposes. Next in Section 2.2.2, it is explained graphically how the dual property in monotone programming emerges when the notion of adjoint transformation is introduced. This enhances an intuitive understanding of the analyses in later sections.

In Section 2.3, the duality theorem and the reciprocity principle in monotone programming, which are studied in Rockafellar (1967) for the case of single-period programs, are extended into multi-period monotone programming with an objective function which is a function only of the final state. Then, two alternative conditions to guarantee the existence of the optimal path in the dual problem are given. Though the existence problem was not investigated in Rockafellar (1967), this may be important for theoretical completeness. After some corollaries to these theorems, the max-min principle in monotone programming is obtained. It is shown that the conventional form of the maximum principle in discrete time optimal control theory can be obtained from our max-min principle when further restrictions on the problem are assumed.

In Section 2.4, the results from the previous section are extended into more general multi-period monotone programming in which the objective is the maximization of the sum of the values of the welfare

---

Malinvaud (1953), Gale (1956, 1967a,b), McKenzie (1963, 1972), Morishima (1961, 1964, 1969)]. Then, one of the most systematic studies on monotone transformations and monotone programming was carried out by Rockafellar (1967, 1970, 1972). Meanwhile, using Rockafellar (1967) as a basis, the author extended a part of his results (i.e., the duality and related theorems) to both multi-period monotone programming and generalized monotone programming in Fujita (1972). This chapter is mostly based on that work. On the other hand, studies similar to those of Rockafellar and Fujita (and including many other results) were independently performed by Makarov and Rubinov, and their recent results were summarized in Makarov and Rubinov (1970). The paper by Makarov and Rubinov represents one of the most advanced studies on monotone programming at present (their works were, unfortunately, not known to the author before completing his dissertation, Fujita (1972)). Hence, for further studies of monotone programming, the interested reader should consult Makarov and Rubinov (1970) as well as Rockafellar (1967, 1970, 1972).

function in each period. These welfare functions are limited to linearly homogeneous functions.

In Section 2.5, the theorems in Section 2.4 are extended into "generalized monotone programming" in which transformations and welfare functions are not necessarily linearly homogeneous but are concave.

In Sections 2.3 to 2.5, it is assumed that the length of the plan period is finite. But, in many dynamic optimization problems in economics, one is led to consider programming problems with an infinite time horizon. Hence, in Section 6, results from the previous sections are extended to programming problems involving infinite horizons.

Purely mathematical (or relatively unimportant) proofs of statements in this chapter are given in Appendix C.

2.2. Monotone transformations and the fundamental duality theorem

In this section, some basic properties of monotone transformations are summarized. In the first subsection, definitions and economic interpretations of two types of monotone transformations are given; economic examples are also included. The emphasis is on the explanation of the dual relationship between the primal and its adjoint transformations. This fundamental dual relationship is summarized in the form of a duality theorem. Then, a two-dimensional geometrical explanation of this duality theorem is given in the second subsection.

2.2.1. A monotone transformation and its adjoint transformation

A *monotone transformation T of concave type* (abbreviated to m.t.cc. T) from $\mathbf{R}^m_+$ into $\mathbf{R}^n_+$ is a multi-valued mapping[2] such that

(i) $T(x^1 + x^2) \supset T(x^1) + T(x^2)$ for all $x^1, x^2 \in \mathbf{R}^m_+$,

(ii) $T(\lambda x) = \lambda T(x)$ for all $x \in \mathbf{R}^m_+$ and $\lambda > 0$,

(iii) $T(0) = \{0\}$,

---

[2]Hereafter, for example, $\mathbf{R}^m_+$ represents the non-negative orthant of the m-dimensional Euclidean space $\mathbf{R}^m$, and $\mathbf{R}^{*m}_+$ is its second copy. The distinction between $\mathbf{R}^m_+$ and $\mathbf{R}^{*m}_+$ is not for a technical necessity, but for convenience in application. That is, in economics, elements of $\mathbf{R}^m_+$ may be vectors of commodities and those of $\mathbf{R}^{*m}_+$ vectors of prices.

(iv)   it is a closed mapping,

(v) $0 \leqq x^1 \leqq x^2$ implies $T(x^1) \subset T(x^2)$ for all $x^1, x^2 \in \mathbf{R}_+^m$,

(vi) $0 \leqq y^1 \leqq y^2 \in T(x)$ implies $y^1 \in T(x)$ for all $y^1, y^2 \in \mathbf{R}_+^n$.

In economic applications, it is convenient to consider that x, y and T represent, respectively, a set of input commodities (= an input vector), a set of output commodities (= an output vector) and the production technology within a system. Then $T(x)$ is the set of all output vectors which are feasible under technology T from input vector x. A pair (x, y) of an input vector x and an output vector y is called an *activity*, and it is called a *feasible activity* when $y \in T(x)$. Thus the above conditions have the following economic interpretations.

Condition (i) means

$$y^1 \in T(x^1) \quad \text{and} \quad y^2 \in T(x^2) \quad \text{implies} \quad y^1 + y^2 \in T(x^1 + x^2).$$

That is, if outputs y^1 and y^2 are feasible from inputs x^1 and x^2, respectively, then $y^1 + y^2$ is feasible from $x^1 + x^2$. This property of *additivity* implies the *absence of external diseconomies* between feasible activities within the system.

Condition (ii) means

$$y \in T(x) \quad \text{implies} \quad \lambda y \in T(\lambda x) \quad \text{for any } \lambda > 0. \tag{1}$$

Namely, if inputs are proportionally increased (decreased) by some percentage, then outputs are proportionally increased (decreased) by the same percentage. This means that *constant returns to scale* prevails in the production technology.

Under condition (ii), we have

$$y^1 \in T(x^1) \quad \text{and} \quad y^2 \in T(x^2) \quad \text{implies}$$
$$\lambda y^1 \in T(\lambda x^1) \quad \text{and} \quad (1-\lambda)y^2 \in T[(1-\lambda)x^2] \quad \text{for all } 0 < \lambda < 1.$$

Hence, from condition (i), we get

$$y^1 \in T(x^1) \quad \text{and} \quad y^2 \in T(x^2) \quad \text{implies}$$
$$\lambda y^1 + (1-\lambda)y^2 \in T[\lambda x^1 + (1-\lambda)x^2] \quad \text{for all } 0 < \lambda < 1. \tag{2}$$

Therefore, conditions (i) and (ii) together assure that, by simultaneously using two inputs x^1 and x^2 with weights λ and $(1-\lambda)$, at least the weighted sum of the original two outputs y^1 and y^2 can be produced. The economic interpretation of this property of *convexity* is that the *law of diminishing marginal returns* (in addition to the absence of external diseconomies) prevails in the technology.

Condition (iii) ensures that it is impossible to get something from nothing; namely, it means the *impossibility of the land of Cockaigne.*

Condition (iv) represents the following "*continuity*" condition:

$$\left.\begin{array}{l} y^j \in T(x^j), j = 1, 2, \ldots, \quad \text{and} \\ \lim_{j \to \infty} x^j = \bar{x}, \lim_{j \to \infty} y^j = \bar{y} \end{array}\right\} \text{ implies } \bar{y} \in T(\bar{x}). \tag{3}$$

Namely, if there are feasible activities "arbitrarily close" to a given activity, then the given activity is also feasible.

Conditions (v) and (vi) mean the *free disposability of inputs and outputs*, respectively.

On the other hand, if we set $x^1 = x^2 = x$ in (2), then we get

$$y^1 \in T(x) \quad \text{and} \quad y^2 \in T(x) \quad \text{implies} \quad \lambda y^1 + (1 - \lambda)y^2 \in T(x)$$
$$\text{for all } 0 < \lambda < 1. \tag{4}$$

From B-definition 3,[3] a non-empty closed bounded convex set C is called a *monotone set of concave type* when $0 \leq y^1 \leq y^2 \in C$ implies $y^1 \in C$. From (4), $T(x)$ is a convex set for any $x \in \mathbf{R}^n_+$, and it is a closed set from (3). In addition, (v) and (vi) imply $0 \in T(x)$; thus $T(x)$ is non-empty for any $x \in \mathbf{R}^n_+$. Finally, condition (iii) together with (ii) and (iv) guarantee that $T(x)$ is bounded for each $x \in \mathbf{R}^n_+$ (refer to B-property 2). Hence, $T(x)$ is a monotone set of concave type for each $x \in \mathbf{R}^m_+$. Figure 2.1 gives several examples of monotone sets of concave type.

An alternative definition of m.t.cc. is given as follows:

A multivalued mapping T from $\mathbf{R}^m_+$ into $\mathbf{R}^n_+$ is a m.t.cc. if and only if the following three conditions are satisfied:

(a) the graph of T (abbreviated to graph T)[4] is a closed convex cone in $\mathbf{R}^m_+ \times \mathbf{R}^n_+$,
(b) $(0, y) \in$ graph T implies $y = 0$.
(c) $(x, y) \in$ graph T, $x' \geq x$ and $y \geq y' \geq 0$ implies $(x', y') \in$ graph T.

The set, graph T, is called the *technology set*, and it represents all feasible activities under T. A pair of input–output vectors (x, y) is a feasible activity under T if and only if $(x, y) \in$ graph T.

---

[3]Throughout this book, "B-definition α" means Definition α in Appendix B.
[4]By definition, graph $T = \{(x, y) | x \in \mathbf{R}^m_+, y \in \mathbf{R}^n_+, y \in T(x)\}$, which is a subset of a Cartesian product, $\mathbf{R}^m_+ \times \mathbf{R}^n_+$.

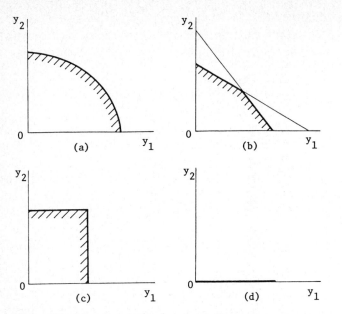

Figure 2.1. Monotone sets of concave type.

The equivalence of the two definitions is clear. From (1), graph T is a cone in $\mathbf{R}_+^m \times \mathbf{R}_+^n$, and it is convex and closed from (2) and (3), respectively. On the other hand, (b) is equivalent to (iii), and (c) is equivalent to (v) plus (vi).

Hence, a m.t.cc. T is simply a restatement of the properties of the technology set assumed in the *closed linear model of production* in the form of a transformation. We could proceed by using the concept of a technology set rather than that of a transformation. But it soon becomes clear that the concept of a transformation is much more convenient than that of a technology set for the analysis of growth planning.

Next, a m.t.cc. T is called *non-singular* if $T(x)$ has a non-empty interior when x is strictly positive. This implies that if x is strictly positive, then there is a strictly positive output vector in $T(x)$. Namely, if the inputs include some amount of every input commodity, then it is possible to produce some amount of every output commodity simultaneously. A singular transformation can always be made non-singular by restricting its range appropriately, namely, by reducing the number of output commodities.

Next, we give several examples of m.t.cc. in economics.

Example 1. Neoclassical production model

$$T(K, L) = \{y|0 \leqq y \leqq F(K, L)\}: \mathbf{R}_+^2 \to \mathbf{R}_+^1,$$

where

K: the amount of capital,
L: the amount of labor,
y: the amount of the output,
F: a neoclassical production function which is a linearly homogeneous, concave and non-decreasing function on $\mathbf{R}_+^2$.

Thus defined T is a m.t.cc. from $\mathbf{R}_+^2$ into $\mathbf{R}_+^1$, and it is non-singular if $F(K, L)$ is not identically equal to zero.

Example 2. von Neumann model of production

$$T(x) = \{y|0 \leqq y \leqq Bv \text{ for some } v \geqq 0 \text{ such that } Av \leqq x\}: \mathbf{R}_+^m \to \mathbf{R}_+^n,$$

where

A: input coefficient matrix of size $m \times k$, of which the jth column represents the input-coefficient-vector of the jth basic activity, $A \geqq 0$,
B: output coefficient matrix of size $n \times k$, of which the jth column represents the output-coefficient-vector of the jth basic activity, $B \geqq 0$,
v: k-dimensional vector, of which the jth component represents the activity level of the jth basic activity,
x: m-dimensional input vector,
y: n-dimensional output vector.

Thus defined T is a m.t.cc. from $\mathbf{R}_+^m$ into $\mathbf{R}_+^n$ if no column vector of A is zero. Moreover, if no row vector of B is zero, then it is non-singular.[5]

Example 3. Technology underlying a linear programming model

$$T(x) = \{y|Ay \leqq x, y \geqq 0\}: \mathbf{R}_+^m \to \mathbf{R}_+^n,$$

[5]For details of the von Neumann model of production, see, for example, von Neumann (1945), Morishima (1969) and Nikaido (1968). Also see Section 5.1 in Chapter 3 of this book.

where

Λ: input coefficient matrix of the size $m \times n$, $A \geqq 0$,

x: m-dimensional resource constraint vector,

y: n-dimensional activity level vector.

Thus defined T is a non-singular m.t.cc. from $\mathbf{R}_+^m$ into $\mathbf{R}_+^n$ if no column vector of matrix A is zero.

Next, given a m.t.cc. T from $\mathbf{R}_+^m$ into $\mathbf{R}_+^n$, the *adjoint transformation* T^* of T is a multivalued mapping from $\mathbf{R}_+^{*n}$ into $\mathbf{R}_+^{*m}$ such that

$$T^*(p) = \{q \in \mathbf{R}_+^{*m} \,|\, q \cdot x \geqq p \cdot y \quad \text{for every } (x, y) \in \text{graph } T\}, \qquad (5)^6$$

for each $p \in \mathbf{R}_+^{*n}$. Thus defined T^* satisfies the following conditions:

(i*) $T^*(p^1 + p^2) \supset T^*(p^1) + T^*(p^2)$ for all $p^1, p^2 \in \mathbf{R}_+^{*n}$,

(ii*) $T^*(\lambda p) = \lambda T^*(p)$ for all $p \in \mathbf{R}_+^{*n}$ and $\lambda > 0$,

(iii*) $T^*(0) = \mathbf{R}_+^{*m}$,

(iv*)   it is a closed mapping,

(v*) $p^1 \geqq p^2 \geqq 0$ implies $T^*(p^1) \subset T^*(p^2)$ for all $p^1, p^2 \in \mathbf{R}_+^{*n}$,

(vi*) $q^1 \geqq q^2 \in T^*(p)$ implies $q^1 \in T^*(p)$ for all $q^1, q^2 \in \mathbf{R}_+^{*m}$.

Condition (i*) is derived as follows: First, observe that

$$q^1 \in T^*(p^1) \quad \text{means} \quad q^1 \cdot x \geqq p^1 \cdot y \quad \text{for any } (x, y) \in \text{graph } T,$$

and

$$q^2 \in T^*(p^2) \quad \text{means} \quad q^2 \cdot x \geqq p^2 \cdot y \quad \text{for any } (x, y) \in \text{graph } T.$$

Therefore, $q^1 \in T^*(p^1)$ and $q^2 \in T^*(p^2)$ implies

$$(q^1 + q^2)x \geqq (p^1 + p^2)y \quad \text{for any } (x, y) \in \text{graph } T,$$

hence

$$(q^1 + q^2) \in T^*(p^1 + p^2).$$

Condition (ii*) means

$$q \in T^*(p) \quad \text{implies} \quad \lambda q \in T^*(\lambda p) \quad \text{for any } \lambda > 0.$$

This is clear from the fact that $q \cdot x \geqq p \cdot y$ implies $\lambda p \cdot x \geqq \lambda p \cdot y$. Similarly, conditions (iii*), (v*) and (vi*) immediately follow from the

---

[6] As defined in footnote 3, graph T is the graph of T. Hence, this is rewritten in detail as follows:

$$T^*(p) = \{q \in \mathbf{R}_+^{*m} \,|\, q \cdot x \geqq p \cdot y \quad \text{for every } (x, y) \text{ such that } y \in T(x)\}.$$

property of the linear inequality in Definition (5). Finally condition (iv*) arises from the fact that graph T is a closed set (for this, refer to the proof of B-property 4^7). Note that (iii*) is not an independent condition since (ii*) and (iv*) imply $0 \in T^*(0)$, and hence it follows from (vi*) that $T^*(0) = \mathbf{R}_+^{*m}$.

From B-definition 1, a transformation which satisfies conditions (i*), (ii*), and (iv*) to (vi*) (hence (iii*) is also satisfied) is called a *monotone transformation of convex type* (abbreviated to m.t.cv.). Therefore, the *adjoint operation* (5) assigns a unique m.t.cv. to each m.t.cc. The economic meaning of the adjoint transformation is clear from the definition itself. If we consider x, y and T in (5) to represent, as before, an input commodity vector, an output commodity vector and the production technology, respectively, then it is appropriate to interpret that p and q represent, respectively, a price vector for output commodities and a price vector for input commodities. Hence $T^*(p)$ is the set of all input price vectors each of which does not allow positive profits for any feasible activity under technology T given output price vector p.

From B-definition 3, a non-empty closed convex set C is called a *monotone set of convex type* when $q^1 \geq q^2 \in C$ implies $q^1 \in C$. It is easy to see, as in the case of m.t.cc., that when T^* is a m.t.cv., $T^*(p)$ is a monotone set of convex type for each $p \in \mathbf{R}_+^{*n}$. Figure 2.2 gives several examples of monotone sets of convex type.

From B-definition 6, a m.t.cv. T^* is called *non-singular* when $0 \in T^*(p)$ only for $p = 0$. In other words, T^* is non-singular if zero pricing for inputs is allowed only when output prices are zero. It is easy to see from (5) that the adjoint transformation T^* is non-singular if and only if T in (5) is non-singular (refer to B-property 9).

Example 4. Adjoint transformation in the neoclassical model

The adjoint transformation of m.t.cc. T defined in Example 1 is given by

$$T^*(p) = \{(q_K, q_L) \in \mathbf{R}_+^{*2} | Kq_K + Lq_L \geq pF(K, L)$$
$$\text{for all } (K, L) \geq 0\}: \mathbf{R}_+^{*1} \to \mathbf{R}_+^{*2},$$

where

p: price of the output,
q_K: rental price on capital,
q_L: wage rate.

[7]Throughout this book, "B-property α" means Property α in Appendix B.

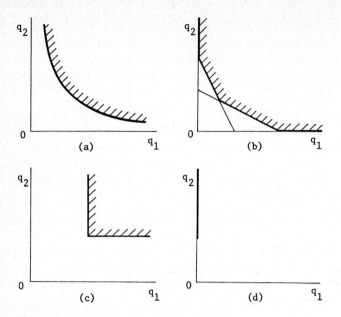

Figure 2.2. Monotone sets of convex type.

Example 5. Adjoint transformation in the von Neumann model

In the case of von Neumann production function T defined in Example 2

$$T^*(p) = \{q \in \mathbf{R}_+^{*m} | v'A'q \geqq v'B'p \quad \text{for all } v' \geqq 0\}$$
$$= \{q \in \mathbf{R}_+^{*m} | A'q \geqq B'p\}: \mathbf{R}_+^{*n} \to \mathbf{R}_+^{*m},$$

where A' and B' are transposed matrices of A and B, respectively, and v' is the transposed vector of v. In terms of basic activities, (a_j, b_j), $j = 1, 2, \ldots, k$, where a_j and b_j are the jth columns of A and B respectively,

$$T^*(p) = \{q \in \mathbf{R}_+^{*m} | q \cdot a_j \geqq p \cdot b_j, \quad j = 1, 2, \ldots, k\}.$$

Hence $T^*(p)$ gives the set of input price vectors each of which does not allow positive profits for any basic activity (a_j, b_j), $j = 1, 2, \ldots, k$, under a given output price vector p.

Example 6. Adjoint transformation in a linear programming model

In Example 3, the adjoint transformation is given by

$$T^*(p) = \{q | A'q \geqq p, q \geqq 0\}: \mathbf{R}_+^{*n} \to \mathbf{R}_+^{*m},$$

where A' is the transposed matrix of A. Vector q gives prices for input resources, and vector p represents prices for outputs.

Next, to explore the dual property between T and T^*, the following definitions are needed. From B-definition 2, a *monotone concave gauge* on $\mathbf{R}_+^m$ is a continuous, real-valued, linearly homogeneous, concave and non-decreasing function on $\mathbf{R}_+^m$. A *monotone convex gauge* is subjected to the same requirements except that concavity is replaced by convexity.

Next, given a monotone set C of concave type in $\mathbf{R}_+^n$ and a monotone set D^* of convex type in $\mathbf{R}_+^{*m}$. We define that

$$\langle p, C \rangle = \sup \{p \cdot y | y \in C\} \qquad \text{for each } p \in \mathbf{R}_+^{*n}, \tag{6}$$

$$\langle D^*, x \rangle = \inf \{q \cdot x | q \in D^*\} \quad \text{for each } x \in \mathbf{R}_+^m. \tag{7}$$

Then, from B-property 3, $\langle p, C \rangle$ is a monotone convex gauge on $\mathbf{R}_+^{*n}$, and $\langle D^*, x \rangle$ is a monotone concave gauge on $\mathbf{R}_+^m$. Conversely, each monotone convex gauge g on $\mathbf{R}_+^{*n}$ is of this form (i.e., $g(p) = \langle p, C \rangle$) for a unique monotone set C of concave type in $\mathbf{R}_+^n$ such that

$$C = \{y \in \mathbf{R}_+^n | p \cdot y \leqq g(p) \quad \text{for all } p \in \mathbf{R}_+^{*n}\}, \tag{8}$$

and each monotone concave gauge f on $\mathbf{R}_+^m$ is of this form (i.e., $f(x) = \langle D^*, x \rangle$) for a unique monotone set D^* of convex type in $\mathbf{R}_+^{*m}$ such that

$$D^* = \{q \in \mathbf{R}_+^{*m} | q \cdot x \geqq f(x) \quad \text{for all } x \in \mathbf{R}_+^m\}. \tag{9}$$

Now given a m.t.cc. T from $\mathbf{R}_+^m$ into $\mathbf{R}_+^n$ and its adjoint transformation T^*, the fundamental dual property between them can be stated as follows:

$$\langle p, T(x) \rangle = \langle T^*(p), x \rangle \quad \text{for all } x \in \mathbf{R}_+^m \text{ and } p \in \mathbf{R}_+^{*n}. \tag{10}$$

Note that since $T(x)$ and $T^*(p)$ are, respectively, a monotone set of concave type in $\mathbf{R}_+^n$ and a monotone set of convex type in $\mathbf{R}_+^{*m}$, each side of (10) is defined as follows:

$$\langle p, T(x) \rangle = \sup \{p \cdot y | y \in T(x)\},$$

$$\langle T^*(p), x \rangle = \inf \{q \cdot x | q \in T^*(p)\}.$$

Therefore, $\langle p, T(x) \rangle$ is a monotone convex gauge on $\mathbf{R}_+^{*n}$ for each fixed $x \in \mathbf{R}_+^m$ and is a monotone concave gauge on $\mathbf{R}_+^m$ for each fixed $p \in \mathbf{R}_+^{*n}$ because of (10). Likewise $\langle T^*(p), x \rangle$ is a monotone concave gauge on

$\mathbf{R}_+^m$ for each fixed $p \in \mathbf{R}_+^{*n}$ and is a monotone convex gauge on $\mathbf{R}_+^{*n}$ for each fixed $x \in \mathbf{R}_+^m$ because of (10).

In economic terms (10) says that, given input vector x and output price vector p, the maximum value of outputs achievable under technology T is equal to the minimum valuation of the inputs under the "market condition" that any feasible activity cannot obtain positive profits. In other words, if the "market" works so as to yield no positive profits to any feasible activities as shown in (5), the maximum revenue becomes equal to the minimum costs for the optimal activity.

We call the dual property of T and T^* expressed by (10) the *fundamental duality theorem* in monotone programming. This duality theorem includes the duality theorem in linear programming as a special case as is shown in the following example.

Example 7. Duality theorem in linear programming
By applying (10) to Examples 3 and 6, we have:

$$\max \{p \cdot y | Ay \leqq x, y \geqq 0\} = \min \{q \cdot x | A'q \geqq p, q \geqq 0\},$$

which expresses the duality theorem in linear programming.

The following question naturally arises. When we define the adjoint transformation by (5), why does such a well behaved property (10) emerge? In other words, how does the adjoint transformation "work"? The mathematically rigorous answer to this question can be obtained by combining B-properties 5 and 6. A more intuitive explanation involving graphical analysis is depicted in Section 2.2.

So far we have started from a m.t.cc. T and obtained a m.t.cv. T^* by adjoint operation (5). But we can go conversely. Namely, given a m.t.cv. S from $\mathbf{R}_+^{*n}$ into $\mathbf{R}_+^{*m}$, we define its adjoint transformation S^* as follows:

$$S^*(x) = \{y \in \mathbf{R}_+^n | p \cdot y \leqq q \cdot x \quad \text{for every } (p, q) \in \text{graph } S\}, \qquad (11)$$

for each $x \in \mathbf{R}_+^m$. Then it is easy to see that thus defined S^* satisfies all the conditions of m.t.cc., and hence it is a m.t.cc. from $\mathbf{R}_+^m$ into $\mathbf{R}_+^n$. Moreover, if S is the adjoint transformation of a m.t.cc. T defined by (5) (i.e., if $S = T^*$), we get $S^* = T$. That is

$$(T^*)^* = T. \qquad (12)$$

This can be proven as follows. From (11),

$$(T^*)^*(x) = \{y \in \mathbf{R}_+^n | p \cdot y \leq q \cdot x \qquad \text{for every } (p, q) \text{ such that}$$
$$q \in T^*(p)\}$$

$$= \{y \in \mathbf{R}_+^n | p \cdot y \leq \langle T^*(p), x\rangle \quad \text{for all } p \in \mathbf{R}_+^{*n}\}$$
$$\text{from Definition (7)},$$

$$= \{y \in \mathbf{R}_+^n | p \cdot y \leq \langle p, T(x)\rangle \quad \text{for all } p \in \mathbf{R}_+^{*n}\} \text{ from (10)},$$

and hence, from (6) and (8), $(T^*)^*(x) = T(x)$ for each $x \in \mathbf{R}_+^m$, which gives (12).

In effect, the adjoint operations (5) and (11) set up, respectively, a one-to-one type-reversing, order-preserving correspondence between the monotone transformation from $\mathbf{R}_+^m$ into $\mathbf{R}_+^n$ and the monotone transformation from $\mathbf{R}_+^{*n}$ into $\mathbf{R}_+^{*m}$. This relation can be depicted as in Figure 2.3.

Finally, to generalize dual property (10), we introduce some notation. Given a monotone set X of concave type in $\mathbf{R}_+^m$ and a monotone set P of convex type in $\mathbf{R}_+^{*n}$, define that

$$\langle P, X\rangle = \sup_{x \in X} \inf_{p \in P} p \cdot x = \inf_{p \in P} \sup_{x \in X} p \cdot x$$
$$= \sup_{x \in X} \langle P, x\rangle = \inf_{p \in P} \langle p, X\rangle. \tag{13}$$

The equalities in the above definition are assured by B-property 15.

Next, given a monotone set C of concave type in $\mathbf{R}_+^m$ and a monotone set D^* of convex type in $\mathbf{R}_+^{*n}$, we define that

$$T(C) = \cup\{T(x)|x \in C\} \quad \text{and} \quad T^*(D^*) = \cup\{T^*(p)|p \in D^*\}, \tag{14}$$

for a m.t.cc. T from $\mathbf{R}_+^m$ into $\mathbf{R}_+^n$ and a non-singular m.t.cv. T^* from $\mathbf{R}_+^{*n}$ into $\mathbf{R}_+^{*m}$, respectively. Then, from B-property 17, $T(C)$ [resp. $T^*(D^*)$] is a monotone set of concave type (convex type) in $\mathbf{R}_+^n$ ($\mathbf{R}_+^{*m}$),

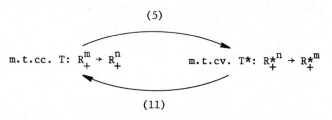

Figure 2.3. Adjoint operations.

and the next important property holds.

$$\langle D^*, T(C)\rangle = \langle T^*(D^*), C\rangle \quad \text{(refer to (13))} \tag{15}$$

which is a generalization of the duality theorem (10).

Note that, in the above, we have defined adjoint transformation T^* of a m.t.cc. T by (5) and considered (10) as one of the properties of T^*. Conversely, we can define that the adjoint transformation T^* of a m.t.cc. T is a m.t.cv. which satisfies relation (10), and consider that (5) is one of the properties of T^*. The equivalence of the two approaches is assured by B-properties 5 and 6. The second approach is taken in the original work by Rockafellar (1967). But for economic interpretation, our approach is more appropriate.

2.2.2. *A geometrical explanation of the fundamental duality theorem*

In this section, a graphical answer is given to the question posed in Section 2.2.1 (How does dual property (10) emerge if we define T^* by (5)?) by using a two-dimensional example. The emphasis here is not on theoretical strictness but rather on an intuitive geometrical explanation.[8]

Given a m.t.cc. T from $\mathbf{R}_+^2$ into $\mathbf{R}_+^2$, the adjoint transformation T^* is defined by (5) as follows:

$$T^*(p) = \{q \in \mathbf{R}_+^{*2} | q \cdot x \geq p \cdot y \quad \text{for every } (x, y) \in \text{graph } T\}, \tag{16}$$

for each $p \in \mathbf{R}_+^{*2}$. Let us fix p in (16) to be an arbitrary vector, $\bar{p} \in \mathbf{R}_+^{*2}$, and examine how to get the set $T^*(\bar{p})$ graphically. Since T has the property of constant returns to scale, we first limit the input vector x in (16) to the set

$$H = \{x | x = (x_1, x_2), x_1 + x_2 = 1, x_1, x_2 \geq 0\}.$$

Here, H is the two-dimensional standard simplex, and it is depicted in Figure 2.4. First, take any point $x^\alpha = (x_1^\alpha, x_2^\alpha)$ in H. Then (16) requires that

$$q \cdot x^\alpha \geq \bar{p} \cdot y \quad \text{for all } y \in T(x^\alpha).$$

---

[8]In particular, it is assumed in (10) that $\langle p, T(x)\rangle = \max\{p \cdot y | y \in T(x)\}$ and $\langle T^*(p), x\rangle = \min\{q \cdot x | q \in T^*(p)\}$. Of course, this assumption will not always be satisfied, and this is the subject of Section 2.3.3.

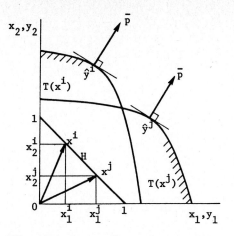

Figure 2.4. Commodity space.

The set of $q \in \mathbf{R}_+^{*2}$ which satisfies the above condition, which is denoted by $T^*(\bar{p})|_{x^\alpha}$, is given by

$$T^*(\bar{p})|_{x^\alpha} = \{q \in \mathbf{R}_+^{*2} | q \cdot x^\alpha \geqq \bar{p} \cdot \hat{y}^\alpha\}, \tag{17}$$

where $\hat{y}^\alpha$ is a point in $T(x^\alpha)$ such that

$$\bar{p} \cdot \hat{y}^\alpha = \max \{\bar{p} \cdot y | y \in T(x^\alpha)\}. \tag{18}$$

For example, two points x^i and x^j are arbitrarily chosen on H in Figure 2.4. Then corresponding output sets $T(x^\alpha)$, $\alpha = i, j$, are depicted in Figure 2.4, and corresponding price sets $T^*(\bar{p})|_{x^\alpha}$, $\alpha = i, j$, are depicted in Figure 2.5, respectively. In this way, the price set $T^*(\bar{p})$ is obtained as the interaction of sets $T^*(\bar{p})|_{x^\alpha}$ while x^α moves all over the set H.[9] That is,

$$T^*(\bar{p}) = \bigcap_{x^\alpha \in H} T^*(\bar{p})|_{x^\alpha}. \tag{19}$$

---

[9]Note that $\mathbf{R}_+^2 = \{x | x = \lambda x', x' \in H, \lambda \geqq 0\}$. Therefore,

$$
\begin{aligned}
T^*(p) &= \{q \in \mathbf{R}_+^{*2} | q \cdot x \geqq p \cdot y \quad \text{for every } y \in T(x), x \in \mathbf{R}_+^2\} \\
&= \{q \in \mathbf{R}_+^{*2} | q \cdot \lambda x' \geqq p \cdot y \quad \text{for every } y \in T(\lambda x'), x' \in H, \lambda \geqq 0\} \\
&= \{q \in \mathbf{R}_+^{*2} | q \cdot \lambda x' \geqq p \cdot y \quad \text{for every } y \in \lambda T(x'), x' \in H, \lambda \geqq 0\} \\
&\qquad\qquad\qquad \text{from condition (ii) of m.t.cc.} \\
&= \{q \in \mathbf{R}_+^{*2} | q \cdot \lambda x' \geqq p \cdot \lambda y' \quad \text{for every } y' \in T(x'), x' \in H, \lambda \geqq 0\} \\
&= \{q \in \mathbf{R}_+^{*2} | q \cdot x' \geqq p \cdot y' \quad \text{for every } y' \in T(x'), x' \in H\}.
\end{aligned}
$$

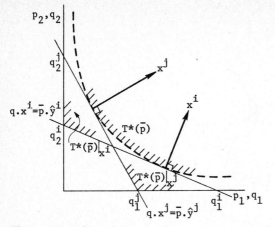

Figure 2.5. Price space.

Hence, if we can show that each line $q \cdot x^{\alpha} = \bar{p} \cdot \hat{y}^{\alpha}$ touches the set $T^*(\bar{p})$ in Figure 2.5 at least at one point, that is,

$$T^*(\bar{p}) \cap \{q \in \mathbf{R}_+^{*2} | q \cdot x^{\alpha} = \bar{p} \cdot \hat{y}^{\alpha}\} \neq \emptyset \quad \text{for each } x^{\alpha} \in H, \tag{20}$$

then, from (17) and (19), surely we have

$$\min \{q \cdot x^{\alpha} | q \in T^*(\bar{p})\} = \bar{p} \cdot \hat{y}^{\alpha} \quad \text{for each } x^{\alpha} \in H. \tag{21}$$

Thus, from (18) and (21) we get

$$\max \{\bar{p} \cdot y | y \in T(x^{\alpha})\} = \min \{q \cdot x^{\alpha} | q \in T^*(\bar{p})\},$$

for each $x^{\alpha} \in H$. Then, taking into account condition (ii) in the definition of m.t.cc., we have

$$\max \{\bar{p} \cdot \lambda y | \lambda y \in T(\lambda x^{\alpha})\} = \min \{q \cdot \lambda x^{\alpha} | q \in T^*(\bar{p})\},$$

for all $x^{\alpha} \in H$ and $\lambda \geqq 0$. This means

$$\max \{\bar{p} \cdot y | y \in T(x)\} = \min \{q \cdot x | q \in T^*(\bar{p})\},$$

for all $x \in \mathbf{R}_+^2$. Then, since $\bar{p}$ was chosen arbitrarily from $\mathbf{R}_+^{*2}$, we obtain the duality theorem, (10).

Condition (20) is satisfied for the following reasons. First, the function

$$z(x) = \langle T(x), \bar{p} \rangle = \max \{\bar{p} \cdot y | y \in T(x)\} \tag{22}$$

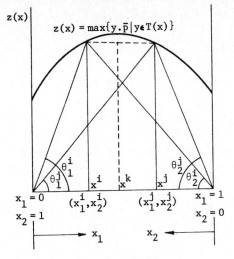

Figure 2.6. Concavity of $z(x)$.

is concave on the set H since

$$z(\beta x^1 + (1 - \beta)x^2) = \langle T[\beta x^1 + (1 - \beta)x^2], \bar{p} \rangle$$
$$\geqq \langle T(\beta x^1) + T[(1 - \beta)x^2], \bar{p} \rangle \quad \text{from (i) of m.t.cc.}$$
$$= \langle T(\beta x^1), \bar{p} \rangle + \langle T[(1 - \beta)x^2], \bar{p} \rangle$$
$$= \beta \langle T(x^1), \bar{p} \rangle$$
$$\qquad + (1 - \beta)\langle T(x^2), \bar{p} \rangle \quad \text{from (ii) of m.t.cc.}$$
$$= \beta z(x^1) + (1 - \beta)z(x^2),$$

where $0 < \beta < 1$. Thus, the graph of function $z(x)$ has a shape like Figure 2.6.

Next, again take any two points, $x^\alpha = (x_1^\alpha, x_2^\alpha)$, $\alpha = i, j$, on H. And suppose, for example, $x_1^i < x_1^j$ (and hence, $x_2^i > x_2^j$) as shown in Figures 2.4 and 2.6. Then, Figure 2.5 depicts the intersection q_1^α between the line $\bar{p} \cdot \hat{y}^\alpha = q \cdot x^\alpha$ and the horizontal axis, and the intersection q_2^α between that line and the vertical axis. They are defined as follows ($\alpha = 1, 2$):

$$q_1^\alpha = \frac{\bar{p} \cdot \hat{y}^\alpha}{x_1^\alpha} = \frac{z(x^\alpha)}{x_1^\alpha}, \qquad q_2^\alpha = \frac{\bar{p} \cdot \hat{y}^\alpha}{x_2^\alpha} = \frac{z(x^\alpha)}{x_2^\alpha} \quad \text{(from (22))}.$$

Thus, in Figure 2.6, we see that $q_1^\alpha = \tan \theta_1^\alpha$ and $q_2^\alpha = \tan \theta_2^\alpha$, $\alpha = 1, 2$, and $\theta_1^i > \theta_1^j$, $\theta_2^i < \theta_2^j$. Therefore, we conclude that

$$x_1^i < x_1^j \text{ (i.e., } x_2^i > x_2^j) \quad \text{implies} \quad q_1^i > q_1^j \quad \text{and} \quad q_2^i < q_2^j. \tag{23}$$

The above relation assures that any given line $q \cdot x^\alpha = \bar{p} \cdot \hat{y}^\alpha$ is not completely dominated by any other line $q \cdot x^{\alpha'} = \bar{p} \cdot \hat{y}^{\alpha'}$ in Figure 2.5. But this does not guarantee that any given line $q \cdot x^\alpha = \bar{p} \cdot \hat{y}^\alpha$ is not completely dominated by any set of other lines, $q \cdot x^{\alpha'} = \bar{p} \cdot \hat{y}^{\alpha'}$, while $x^{\alpha'}$ moves over H. However, the latter condition is satisfied for the following reason.

Given two points x^i and x^j on H, choose any third point x^k between x^i and x^j. In this case, $x_1^i < x_1^k < x_1^j$ (and hence, $x_2^i > x_2^k > x_2^j$) as shown in Figure 2.6. If we denote by $(q_1^{\alpha\alpha'}, q_2^{\alpha\alpha'})$ the intersection between the lines $\bar{p} \cdot \hat{y}^\alpha = q \cdot x^\alpha$ and $\bar{p} \cdot \hat{y}^{\alpha'} = q \cdot x^{\alpha'}$ $(\alpha, \alpha' = i, j$ or $k)$, then, by simple calculations, we get

$$q_1^{ik} - q_1^{kj} = (x_2^i - x_2^j)x_2^k[z(x^k) - (\beta z(x^i) + (1 - \beta)z(x^j))]/\gamma,$$

$$q_2^{ik} - q_2^{kj} = (x_1^i - x_1^j)x_1^k[z(x^k) - (\beta z(x^i) + (1 - \beta)z(x^j))]/\gamma,$$

where

$$\gamma = (x_1^k x_2^i - x_1^i x_2^k)(x_1^j x_2^k - x_1^k x_2^j) > 0, \quad \beta = \frac{x_1^k - x_1^j}{x_1^i - x_1^j} = \frac{x_2^k - x_2^j}{x_2^i - x_2^j} > 0.$$

Therefore, since we see in Figure 2.6 that $z(x^k) \geqq (\beta z(x^i) + (1 - \beta)z(x^j))$, we conclude that

$$x_1^i < x_1^k < x_1^j \text{ (i.e., } x_2^i > x_2^k > x_2^j) \quad \text{implies that}$$

$$q_1^{ik} \geqq q_1^{kj} \quad \text{and} \quad q_2^{ik} \leqq q_2^{kj}. \tag{24}$$

This relation is depicted in Figure 2.7.

One can easily see in Figure 2.5 that when conditions (23) and (24) are satisfied, any line $\bar{p} \cdot \hat{y}^\alpha = q \cdot x^\alpha$ is not completely dominated by

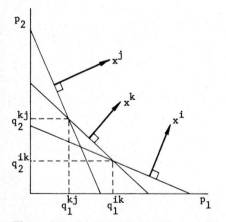

Figure 2.7. General relationship in price space.

any set of other lines, $\bar{p} \cdot \hat{y}^{\alpha'} = q \cdot x^{\alpha'}$, while $x^{\alpha'}$ moves over H. Hence (20) is true.

From the above discussion, it is apparent that the sole source of duality between T and T^* are conditions (i) and (ii) in the definition of m.t.cc. For example, if T does not satisfy condition (i), function $z(k)$ is not concave. In this case, one or both of (23) and (24) would not be satisfied, and hence we would not have the dual property.

2.3. Duality and related theorems in final state problems

2.3.1. Notation and problem formulation

In Part 4 of Rockafellar (1967), the duality theorem and the reciprocity principle in monotone programming are examined for the case of single period programs. In this section, these theorems are extended to multi-period monotone programming with objective functions involving only the final state. The existence theorem and the max-min principle are developed. Results of this section constitute the base of more general multi-period monotone programming.

The state of the system at the beginning of period t is denoted by x_t which is an m-dimensional vector in $\mathbf{R}_+^m$. Each component, x_i^t, may be considered as the stock of the ith commodity at that time ($i = 1, 2, \ldots, i, \ldots, m$). The set of all feasible states of the system at the beginning of the next period $t + 1$ is denoted by the set $T_t(x_t)$ which is a subset of $\mathbf{R}_+^m$.[10] Hence, (x_t, x_{t+1}) is a *feasible activity* in period t if and only if $x_{t+1} \in T_t(x_t)$, that is, if and only if $(x_t, x_{t+1}) \in$ graph T_t.

Throughout this section it is assumed that T_t is a non-singular m.t.cc. from $\mathbf{R}_+^m$ into $\mathbf{R}_+^m$ ($t = 0, 1, \ldots, N - 1$). This condition is assumed because if T_t is singular most of the necessary conditions of optimality expressed in terms of auxiliary variables give little information. Recall that a singular transformation can be made non-singular just by reducing the number of output commodities appropriately.

From (5), the adjoint transformation T_t^* for each T_t is defined as

---

[10]In this chapter we assume that the number of commodities does not change over time. This assumption is just for simplicity of notation, and one should notice that all the results in this chapter hold to be true with appropriate notational modifications even when the number of commodities changes over time.

follows:

$$T_t^*(p_{t+1}) = \{p_t \in \mathbf{R}_+^{*m} | p_t \cdot x_t \geqq p_{t+1} \cdot x_{t+1} \quad \text{for all } (x_t, x_{t+1}) \in \text{graph } T_t\},$$
(25)

for each $p_{t+1} \in \mathbf{R}_+^{*m}$ $(t = 0, 1, \ldots, N-1)$. When T_t is non-singular m.t.cc., T_t^* is a non-singular m.t.cv. from $\mathbf{R}_+^{*m}$ into $\mathbf{R}_+^{*m}$.

Using these definitions and notation, the pair of problems which are analyzed in this section are stated as follows.

Problem I. Given a monotone concave gauge f on $\mathbf{R}_+^m$, a non-singular m.t.cc. T_t $(t = 0, 1, \ldots, N-1)$ from $\mathbf{R}_+^m$ into $\mathbf{R}_+^m$ and a monotone set C of concave type in $\mathbf{R}_+^m$:

I(a)   Maximize $f(x_N)$,

   subject to $x_{t+1} \in T_t(x_t)$, $\quad t = 0, 1, \ldots, N-1$,

   and $x_0 \in C$.

I(b)   Minimize $g(p_0) \equiv \langle p_0, C \rangle$,

   subject to $p_t \in T_t^*(p_{t+1})$, $\quad t = 0, 1, \ldots, N-1$,

   and $p_N \in D^*$.

Here T_t^* is the adjoint transformation of T_t $(t = 0, 1, \ldots, N-1)$ and D^* is the monotone set of convex type in $\mathbf{R}_+^{*m}$ corresponding to f by (9) (i.e., $f(x_N) = \langle D^*, x_N \rangle$).

On the other hand, for a given vector $\bar{p}_N \in \mathbf{R}_+^{*m}$, $f(x_N) = \bar{p}_N \cdot x_N$ is a monotone concave gauge on $\mathbf{R}_+^m$, and the corresponding monotone set is given by $D^* = \{p \in \mathbf{R}_+^{*m} | p \geqq \bar{p}_N\}$ from (9). In addition, when the initial position is given by a point $\bar{x}_0 \in \mathbf{R}_+^m$, the set $C = \{x | 0 \leqq x \leqq \bar{x}_0\}$ is a monotone set of concave type in $\mathbf{R}_+^m$, and $\langle p_0, C \rangle = p_0 \cdot \bar{x}_0$. Then from condition (v) (resp. (v*)) of the definition of m.t.cc. (m.t.cv.), $T_0(C) = T_0(\bar{x}_0)$ and $T_{N-1}^*(D^*) = T_{N-1}^*(\bar{p}_N)$. Hence, the above problem includes the next problem as a special case.

Problem I'. Given a final price vector $\bar{p}_N \in \mathbf{R}_+^{*m}$, an initial point $\bar{x}_0 \in \mathbf{R}_+^m$ and a non-singular m.t.cc. T_t $(t = 0, 1, \ldots, N-1)$ from $\mathbf{R}_+^m$ into $\mathbf{R}_+^m$:

I'(a)   Maximize $\bar{p}_N \cdot x_N$,

   subject to $x_{t+1} \in T_t(x_t)$, $\quad t = 0, 1, \ldots, N-1$,

   and $x_0 = \bar{x}_0$.

I'(b)   Minimize $p_0 \cdot \bar{x}_0$,

   subject to $p_t \in T_t^*(p_{t+1})$, $t = 0, 1, \ldots, N - 1$,

   and $p_N = \bar{p}_N$.

Here T_t^* is the adjoint transformation of T_t $(t = 0, 1, \ldots, N - 1)$.

To simplify the representation of the problem, we introduce the ordinary definition of the product of transformations. That is, given two monotone transformations of the same type, T and S, from $\mathbf{R}_+^m$ to $\mathbf{R}_+^n$ and from $\mathbf{R}_+^n$ to $\mathbf{R}_+^r$, respectively, the *binary multiplication* $S \cdot T$ is defined by

$$(S \cdot T)(x) = S(T(x)) = \cup\{S(y) | y \in T(x)\} \quad \text{(refer to (14)).} \tag{26}$$

Then, when each of T and S is non-singular, from B-property 23, $S \cdot T$ is also a non-singular monotone transformation of the same type with T (and S) from $\mathbf{R}_+^m$ into $\mathbf{R}_+^r$; and the adjoint transformation $(S \cdot T)^*$ of $S \cdot T$ is given by

$$(S \cdot T)^* = T^* \cdot S^* \tag{27}$$

where T^* is the adjoint transformation of T and S^* is the adjoint transformation of S. Note that binary multiplication is the same as the ordinary definition of composite product except that we define it only for the same type of monotone transformation. Binary multiplication is also associative.

Hence, in the context of Problem I, we set

$$\begin{aligned} A_t &= T_{t-1} \cdot T_{t-2} \cdot \ldots \cdot T_1 \cdot T_0, & t = 1, 2, \ldots, N, \\ B_t &= T_{N-1} \cdot T_{N-2} \cdot \ldots \cdot T_{t+1} \cdot T_t, & t = 0, 1, \ldots, N - 1. \end{aligned} \tag{28a}^{11}$$

Then, A_t is a non-singular m.t.cc. from $\mathbf{R}_+^m$ in period 0 into $\mathbf{R}_+^m$ in period t, and B_t is a non-singular m.t.cc. from $\mathbf{R}_+^m$ in period t into $\mathbf{R}_+^m$ in period N. Hence, by using (27), we represent the adjoint trans-

---

[11]Since binary multiplication is associative, for example we have

$$T_2(T_1(T_0(x))) = T_2((T_1 \cdot T_0)(x)) = (T_2 \cdot T_1)(T_0(x)).$$

Therefore we represent all of them commonly by $(T_2 \cdot T_1 \cdot T_0)(x)$, and denote it simply by $A_3(x)$. In general,

$$T_{t-1}(T_{t-2}(\ldots(T_1(T_0(x)))\ldots)) = \cdots = (T_{t-1} \cdot T_{t-2})(T_{t-3}(\ldots(T_1(T_0(x)))\ldots)),$$

and hence all of them are commonly represented by $(T_{t-1} \cdot T_{t-2} \cdot \cdots \cdot T_1 \cdot T_0)(x)$, which we denote by $A_t(x)$.

formation A_t^* of A_t and the adjoint transformation B_t^* of B_t, respectively, by

$$A_t^* = T_0^* \cdot T_1^* \cdot \ldots \cdot T_{t-2}^* \cdot T_{t-1}^*, \qquad t = 1, 2, \ldots, N,$$
$$B_t^* = T_t^* \cdot T_{t+1}^* \cdot \ldots \cdot T_{N-2}^* \cdot T_{N-1}^*, \quad t = 0, 1, \ldots, N - 1. \qquad (28b)$$

Then, A_t^* is a non-singular m.t.cv. from $\mathbf{R}_+^{*m}$ in period t into $\mathbf{R}_+^{*m}$ in period 0, and B_t^* is a non-singular m.t.cv. from $\mathbf{R}_+^{*m}$ in period N into $\mathbf{R}_+^{*m}$ in period t.

In addition to (28a) and (28b), for convenience we define the following *identity transformations*.

$$A_0(x_0) = x_0 \text{ for each } x_0 \in \mathbf{R}_+^m, \qquad B_N(x_N) = x_N \text{ for each } x_N \in \mathbf{R}_+^m,$$
$$A_0^*(p_0) = p_0 \text{ for each } p_0 \in \mathbf{R}_+^{*m}, \qquad B_N^*(p_N) = p_N \text{ for each } p_N \in \mathbf{R}_+^{*m}. \qquad (28c)$$

According to the above definitions, we have the following relations.

$$B_t \cdot A_t = A_N = B_0, \qquad A_t^* \cdot B_t^* = A_N^* = B_0^*, \quad t = 0, 1, \ldots, N. \qquad (29)$$

Note also that according to (28a),

$$A_t(x_0) = \{x_t | x_t \in T_{t-1}(x_{t-1}), x_{t-1} \in T_{t-2}(x_{t-2}), \ldots, x_1 \in T_0(x_0),$$
$$\text{for some } (x_1, x_2, \ldots, x_{t-2}, x_{t-1}) \in (\mathbf{R}_+^m)^{t-1}\}, \qquad (30)$$

for each $x_0 \in \mathbf{R}_+^m$. Hence $A_t(x_0)$ gives the set of all output vectors at the beginning of period t which are feasible in t periods starting from the initial vector x_0. Similarly according to (28b),

$$B_t^*(p_N) = \{p_t | p_t \in T_t^*(p_{t+1}), p_{t+1} \in T_{t+1}^*(p_{t+2}), \ldots, p_{N-1} \in T_{N-1}^*(p_N)$$
$$\text{for some } (p_{t+1}, p_{t+2}, \ldots, p_{N-1}) \in (\mathbf{R}_+^{*m})^{N-t+1}\}, \qquad (31)$$

for each $p_N \in \mathbf{R}_+^{*m}$.

Throughout this book, a sequence of vectors is called a *feasible path* for a certain problem when it satisfies all the restrictions on that problem. And, if a feasible path for a problem maximizes (or, minimizes) the objective function among all feasible paths for that problem, it is called an *optimal path* for that problem. Then, from (30), one easily sees that if a sequence of vectors $\{x_t\}_0^N$ is a feasible path for I(a), then $x_N \in A_N(C)$. Conversely, if a vector x_N is in $A_N(C)$, from (30) there exists a feasible path $\{x_t'\}_0^N$ for I(a) such that $x_N' = x_N$. Similarly, from (31) if $\{p_t\}_0^N$ is a feasible path for I(b), then $p_0 \in B_0^*(D^*)$. Conversely, if a point p_0 is in $B_0^*(D^*)$, there exists a feasible path $\{p_t'\}_0^N$ for I(b) such that $p_0' = p_0$. Hence, by using Definition (28), Problems I and I' can simply be rewritten as follows:

I(a) Maximize $f(x_N)$, subject to $x_N \in A_N(C)$.

I(b) Minimize $g(p_0) = \langle p_0, C \rangle$, subject to $p_0 \in B_0^*(D^*)$.

$$(32)$$

I'(a) Maximize $\bar{p}_N \cdot x_N$, subject to $x_N \in A_N(\bar{x}_0)$.

I'(b) Minimize $p_0 \cdot \bar{x}_0$, subject to $p_0 \in B_0^*(\bar{p}_N)$.

$$(33)$$

The economic meanings of I(a) and I'(a) are clear, while the economic meanings of I(b) and I'(b) become apparent if we use the fact that B_0^* is the adjoint transformation of A_N. Namely,

$$B_0^*(\bar{p}_N) = \{p_0 | p_0 \cdot x_0 \geqq \bar{p}_N \cdot x_N \quad \text{for all } (x_0, x_N) \in \text{graph } A_N\}$$

and

$$B_0^*(D^*) = \bigcup_{p_N \in D^*} \{p_0 | p_0 \cdot x_0 \geqq p_N \cdot x_N \quad \text{for all } (x_0, x_N) \in \text{graph } A_N\}$$

Hence, I'(b) says that given a final price vector $\bar{p}_N$, find the price vector p_0 for the initial stock of commodities $\bar{x}_0$ that minimizes the value $p_0 \cdot \bar{x}_0$ among the set of initial price vectors $B_0^*(\bar{p}_N)$, each of which does not allow positive profits for any feasible activity of length N. I(b) also has a similar interpretation.

Finally we define transformation $A_{\tau t}$ by

$$A_{\tau t}(x_\tau) = (T_{t-1} \cdot T_{t-2} \cdot \ldots \cdot T_{\tau+1} \cdot T_\tau)(x_\tau), \qquad (34a)$$

for each $x_\tau \in \mathbf{R}_+^m$ and $N \geqq t > \tau \geqq 0$.

Then in the context of Problem I, $A_{\tau t}$ is a non-singular m.t.cc. from $\mathbf{R}_+^m$ into $\mathbf{R}_+^m$, and its adjoint transformation $A_{\tau t}^*$ is given by

$$A_{\tau t}^*(p_t) = (T_\tau^* \cdot T_{\tau+1}^* \cdot \ldots \cdot T_{t-2}^* \cdot T_{t-1}^*)(p_t). \qquad (34b)$$

The meanings of each set $A_{\tau t}(x_\tau)$ and $A_{\tau t}^*(p_t)$ are clear. In particular, when $\tau = t - 1$, $A_{\tau t}(x_\tau) = T_{t-1}(x_{t-1})$ and $A_{\tau t}^*(p_t) = T_{t-1}^*(p_t)$. For convenience, we define

$$A_{00}(x_0) = x_0 \quad \text{for each } x_0 \in \mathbf{R}_+^m,$$

$$A_{NN}^*(p_N) = p_N \quad \text{for each } p_N \in \mathbf{R}_+^{*m}. \qquad (34c)$$

Then, from Definitions (28) and (34), we have

$$A_{0t} = A_t \quad \text{for } t = 0, 1, \ldots, N, \qquad A_{tN}^* = B_t^* \quad \text{for } t = 0, 1, \ldots, N.$$

$$(35)$$

2.3.2. Duality theorem

In this section, the part of the theorems in Rockafellar (1967) which relate to duality, is extended to multi-period monotone programming. First, we need the next lemma.

Lemma 2.1. In the context of Problem I, $\langle D^*, A_N(C)\rangle = \langle B^*_{N-1}(D^*), A_{N-1}(C)\rangle = \cdots = \langle B^*_t(D^*), A_t(C)\rangle = \cdots = \langle B^*_1(D^*), A_1(C)\rangle = \langle B^*_0(D^*), C\rangle$. That is $\langle B^*_t(D^*), A_t(C)\rangle$ is constant for $t = 0, 1, \ldots, N$.

Proof. From Definition (28), $\langle B^*_t(D^*), A_t(C)\rangle = \langle T^*_t(B^*_{t+1}(D^*)), T_{t-1}(A_{t-1}(C))\rangle$ for $t = 1, 2, \ldots, N - 1$. On the other hand, in the context of Problem I, $B^*_{t+1}(D^*)$ (resp. $A_{t-1}(C)$) is a monotone set of convex type (resp. concave type). Hence from property (15), $\langle B^*_{t+1}(D^*), T_t(T_{t-1}(A_{t-1}(C)))\rangle = \langle T^*_t(B^*_{t+1}(D^*)), T_{t-1}(A_{t-1}(C))\rangle = \langle T^*_{t-1}(T^*_t(B^*_{t+1}(D^*))), A_{t-1}(C)\rangle$. Hence by Definition (28), $\langle B^*_{t+1}(D^*), A_{t+1}(C)\rangle = \langle B^*_t(D^*), A_t(C)\rangle = \langle B^*_{t-1}(D^*), A_{t-1}(C)\rangle$ for $t = 1, 2, \ldots, N - 1$. Q.E.D.

By using the above lemma, the next two theorems are obtained. Though Theorem 2.1′ is a special case of Theorem 2.1, it is stated separately because of its simplicity and importance.

Theorem 2.1 (duality theorem). Suppose $\{\hat{x}_t\}_0^N$ is an optimal path for I(a) and $\{\hat{p}_t\}_0^N$ is an optimal path for I(b). Then we have:

(i) $\mu = f(\hat{x}_N) = \hat{p}_N \cdot \hat{x}_N = \cdots = \hat{p}_t \cdot \hat{x}_t = \cdots = \hat{p}_0 \cdot \hat{x}_0 = g(\hat{p}_0)$,
      where μ is a non-negative real number.

(ii′) Max $\{\hat{p}_t \cdot x_t | x_t \in A_t(C)\} = \hat{p}_t \cdot \hat{x}_t = $ Min $\{p_t \cdot \hat{x}_t | p_t \in B^*_t(D^*)\}$,
      $t = 0, 1, \ldots, N$.

(ii″) Max $\{\hat{p}_t \cdot x_t | x_t \in A_{\tau t}(\hat{x}_\tau)\} = \hat{p}_t \cdot \hat{x}_t = $ Min $\{p_t \cdot \hat{x}_t | p_t \in A^*_{\tau't}(\hat{p}_{\tau'})\}$
      for any $N \geq \tau' > t > \tau \geq 0$.

(iii) If $f(x_N) \neq 0$, then $\hat{p}_t \geq 0$ for $t = 0, 1, \ldots, N$.[12]

Conversely, if a pair of feasible paths $\{\hat{x}_t\}_0^N$ for I(a) and $\{\hat{p}_t\}_0^N$ for I(b) satisfy the condition, $f(\hat{x}_N) = g(\hat{p}_0)$, then they are optimal paths for I(a) and I(b), respectively.

---

[12]Given two appropriate vectors $x = (x_i)$ and $x' = (x'_i)$ in this book, $x > x'$, $x \geq x'$ and $x' \geq x'$ mean, respectively, $x_i > x'_i$ for all i, $x_i \geq x'_i$ for all i, and $x_i > x'_i$ for at least one i and $x_i \geq x'_i$ for all i.

Theorem 2.1′ (duality theorem). Suppose $\{\hat{x}_t\}_0^N$ is an optimal path for I′(a) and $\{\hat{p}_t\}_0^N$ is an optimal path for I′(b). Then we have:

(i) $\mu = \hat{p}_t \cdot \hat{x}_N = \cdots = \hat{p}_t \cdot \hat{x}_t = \cdots = \hat{p}_0 \cdot \hat{x}_0$,
  where μ is a non-negative real number.

(ii) Max $\{\hat{p}_t \cdot x_t | x_t \in A_{\tau t}(\hat{x}_\tau)\} = \hat{p}_t \cdot \hat{x}_t = $ Min $\{p_t \cdot \hat{x}_t | p_t \in A_{t\tau}^*(\hat{p}_\tau)\}$
  for any $N \geq \tau' > t > \tau \geq 0$.

(iii) If $\bar{p}_N \geq 0$, then $\hat{p}_t \geq 0$ for $t = 0, 1, \ldots, N - 1$.

Conversely, if a pair of feasible paths $\{\hat{x}_t\}_0^N$ for I′(a) and $\{\hat{p}_t\}_0^N$ for I′(b) satisfy the condition, $\bar{p}_N \cdot \hat{x}_N = \hat{p}_0 \cdot \bar{x}_0$, then they are optimal paths for I′(a) and I′(b), respectively.

Proof. Since Theorem 2.1′ is a special case of Theorem 2.1, it is sufficient to prove Theorem 2.1.

(i) From the assumptions of Problem I, $f(\hat{x}_N) = \langle D^*, \hat{x}_N \rangle$ and $g(\hat{p}_0) = \langle \hat{p}_0, C \rangle$. On the other hand, $A_N(C)$ (resp. $B_0^*(D^*)$) is a monotone set of concave type (convex type), and hence from (13) and (30) and by the definition of optimal paths, $f(\hat{x}_N) = \langle D^*, A_N(C) \rangle$ and $g(\hat{p}_0) = \langle B_0^*(D^*), C \rangle$. Thus, from Lemma 2.1, we get

$$\langle D^*, \hat{x}_N \rangle = f(\hat{x}_N) = \langle D^*, A_N(C) \rangle = \langle B_t^*(D^*), A_t(C) \rangle$$
$$= \langle B_0^*(D^*), C \rangle = g(\hat{p}_0) = \langle \hat{p}_0, C \rangle, \quad t = 0, 1, \ldots, N. \quad (36)$$

Hence, $\langle D^*, \hat{x}_N \rangle \geq \langle B_t^*(D^*), \hat{x}_t \rangle$, $t = 0, 1, \ldots, N$, since $\hat{x}_t \in A_t(C)$ and $A_t(C)$ is a monotone set of concave type.

Suppose $\langle D^*, \hat{x}_N \rangle > \langle B_t^*(D^*), \hat{x}_t \rangle$. From (35) and (12), $(B_t^*)^* = (A_{tN}^*)^* = A_{tN}$, and thus from property (15), $\langle B_t^*(D^*), \hat{x}_t \rangle = \langle D^*, A_{tN}(\hat{x}_t) \rangle$. Hence, we get $\langle D^*, \hat{x}_N \rangle > \langle D^*, A_{tN}(\hat{x}_t) \rangle$. But this is a contradiction since $\hat{x}_N \in A_{tN}(\hat{x}_t)$ and since $A_{tN}(\hat{x}_t)$ is a monotone set of concave type. Therefore, it should be true that

$$\langle D^*, \hat{x}_N \rangle = \langle B_t^*(D^*), \hat{x}_t \rangle, \quad t = 0, 1, \ldots, N.$$

Next, from Definition (13), $\langle \hat{p}_t, A_t(C) \rangle \geq \langle B_t^*(D^*), A_t(C) \rangle$, $t = 0, 1, \ldots, N$, since $\hat{p}_t \in B_t^*(D^*)$ and $B_t^*(D^*)$ is a monotone set of convex type. Thus from (36), $\langle \hat{p}_t, A_t(C) \rangle \geq \langle \hat{p}_0, C \rangle$. Suppose $\langle \hat{p}_t, A_t(C) \rangle > \langle \hat{p}_0, C \rangle$. Again from (15) and (35), we obtain $\langle \hat{p}_t, A_t(C) \rangle = \langle A_t^*(\hat{p}_t), C \rangle = \langle A_{0t}^*(\hat{p}_t), C \rangle$. Thus $\langle A_{0t}^*(\hat{p}_t), C \rangle > \langle \hat{p}_0, C \rangle$. This is a contradiction since $\hat{p}_0 \in A_{0t}^*(\hat{p}_t)$ and $A_{0t}^*(\hat{p}_t)$ is a monotone set of convex type. Hence, it should be true that

$$\langle \hat{p}_0, C \rangle = \langle \hat{p}_t, A_t(C) \rangle, \quad t = 0, 1, \ldots, N.$$

Finally, from Definitions (6) and (7)

$$\left.\begin{array}{ll}\hat{p}_t \cdot \hat{x}_t \geqq \langle B_t^*(D^*), \hat{x}_t \rangle & \text{since } \hat{p}_t \in B_t^*(D^*) \\ \hat{p}_t \cdot \hat{x}_t \leqq \langle \hat{p}_t, A_t(C) \rangle & \text{since } \hat{x}_t \in A_t(C) \end{array}\right\} \quad t = 0, 1, \ldots, N.$$

Combining the above four results, we have

$$f(\hat{x}_N) = \langle \hat{p}_t, A_t(C) \rangle = \hat{p}_t \cdot \hat{x}_t = \langle B_t^*(D^*), \hat{x}_t \rangle = g(\hat{p}_0) = \mu,$$
$$t = 0, 1, \ldots, N, \qquad (37)$$

where μ is a non-negative real number since $g(\hat{p}_0) = \langle \hat{p}_0, C \rangle$ and C is a compact set in $\mathbf{R}_+^m$.

(ii) Since $\hat{x}_t \in A_t(C)$ and $\hat{p}_t \in B_t^*(D^*)$, (ii′) comes immediately from (37). In addition, since optimal paths are feasible paths, $\hat{x}_t \in A_{\tau t}(\hat{x}_\tau) \subset A_t(C)$ and $\hat{p}_t \in A_{t\tau}^*(\hat{p}_\tau) \subset B_t^*(D^*)$ from Definitions (28) and (34). Hence (ii″) is also true from (ii′).

(iii) Suppose $\hat{p}_t = 0$ for some $t = 0, 1, \ldots, N-1$. By the definition of optimal paths for I(b), $\hat{p}_t \in T_t^*(\hat{p}_{t+1})$, $t = 0, 1, \ldots, N-1$ and $\hat{p}_N \in D^*$. Hence from Definition (25),

$$\hat{p}_t \cdot x_t \geqq \hat{p}_{t+1} \cdot x_{t+1} \quad \text{for all } (x_t, x_{t+1}) \in \text{graph } T_t.$$

On the other hand, from the assumption of the non-singularity of T_t, there exists $(x_t, x_{t+1}) \in \text{graph } T_t$ with $x_{t+1} > 0$. Therefore, $\hat{p}_t = 0$ implies $\hat{p}_{t+1} = 0$. Then $\hat{p}_{t+2} = 0$ by the same reason. Similarly, $\hat{p}_t = 0$ implies $\hat{p}_N = 0$, which implies $0 \in D^*$. Then $f(x_N) = \langle D^*, x_N \rangle = 0$ for any $x_N \in \mathbf{R}_+^m$. This contradicts the condition $f(x_N) \not\equiv 0$. Thus (iii) must be true.

Finally, from Definition (13) of operation $\langle \cdot, \cdot \rangle$, and since $f(x_N) = \langle D^*, x_N \rangle$ and $f^*(p_0) = \langle p_0, C \rangle$,

$$\langle D^*, A_N(C) \rangle \geqq \langle D^*, \hat{x}_N \rangle = f(\hat{x}_N) \quad \text{for any } \hat{x}_N \in A_N(C),$$

$$\langle A_0^*(D^*), C \rangle \leqq \langle \hat{p}_0, C \rangle = g(\hat{p}_0) \quad \text{for any } \hat{p}_0 \in B_0^*(D^*).$$

In addition, from Lemma 2.1, $\langle D^*, A_N(C) \rangle = \langle A_0^*(D^*), C \rangle$, and it is assumed that $f(\hat{x}_N) = g(\hat{p}_0)$. Therefore, combining these relations, we get

$$f(\hat{x}_N) = \langle D^*, A_N(C) \rangle = \langle B_0^*(D^*), C \rangle = g(\hat{p}_0).$$

This implies, since $\hat{x}_N \in A_N(C)$ and $\hat{p}_0 \in B_0^*(D^*)$, that

$$f(\hat{x}_N) = \max \{f(x_N) | x_N \in A_N(C)\}, \quad g(\hat{p}_0) = \min \{g(p_0) | p_0 \in B_0^*(D^*)\}.$$

$$\text{Q.E.D.}$$

The most important characteristic of the above duality theorem is that each dual problem does not include any variables from its primal problem. Therefore, the primal problem and dual problem can be solved independently of each other; and this duality theorem states the relation of the solutions for two independent programs.

To date, only linear programming was known to possess this property. For example, the dual problem in any nonlinear programming includes variables from both the primal problem and the dual problem, and hence the duality theorem in nonlinear programming does not state the relation of two independent programs. Similarly, the maximum principle in optimal control theory is not a statement of the relation between two independent programs.

The value $\hat{p}_{it}$ in the above theorems has, as usual, the interpretation of the (efficiency, or imputed) price of commodity i in period t. Though the exact meaning is given by (40) in Section 2.3.3, roughly speaking, $\hat{p}_{it}$ measures how much the maximum possible value of the objective function $f(x_N)$ (or, $\bar{p}_N \cdot x_N$) can be increased if a unit of commodity i is increased exogenously in period t on the optimal path. With this interpretation of $\hat{p}_{it}$, (i) in the above theorems may be considered as the proof of the following statement, which is suggested in Dorfman, Samuelson and Solow (1958, p. 332).

> Consider any efficient capital program and its corresponding profile of prices and own rates. At every point of time the value of the capital stock at current efficiency prices, discounted back to the initial time, is a constant, equal to the initial value. This law of conservation of discounted value of capital (or discounted Net National Product) reflects, as do the grand laws of conservation of energy of physics, the maximizing nature of the path.

But, it must be noted that the constancy of the value, $\hat{p}_t \cdot \hat{x}_t$, is true only when T_t ($t = 0, 1, \ldots, N-1$) is linearly homogeneous as will be shown by Theorem 2.14.

Interpretations of (ii) in the above theorems will be given after Theorem 2.4. Condition (iii) says that price vector p_t is semi-positive for each t ($t = 0, 1, \ldots, N-1$) in any non-trivial problem. On the other hand, (i) and the last statement in Theorem 2.1 together say that

Corollary 2.1. A pair of feasible paths $\{\hat{x}_t\}_0^N$ for I(a) and $\{\hat{p}_t\}_0^N$ for I(b) are optimal paths for I(a) and I(b), respectively, if and only if $f(\hat{x}_N) = g(\hat{p}_0)$.

This can be restated as follows:

Corollary 2.2.   A pair of feasible paths $\{\hat{x}_t\}_0^N$ for I(a) and $\{\hat{p}_t\}_0^N$ for I(b) are optimal paths for I(a) and I(b), respectively, if and only if $\hat{p}_0 \cdot \hat{x}_0 = \hat{p}_N \cdot \hat{x}_N$, and the hyperplane $\hat{p}_0 \cdot x = \hat{p}_0 \cdot \hat{x}_0$ supports the set C at $\hat{x}_0$ (i.e., $\hat{p}_0 \cdot x \leqq \hat{p}_0 \cdot \hat{x}_0$ for every $x \in C$), and the hyperplane $p \cdot \hat{x}_N = \hat{p}_N \cdot \hat{x}_N$ supports the set D^* at p_N (i.e., $p \cdot \hat{x}_N \geqq \hat{p}_N \cdot \hat{x}_N$ for every $p \in D^*$).

For the proof, see Appendix C.1. With respect to Problem I', the above corollary is simply rewritten as follows:

Corollary 2.1'.   A pair of feasible paths $\{\hat{x}_t\}_0^N$ for I'(a) and $\{\hat{p}_t\}_0^N$ for I'(b) are optimal paths for I'(a) and I'(b), respectively, if and only if $\hat{p}_0 \cdot \bar{x}_0 = \bar{p}_N \cdot \hat{x}_N$.

2.3.3. Existence theorem

Though the above theorems are interesting, they report nothing about the existence of optimal paths. And, without the guarantee of the existence of optimal paths, the usefulness of these theorems is limited. Hence we now turn to an examination of the existence of such paths.

The existence of optimal paths for I(a) is easily demonstrated since I(a) is concerned with the maximization of a continuous function over a compact set $A_N(C)$. But unfortunately I(b) does not always have an optimal path. First, the necessary and sufficient condition for the existence of optimal paths for I(b) is obtained by using the Kuhn–Tucker Theorem from Gale (1967a,b). Then, two different sufficient conditions are given to guarantee the existence of the dual optimal path. One involves restrictions on the objective function and transformations, and the other involves a restriction on the initial set.

Let h be a concave function defined on some convex set X in $\mathbf{R}_+^m$. Then the *steepness* $\sigma(\bar{x})$ of this function at a point $\bar{x} \in X$ is defined by

$$\sigma(\bar{x}) = \sup\{(h(x) - h(\bar{x}))/\|x - \bar{x}\| \,|\, x \in X\},$$

where $\|x - \bar{x}\|$ may be any convenient norm on $\mathbf{R}^m$.

Next, in the context of Problem I, we define function $\mu_{tN}(x_t)$ on $\mathbf{R}_+^m$ by

$$\mu_{tN}(x_t) = \sup\{f(x_N) \,|\, x_N \in A_{tN}(x_t)\}$$
$$= \langle D^*, A_{tN}(x_t) \rangle, \quad t = 0, 1, \ldots, N. \tag{38a}$$

Similarly, in the context of Problem I',

$$\mu_{tN}(x_t) = \sup\{\bar{p}_N \cdot x_N \,|\, x_N \in A_{tN}(x_t)\}$$
$$= \langle \bar{p}_N, A_{tN}(x_t)\rangle, \quad t = 0, 1, \ldots, N. \tag{38b}$$

Using these definitions, we first obtain the following lemma.

Lemma 2.2. Suppose $\{\hat{x}_t\}_0^N$ is an optimal path for I(a). Then the following three statements are equivalent to each other.

(a) Function μ_{0N} defined by (38a) has a finite steepness at $\hat{x}_0$.
(b) The next *Kuhn–Tucker Theorem* holds at $\hat{x}_0$ for a non-negative vector $\hat{p}_0 \in \boldsymbol{R}_{\ddagger}^{*m}$:

$$\langle D^*, A_N(x_0)\rangle - \langle D^*, A_N(\hat{x}_0)\rangle \leqq \hat{p}_0 \cdot (x_0 - \hat{x}_0) \quad \text{for all } x_0 \in \boldsymbol{R}_+^m. \tag{39}$$

(c) There exists an optimal path for I(b).

For the proof, see Appendix C.2. From this lemma, the next theorem immediately follows, and it gives the necessary and sufficient condition for the existence of an optimal path for I(b) and I'(b), respectively.

Theorem 2.2 (Existence theorem: the necessary and sufficient condition)

 (i) There always exists an optimal path for I(a) (hence, also for I'(a)).
(ii') There exists an optimal path for I'(b) if and only if function μ_{0N} defined by (38b) has a finite steepness at $\bar{x}_0$.
(ii") There exists an optimal path for I(b) if and only if there exists an optimal path $\{\hat{x}_t\}_0^N$ for I(a) such that function μ_{0N} defined by (38a) has a finite steepness at $\hat{x}_0$. (Note that if μ_{0N} has a finite steepness for a particular optimal path for I(a), then from Lemma 2.2 it is true for any optimal path for I(a).)

Proof. (i) In the context of Problem I(a) (resp. I'(a)), $A_N(C)$ (resp. $A_N(\bar{x})$) is a monotone set of concave type and hence it is compact. Then, since I(a) (resp. I'(a)) involves the maximization of a continuous function $f(x_N)$ (resp. $\bar{p}_N \cdot x_N$) on a compact set, there always exists a solution.

(ii) Since (ii') is a special case of (ii"), it is sufficient to give a proof for (ii"). We see that (ii") comes directly from the equivalence of (a) and (c) in Lemma 2.2. Note that $\{\hat{x}_t\}_0^N$ in Lemma 2.2 is an arbitrary optimal path for I(a). Then, again from the equivalence between (a) and (c), if μ_{0N} has a finite steepness for a particular optimal path for I(a), it must be true for any optimal path in Problem I(a).   Q.E.D.

It is convenient to have some sufficient conditions for the existence of the optimal path for I(b) which are easily identifiable. For this purpose, we introduce the *Hausdolf metric* to define the distance $\|X - Y\|$ between non-empty closed sets X and Y in $\mathbf{R}^m_+$, defined by

$$\|X - Y\| = \max\{\sup_{x \in X} \inf_{y \in Y} \|x - y\|, \sup_{y \in Y} \inf_{x \in X} \|x - y\|\},$$

where $\|x - y\|$ may be any convenient norm on $\mathbf{R}^m$. Then, given any m.t.cc. T from $\mathbf{R}^m_+$ into $\mathbf{R}^m_+$, we say that T satisfies the *Lipschitz condition* if there is a constant k such that

$$\|T(x^1) - T(x^2)\| \leq k\|x^1 - x^2\| \quad \text{for all } x^1, x^2 \in \mathbf{R}^m_+.$$

Intuitively speaking, satisfaction of the Lipschitz condition by T means that the marginal productivity of each input, if defined, is always finite.

First, we obtain the next lemma.

Lemma 2.3. In the context of Problem I, if each T_t $(t = 0, 1, \ldots, N - 1)$ satisfies the Lipschitz condition, then A_t also satisfies this condition for any $t = 0, 1, \ldots, N - 1$.

For the proof, see Appendix C.3. By using this lemma, the next theorem is obtained.

Theorem 2.3 (Existence theorem: sufficient conditions). There exists an optimal path for Problem I(b) under either one of the next two conditions:

(a) The steepness of function f has an upper bound on $\mathbf{R}^m_+$, and each T_t
    $(t = 0, 1, \ldots, N - 1)$ satisfies the Lipschitz condition.
(b) There exists a point $x_0 \in C$ with $x_0 > 0$.

Proof. (a) Following the proof of Lemma 1 in Gale (1967a) (see Appendix C-13), this can be proved as follows. Suppose $\{\hat{x}_t\}_0^N$ is an optimal path for I(a). Take another initial point $x_0' \in \mathbf{R}^m_+$ such that $\mu_{0N}(x_0') > \mu_{0N}(\hat{x}_0)$ (if there is no such x_0', $f(x_N) \equiv 0$ because f is linearly homogeneous, and hence $\mu_{0N}(x_N) \equiv 0$ and $\{\hat{p}_t = 0\}_0^N$ is an optimal path for I(b)). Then, $\mu_{0N}(\hat{x}_0) = f(\hat{x}_N)$ and $\mu_{0N}(x_0') = f(x_N')$ for some $x_N' \in A_N(x_0')$. Further, if $x_N' \in A_N(\hat{x}_0)$ this would contradict $\mu_{0N}(x_0') > \mu_{0N}(\hat{x}_0)$. Let x_N'' be the closest point to x_N' in $A_N(\hat{x}_0)$. Then since each T_t satisfies the Lipschitz condition, from Lemma 2.3 there exists a constant k such that

$$\|x_N' - x_N''\| \leq \|A_N(x_0') - A_N(\hat{x}_0)\| \leq k\|x_0' - \hat{x}_0\|.$$

But if σ is an upper bound on the steepness of f on $\mathbf{R}_+^m$, then

$$f(x_N') - f(\hat{x}_N) \leq f(x_N') - f(x_N'') \leq \sigma\|x_N' - x_N''\| \leq \sigma k\|x_0' - \hat{x}_0\|$$

so $\mu_{0N}(x_0') - \mu_{0N}(\hat{x}_0) \leq \sigma k\|x_0' - \hat{x}_0\|$, and hence the steepness of μ_{0N} at $\hat{x}_0$ is at most σk. Therefore, from Theorem 2.2 (ii'') there exists an optimal path for I(b).

(b) We first show that function μ_{0N} defined by (38a) has a finite steepness at any point $x' > 0$. Recall that μ_{0N} is a concave function on $\mathbf{R}_+^m$. Hence the set

$$\{(\mu_{0N}(x) - \mu(x'), x' - x) | x \in \mathbf{R}_+^m\},$$

is a convex set containing no positive point. Therefore, from Theorem A.9, there is a semipositive vector $(p_0, p) \geq 0$ such that

$$p_0(\mu_{0N}(x) - \mu(x')) + p \cdot (x - x') \leq 0 \quad \text{for all } x \in \mathbf{R}_+^m.$$

Suppose $p_0 = 0$. Then, the above relation implies that $p \cdot (x' - 0) \leq 0$. Thus, since $x' > 0$, $p \leq 0$, and hence $(p_0, p) \leq 0$, which is a contradiction. Therefore, it should be true that $p_0 > 0$. Hence, from Theorem 3 of Gale (1967b), function μ_{0N} has a finite steepness at x'.

Next, suppose $\{\hat{x}_t\}_0^N$ is an optimal path for I(a), which surely exists from (i) of Theorem 2.2. If $\hat{x}_0 > 0$, then from the above result and (ii') of Theorem 2.2 there is an optimal path for I(b). If $\hat{x}_0$ is a boundary point of $\mathbf{R}_+^m$, recalling that $\hat{x}_0 \in C$ and C is a monotone set of concave type and by using the above result, it can be easily shown that function μ_{0N} should have a finite steepness at $\hat{x}_0$. Hence, from (ii') of Theorem 2.2, there is an optimal path for I(b).   Q.E.D.

Recall that any linear function has a finite steepness at each point. Hence, in terms of Problem I'(b), each condition in Theorem 2.3 is restated as follows.

(a') Each T_t $(t = 0, 1, \ldots, N-1)$ satisfies the Lipschitz condition.
(b') $\bar{x}_0 > 0$.

If a transformation T_t consists, for example, of a neoclassical production function (see Chapter 5, Section 5.2.1), then condition (a) may not be satisfied at some boundary points of $\mathbf{R}_+^m$. However, it is not unreasonable to assume that condition (b) (or, (b')) is satisfied since problems are usually highly aggregated in neoclassical production analyses. On the other hand, problems are often highly disaggregated in von Neumann type production models, and hence it would be too restrictive to assume that condition (b) holds. But the von Neumann technology does satisfy the Lipschitz condition. Thus conditions (a) and (b) (or, (a') and (b')) are mutually complementary.

We conclude this section by stating relations between price vectors and partial derivatives. Suppose that $\{\hat{x}_t\}_0^N$ is an optimal path for I(a), and suppose that an optimal path $\{\hat{p}_t\}$ exists for I(b). Let $\partial \mu_{iN}^+(\hat{x}_t)/\partial x_i^t$ and $\partial \mu_{iN}^-(\hat{x}_t)/\partial x_i^t$ denote the partial derivative of μ_{iN} (defined by (38)) with respect to x_i^t at $\hat{x}_t$ from the right and the left respectively. Then, in the context of Problem I;

(a) If $\hat{x}_i^t > 0$, then both partial derivatives exist and

$$\frac{\partial \mu_{iN}^+(\hat{x}_t)}{\partial x_i^t} \leqq \hat{p}_i^t \leqq \frac{\partial \mu_{iN}^-(\hat{x}_t)}{\partial x_i^t}, \quad t = 0, 1, \ldots, N, \tag{40a}$$
$$i = 1, 2, \ldots, m.$$

(b) If $\hat{x}_i^t = 0$, then the partial derivative from the right exists and

$$\frac{\partial \mu_{iN}^+(\hat{x}_t)}{\partial x_i^t} \leqq \hat{p}_i^t, \quad t = 0, 1, \ldots, N, \tag{40b}$$
$$i = 1, 2, \ldots, m.$$

For the proof of the above statements, see the proof of Theorem 5 in Gale (1967b). Because of relations (a) and (b), it is reasonable to call $\hat{p}_{it}$ the (*efficiency*) *price* of commodity i in period t on the optimal path.

2.3.4. *Reciprocity principle*

In this section, the reciprocity principle for single-period programming (Rockafellar, 1967) is extended to the multi-period programming under consideration. Then the character of optimal paths is further clarified.

First, let us define the inverse and polar transformations of a monotone transformation. Given a non-singular monotone transformation S, which may be of concave type or of convex type, from

$\mathbf{R}^m_+$ into $\mathbf{R}^n_+$, the *inverse transformation* S^{-1} of S is defined as follows:

$$S^{-1}(y) = \{x \mid y \in S(x)\} \quad \text{for each } y \in \mathbf{R}^n_+. \tag{41}$$

From B-property 8, S^{-1} is a monotone transformation from $\mathbf{R}^n_+$ into $\mathbf{R}^m_+$ of the opposite type of S and itself non-singular, with

$$(S^{-1})^{-1} = S. \tag{42}$$

The *polar transformation* $(S^{-1})^*$ of a non-singular monotone transformation S from $\mathbf{R}^m_+$ into $\mathbf{R}^n_+$ is defined as the adjoint transformation of S^{-1}. Then, from B-property 9, $(S^{-1})^*$ is a non-singular monotone transformation from $\mathbf{R}^{*m}_{\ddagger}$ into $\mathbf{R}^{*n}_{\ddagger}$ of the same type as S, with

$$(S^{-1})^* = (S^*)^{-1}, \tag{43}$$

and hence hereafter abbreviated to S^{*-1}.

By using these two transformations, we define a pair of problems which are "reciprocal" to Problem I and Problem I′.

Problem I (reciprocal). Given two vectors $\bar{x}_N \in \mathbf{R}^m_+$ and $\bar{p}_0 \in \mathbf{R}^{*m}_{\ddagger}$:

I(c) Minimize $\bar{p}_0 \cdot x_0$,
   subject to $x_t \in T_t^{-1}(x_{t+1}), \quad t = 0, 1, \ldots, N-1,$
   and $x_N = \bar{x}_N$.

I(d) Maximize $p_N \cdot \bar{x}_N$
   subject to $p_{t+1} \in T_t^{*-1}(p_t), \quad t = 0, 1, \ldots, N-1,$
   and $p_0 = \bar{p}_0$.

Here T_t^{-1} and T_t^{*-1} are, respectively, the inverse transformation and the polar transformation of T_t ($t = 0, 1, \ldots, N-1$) in Problem I.

We denote by A_t^{-1} and B_t^{-1} the inverse transformation of A_t and the inverse transformation of B_t, respectively. Then, from B-property 24(c),

$$\begin{aligned}
A_t^{-1} &= T_0^{-1} \cdot T_1^{-1} \cdot \ldots \cdot T_{t-2}^{-1} \cdot T_{t-1}^{-1}, \quad t = 1, 2, \ldots, N \\
B_t^{-1} &= T_t^{-1} \cdot T_{t+1}^{-1} \cdot \ldots \cdot T_{N-2}^{-1} \cdot T_{N-1}^{-1}, \quad t = 0, 1, \ldots, N-1.
\end{aligned} \tag{44a}$$

And, in the context of Problem I, A_t^{-1} (resp. B_t^{-1}) is a non-singular m.t.cv. from $\mathbf{R}^m_+$ in period t (in period N) into $\mathbf{R}^m_+$ in period 0 (in period t). Similarly, inverse transformations A_t^{*-1} and B_t^{*-1} of A_t^* and B_t^* are given by:

$$\begin{aligned}
A_t^{*-1} &= T_{t-1}^{*-1} \cdot T_{t-2}^{*-1} \cdot \ldots \cdot T_1^{*-1} \cdot T_0^{*-1}, \quad t = 1, 2, \ldots, N, \\
B_t^{*-1} &= T_{N-1}^{*-1} \cdot T_{N-2}^{*-1} \cdot \ldots \cdot T_{t+1}^{*-1} \cdot T_t^{*-1}, \quad t = 0, 1, \ldots, N-1.
\end{aligned} \tag{44b}$$

From (43), A_t^{*-1} and B_t^{*-1} are also polar transformations of A_t and B_t, respectively. In the context of Problem I, A_t^{*-1} (resp. B_t^{*-1}) is a non-singular m.t.cc. from $\mathbf{R}_+^{*m}$ in period 0 (in period t) into $\mathbf{R}_+^{*m}$ in period t (in period N). In addition, for convenience we define

$$A_0^{-1}(x_0) = x_0, \qquad B_N^{-1}(x_N) = x_N$$
$$A_0^{*-1}(p_0) = p_0, \qquad B_N^{*-1}(p_N) = p_N. \tag{44c}$$

According to the above definitions,

$$A_t^{-1} \cdot B_t^{-1} = A_N^{-1} = B_0^{-1}, \qquad B_t^{*-1} \cdot A_t^{*-1} = A_N^{*-1} = B_0^{*-1},$$
$$t = 0, 1, \ldots, N. \tag{45}$$

For example, according to (44a),

$$B_t^{-1}(x_N) = \{x_t | x_t \in T_t^{-1}(x_{t+1}), x_{t+1} \in T_{t+1}^{-1}(x_{t+2}), \ldots, x_{N-1} \in T_{N-1}^{-1}(x_N)$$
$$\text{for some } (x_{t+1}, x_{t+2}, \ldots, x_{N-1}) \in (\mathbf{R}_+^m)^{N-t+1}\}, \tag{46}$$

for each $x_N \in \mathbf{R}_+^m$. Similarly, according to (44b)

$$A_t^{*-1}(p_0) = \{p_t | p_t \in T_{t-1}^{*-1}(p_{t-1}), p_{t-1} \in T_{t-2}^{*-1}(p_{t-2}), \ldots, p_1 \in T_0^{*-1}(p_0)$$
$$\text{for some } (p_1, p_2, \ldots, p_{t-1}) \in (\mathbf{R}_+^{*m})^{t-1}\}, \tag{47}$$

for each $p_0 \in \mathbf{R}_+^{*m}$.

By using the above definitions, I(c) and I(d) can be rewritten as follows:

I(c) Minimize $\bar{p}_0 \cdot x_0$, subject to $x_0 \in B_0^{-1}(\bar{x}_N)$.

I(d) Maximize $p_N \cdot \bar{x}_N$, subject to $p_N \in A_N^{*-1}(\bar{p}_0)$. $\qquad$ (48)

I(c) and I(d) are just reciprocals of I(a) and I(b), respectively. And I(c) and I(d) are dual to each other.

Let us next examine the economic meanings of I(c). Given a final commodity vector $\bar{x}_N \in \mathbf{R}_+^m$, from (46) $B_t^{-1}(\bar{x}_N)$ is the set of all input vectors in period t from which $\bar{x}_N$ can be reached in $N - t$ periods. Therefore, I(c) simply says that given a price vector $\bar{p}_0$ for the initial input commodities find an initial input vector x_0 that minimizes the value of initial inputs $\bar{p}_0 \cdot x_0$ among the set of all initial input vectors $B_0^{-1}(\bar{x}_N)$ from which $\bar{x}_N$ can be reached in N periods. If we find such a vector $\hat{x}_0$, then from Definition (46) there is at least one feasible path $\{\hat{x}_t\}_0^N$ for I(c) with $\hat{x}_N = \bar{x}_N$, and this path is an optimal path for the original maximization problem I(c).

The economic interpretation of I(d) becomes clear if we use the fact that A_t^{*-1} is the adjoint transformation of A_t^{-1} which is a m.t.cv. That is, by applying the second equation in B-property 6 (equivalently, definition (11))

$$A_t^{*-1}(p_0) = \{p_t \in \mathbf{R}_+^{*m} | p_t \cdot x_t \le p_0 \cdot x_0 \quad \text{for all } (x_t, x_0) \in \text{graph } A_t^{-1}\}$$
$$= \{p_t \in \mathbf{R}_+^{*m} | p_t \cdot x_t \le p_0 \cdot x_0 \quad \text{for all } (x_0, x_t) \in \text{graph } A_t\}.$$
(49)

Therefore, Problem I(d) says that given the final commodity vector $\bar{x}_N$ find a final price vector $\hat{p}_N$ that maximizes the value $p_N \cdot \bar{x}_N$ among the set $A_N^{*-1}(\bar{p}_0)$ which is the set of all output prices in the final period N under which no feasible activity of length N can yield positive profits with given price vector $\bar{p}_0$ on the initial input commodities. For such $\hat{p}_N$, there corresponds at least one sequence of vectors for which the feasibility conditions $p_{t+1} \in T_t^{*-1}(p_t)$, $t = 0, 1, \ldots, N-1$, and $p_0 = \bar{p}_0$ are satisfied. Such a sequence of vectors $\{\hat{p}_t\}_0^N$ is an optimal path for I(d).

The reciprocity principle is essentially a relation between Problem I' and Problem I (reciprocal), and it is stated as follows:

Theorem 2.4 (reciprocity principle). A pair of paths $\{\hat{x}_t\}_0^N$ and $\{\hat{p}_t\}_0^N$ are optimal paths for I'(a) and I'(b) associated with $\bar{x}_0 = \hat{x}_0$ and $\bar{p}_N = \hat{p}_N$ if and only if they are optimal paths for I(c) and I(d) associated with $\bar{x}_N = \hat{x}_N$ and $\bar{p}_0 = \hat{p}_0$.

Proof. First, comparing the feasibility conditions for Problems I'(a) and I'(b) with those for Problems I(c) and I(d), one easily sees that two paths $\{\hat{x}_t\}_0^N$ and $\{\hat{p}_t\}_0^N$ are feasible paths for I'(a) and I'(b) associated with $\bar{x}_0 = \hat{x}_0$ and $\bar{p}_N = \hat{p}_N$ if and only if they are feasible paths for I(c) and I(d) associated with $\bar{x}_N = \hat{x}_N$ and $\bar{p}_0 = \hat{p}_0$. Second, from Corollary 2.1', these paths are optimal paths for I'(a) and I'(b) associated with $\bar{x}_0 = \hat{x}_0$ and $\bar{p}_N = \hat{p}_N$ if and only if $\hat{p}_0 \cdot \hat{x}_0 = \hat{p}_N \cdot \hat{x}_N$. Hence, the proof is completed if we show that $\{\hat{x}_t\}_0^N$ and $\{\hat{p}_t\}_0^N$ are optimal paths for I(c) and I(d) associated with $\bar{x}_N = \hat{x}_N$ and $\bar{p}_0 = \hat{p}_0$ if and only if $\hat{p}_0 \cdot \hat{x}_0 = \hat{p}_N \cdot \hat{x}_N$.

From Definition (44), B_0^{-1} is a m.t.cv. and A_N^{*-1} is its adjoint transformation (i.e., A_N^{*-1} is a m.t.cc. and B_0^{-1} is its adjoint transformation). Consequently, from property (10) (i.e., B-property 5), $\langle A_N^{*-1}(\hat{p}_0), \hat{x}_N \rangle = \langle \hat{p}_0, B_0^{-1}(\hat{x}_N) \rangle$ for any $\hat{x}_N \in \mathbf{R}_+^m$ and $\hat{p}_0 \in \mathbf{R}_+^{*m}$. In other words

$$\sup\{p_N \cdot \hat{x}_N | p_N \in A_N^{*-1}(\hat{p}_0)\} = \inf\{\hat{p}_0 \cdot x_0 | x_0 \in B_0^{-1}(\hat{x}_N)\}.$$

From (48) and the above relation, if $\{\hat{x}_t\}_0^N$ and $\{\hat{p}_t\}_0^N$ are optimal paths for I(c) and I(d) associated with $\bar{x}_N = \hat{x}_N$ and $\bar{p}_0 = \hat{p}_0$, it should be true that $\hat{p}_N \cdot \hat{x}_N = \hat{p}_0 \cdot \hat{x}_0$.

Conversely, if a pair of feasible paths $\{\hat{x}_t\}_0^N$ and $\{\hat{p}_t\}_0^N$ for I(c) and I(d) associated with $\bar{x}_N = \hat{x}_N$ and $\bar{p}_0 = \hat{p}_0$ satisfy the condition, $\hat{p}_N \cdot \hat{x}_N = \hat{p}_0 \cdot \hat{x}_0$, then from the above relation and (48) they are optimal paths for Problems I(c) and I(d).   Q.E.D.

In other words, Theorem 2.4 says the following. Suppose $\{\hat{x}_t\}_0^N$ and $\{\hat{p}_t\}_0^N$ are optimal paths for Problems I'(a) and I'(b), respectively. Then take one pair of points $(\hat{x}_N, \hat{p}_0)$ from these paths, and solve Problems I(c) and I(d) by equating $\bar{x}_N = \hat{x}_N$ and $\bar{p}_0 = \hat{p}_0$. Then original paths $\{\hat{x}_t\}_0^N$ and $\{\hat{p}_t\}_0^N$ are also optimal paths for these reciprocal problems, respectively. The converse also holds.

Though the complete symmetry is lost, the next theorem gives the reciprocal relation between Problem I and Problem I(reciprocal).

Theorem 2.4' (reciprocity principle). Suppose $\{\hat{x}_t\}_0^N$ and $\{\hat{p}_t\}_0^N$ are optimal paths for, respectively, I(a) and I(b). Then they are optimal paths for, respectively, I(c) and I(d) with $\bar{x}_N = \hat{x}_N$ and $\bar{p}_0 = \hat{p}_0$. On the other hand, suppose $\{\hat{x}_t\}_0^N$ with $\hat{x}_N = \bar{x}_N$ and $\{\hat{p}_t\}_0^N$ with $\hat{p}_0 = \bar{p}_0$ are optimal paths for I(c) and I(d), respectively, and $\hat{x}_0 \in C$ and $\hat{p}_N \in D^*$ where C and D^* are the sets specified in Problem I. Then, if the hyperplane $\bar{p}_0 \cdot x = \bar{p}_0 \cdot \hat{x}_0$ supports the set C at $\hat{x}_0$ and the hyperplane $p \cdot \bar{x}_N = \hat{p}_N \cdot \bar{x}_N$ supports the set D^* at $\hat{p}_N$, they are optimal paths for I(a) and I(b), respectively.

The proof of this theorem, by using Corollary 2.2, is quite similar to the proof of Theorem 2.4, and hence is omitted. The usefulness of Theorems 2.4 and 2.4' will become apparent in the next section.

2.3.5. *Some corollaries to Theorem 2.1 through Theorem 2.4*

In this subsection, some corollaries are derived by combining the preceeding theorems to further clarify the character of the optimal paths for Problems I(a) and I'(a). Note first that inverse transformations of $A_{\tau t}$ and $A_{\tau t}^*$ (see (34)) are given, from B-property 24(c), as follows.

$$A_{\tau t}^{-1}(x_t) = (T_\tau^{-1} \cdot T_{\tau+1}^{-1} \cdot \ldots \cdot T_{t-2}^{-1} \cdot T_{t-1}^{-1})(x_t) \tag{50a}$$

for each $x_t \in \mathbf{R}_+^m$ and $N \geq t > \tau \geq 0$, and

$$A_{\tau t}^{*-1}(p_\tau) = (T_{t-1}^{*-1} \cdot T_{t-2}^{*-2} \cdot \ldots \cdot T_{\tau+1}^{*-1} \cdot T_\tau^{*-1})(p_\tau) \tag{50b}$$

for each $p_\tau \in \mathbf{R}_+^{*m}$ and $N \geq t > \tau \geq 0$. In particular, when $t = \tau + 1$, $A_{\tau t}^{-1}(x_t) = T_\tau^{-1}(x_{\tau+1})$, and when $\tau = t - 1$, $A_{\tau t}^{*-1}(p_\tau) = T_{t-1}^{*-1}(p_{t-1})$. Moreover, from (44) and (50),

$$A_{\tau t}^{*-1} = A_t^{*-1} \quad \text{when } \tau = 0, \quad \text{and} \quad A_{\tau t}^{-1} = B_\tau^{-1} \quad \text{when } t = N.$$

The economic meaning of set $A_{\tau t}^{-1}(x_t)$ is clear, while the meaning of $A_{\tau t}^{*-1}(p_\tau)$ becomes clear if we make use of the second equation in B-property 6 (equivalently, Definition (11)). That is,

$$\begin{aligned} A_{\tau t}^{*-1}(p_\tau) &= \{p_t \in \mathbf{R}_+^{*m} | p_t \cdot x_t \leq p_\tau \cdot x_\tau \quad \text{for all } (x_t, x_\tau) \in \text{graph } A_{\tau t}^{-1}\} \\ &= \{p_t \in \mathbf{R}_+^{*m} | p_t \cdot x_t \leq p_\tau \cdot x_\tau \quad \text{for all } (x_\tau, x_t) \in \text{graph } A_{\tau t}\} \end{aligned} \tag{51}$$

Hence $A_{\tau t}^{*-1}(p_\tau)$ gives the set of all output prices in period t under which no feasible activity of length $t - \tau$ can earn positive profits under given price vector p_τ for the input commodities in period τ.

Next, by combining the duality theorem and the reciprocity principle, we derive:

Corollary 2.3. A pair of optimal paths $\{\hat{x}_t\}_0^N$ and $\{\hat{p}_t\}_0^N$ for I(a) and I(b) (resp. I'(a) and I'(b)) satisfy not only (i), (ii) and (iii) in Theorem 2.1 (Theorem 2.1') but also the following:

(iv) $\max\{p_t \cdot \hat{x}_t | p_t \in A_{t\tau}^{*-1}(\hat{p}_\tau)\} = \hat{p}_t \cdot \hat{x}_t = \min\{\hat{p}_t \cdot x_t | x_t \in A_{t\tau'}^{-1}(\hat{x}_{\tau'})\}$

for any $N \geq \tau' > t > \tau \geq 0$.

For the proof, see Appendix C.4.

By summarizing (ii) in Theorem 2.1 and the above corollary, we obtain:

Corollary 2.4. Suppose $\{\hat{x}_t\}_0^N$ and $\{\hat{p}_t\}_0^N$ are optimal paths for I(a) and I(b), respectively. Then, in the commodity space, the hyperplane $\hat{p}_t \cdot x = \hat{p}_t \cdot \hat{x}_t$ separates the sets $A_t(C)$ and $A_{\tau t}(\hat{x}_\tau)$ $(\tau = 1, 2, \ldots, t - 1)$ from the sets $A_{t\tau'}^{-1}(\hat{x}_{\tau'})$ $(\tau' = t + 1, t + 2, \ldots, N)$ for each $t = 0, 1, \ldots, N$; and all these sets are supported at $\hat{x}_t$. That is,

$$\max_{x_t \in A_t(C)} \hat{p}_t \cdot x_t = \max_{x_t \in A_{\tau t}(\hat{x}_\tau)} \hat{p}_t \cdot x_t = \hat{p}_t \cdot \hat{x}_t = \min_{x_t \in A_{t\tau'}^{-1}(\hat{x}_{\tau'})} \hat{p}_t \cdot x_t$$

and

$$\hat{x}_t \in A_{\tau t}(\hat{x}_\tau) \subset A_t(C), \qquad \hat{x}_t \in A_{t\tau'}^{-1}(\hat{x}_{\tau'}),$$

for $t = 0, 1, \ldots, N$, $\tau = 1, 2, \ldots, t - 1$, $\tau' = t + 1$, $t + 2, \ldots, N$. On the other hand, in the price space, the hyperplane $p \cdot \hat{x}_t = \hat{p}_t \cdot \hat{x}_t$ separates the sets $B_t^*(D^*)$ and $A_{t\tau'}^*(\hat{p}_\tau)$ $(\tau' = t + 1, t + 2, \ldots, N)$ from the sets $A_{\tau t}^{*-1}(\hat{p}_\tau)$ $(\tau = 1, 2, \ldots, t - 1)$ for each $t = 0, 1, \ldots, N$; and these sets are supported by that hyperplane at the point $\hat{p}_t$. That is,

$$\max_{p_t \in A_{\tau t}^{*-1}(\hat{p}_\tau)} p_t \cdot \hat{x}_t = \hat{p}_t \cdot \hat{x}_t = \min_{p_t \in A_{t\tau'}^*(\hat{p}_{\tau'})} p_t \cdot \hat{x}_t = \min_{p_t \in B_t^*(D^*)} p_t \cdot \hat{x}_t$$

and

$$\hat{p}_t \in A_{\tau t}^{*-1}(\hat{p}_\tau), \qquad \hat{p}_t \in A_{t\tau'}^*(\hat{p}_{\tau'}) \subset B_t^*(D^*),$$

for $t = 0, 1, \ldots, N$, $\tau = 1, 2, \ldots, t - 1$, $\tau' = t + 1$, $t + 2, \ldots, N$.

Thus, when $m = 2$, these optimal paths may be depicted as in Figure 2.8(a) and (b). Roughly speaking, this corollary says the following. In the commodity space, the optimal path $\{\hat{x}_t\}_0^N$ for I(a) should follow the tangential point between the set $A_t(C)$ and the set $B_t^{-1}(\hat{x}_N)$ $(\equiv A_{tN}^{-1}(\hat{x}_N))$ in each period t, where the set $A_t(C)$ is the set of all output vectors in period t which are feasible from initial set C, and the set $B_t^{-1}(\hat{x}_N)$ is the set of all input vectors in period t from which $\hat{x}_N$ can be reached in $(N - t)$ periods. And $\hat{p}_t$ is the normal vector of the separating hyperplane between these two sets. "Microscopically," $\hat{x}_t$ should be the tangential point between the set $T_{t-1}(\hat{x}_{t-1})$ and the set $T_t^{-1}(\hat{x}_{t+1})$ in each period t where $T_{t-1}(\hat{x}_{t-1})$ is the set of all output vectors in period t which are feasible from $\hat{x}_{t-1}$ in one period, and $T_t^{-1}(\hat{x}_{t+1})$ is the set of all input vectors in period t from which $\hat{x}_{t+1}$ can be reached in one period. This was pointed out to be the intertemporal efficiency condition of optimal paths by Dorfman, Samuelson and Solow (1958, Ch. 3). Furthermore, though they are not depicted in Figure 2.8, any two sets $A_{\tau t}(\hat{x}_\tau)$ and $A_{t\tau'}^{-1}(\hat{x}_{\tau'})$ $(N \geqq \tau' > t > \tau \geqq 0)$ are tangent to each other at point $\hat{x}_t$. Thus, this fact reflects the *principle of optimality in dynamic programming* which says that "any portion of an optimal trajectory is also an optimal trajectory."[13] Dual relationships hold in the price space as illustrated in Figure 2.8(b).

The next statement is obtained from Theorem 2.3 and Corollary 2.3.

---

[13]See, Pontryagin and others (1962, p. 16), and also Bellman (1957, p. 83).

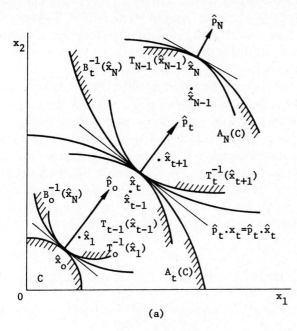

(a)

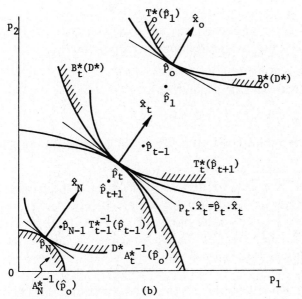

(b)

Figure 2.8. Optimal paths for Problem I; (a) optimal commodity path, (b) optimal price path.

Corollary 2.5 Suppose $\{\hat{x}_t\}_0^N$ is an optimal path for I(a). Then, if either one of the two conditions, (a) or (b), in Theorem 2.3 is satisfied, there is a sequence of vectors $\{\hat{p}_t\}_0^N$ under which (i) to (iii) in Theorem 2.1 and (iv) in Corollary 2.3 are satisfied. And such a sequence of vectors is obtained as an optimal path for I(b).

Next, we examine the linkage between duality theorems developed in this chapter and the theory of efficient price vectors developed in activity analysis. For this purpose, let us introduce several terms.

Given a set X in $\mathbf{R}_+^m$, we say a point $x \in X$ is a *frontier point* of X if $\lambda x \notin X$ for any $\lambda > 1$, and a point $x \in X$ is an *efficient point* of X if there is no point x' in X such that $x' \geq x$. We denote by $F(X)$ the set of all frontier points of set X, and call $F(X)$ the *frontier set* of X. Similarly, we denote by $E(X)$ the set of all efficient points of set X, and call $E(X)$ the *efficient set* of X. Then, one easily sees that

$$E(X) \subset F(X).$$

That is, any efficient point of a set is a frontier point of that set. But the opposite is not always true.

The next lemma provides a linkage between frontier points and efficient price vectors.

Lemma 2.4. If X is a monotone set of concave type in $\mathbf{R}_+^m$ and $X \neq \{0\}$, then the following two statements are equivalent:

(a) $\hat{x}$ is a frontier point of X.
(b) There exists a vector $\hat{p} \in \mathbf{R}_+^{*m}$ such that

$$\hat{p} \cdot \hat{x} = \max\{\hat{p} \cdot x \,|\, x \in X\} > 0, \quad \hat{p} \geq 0.$$

This lemma is intuitively obvious, and the proof is given in Appendix C.5.

In the context of Problem I, we denote by $F(A_N(C))$ the frontier set of $A_N(C)$, and by $E(A_N(C))$ the efficient set of $A_N(C)$. And we call a feasible path $\{\hat{x}_t\}_0^N$ for I(a) a *frontier path* if $\hat{x}_N \in F(A_N(C))$, and call it an *efficient path* if $\hat{x}_N \in E(A_N(C))$. Then, by using the above lemma, we get:

Theorem 2.5. Suppose either one of the two conditions in Theorem 2.3 is satisfied, where the condition on f is omitted in (a) (i.e., the condition on f is not necessary here). Then if $\{\hat{x}_t\}_0^N$ is a frontier path for I(a), there exists a sequence of vectors $\{\hat{p}_t\}_0^N$ with $p_t \geq 0$ ($t = 0, 1, \ldots, N$) such that $\{\hat{x}_t\}_0^N$ and $\{\hat{p}_t\}_0^N$ together satisfy (i) and (ii) in Theorems 2.1 and (iv) in Corollary 2.3 where $f(\hat{x}_N)$ and D^* are replaced by $\hat{p}_N \cdot \hat{x}_N$ and $\hat{p}_N$, respectively. Such a sequence of vectors $\{\hat{p}_t\}_0^N$ can be obtained as the optimal path for I(b) with $f(x_N) = \hat{p}_N \cdot x_N$ and $D^* = \hat{p}_N$ for an appropriate semipositive vector $\hat{p}_N$ in $\mathbf{R}_+^{*m}$. Conversely, if a feasible path $\{\hat{x}_t\}_0^N$ for I(a) has a corresponding feasible path $\{\hat{p}_t\}_0^N$ for I(b), where D^* is replaced by $\hat{p}_N$, such that they together satisfy $\hat{p}_N \cdot \hat{x}_N = g(\hat{p}_0)$, then $\{\hat{x}_t\}_0^N$ is a frontier path for I(a).

Proof. Suppose $\{\hat{x}_t\}_0^N$ is a frontier path for I(a). Then it is true that $\hat{x}_N \in F(A_N(C))$, and hence from Lemma 2.4 there exists a semipositive vector $\hat{p}_N \in \mathbf{R}_+^{*m}$ such that

$$\hat{p}_N \cdot \hat{x}_N = \max\{\hat{p}_N \cdot x_N \mid x_N \in A_N(C)\} > 0.$$

Hence, solving I(b) with $D^* = \hat{p}_N$, from Theorem 2.1 and Corollary 2.3, we obtain the sequence of desired vectors $\{\hat{p}_t\}_0^N$. The converse is also true from the last part of Theorem 2.1 and Lemma 2.4. Q.E.D.

Note that only the first half of the above theorem is true for an efficient path for problem I(a).

Finally, the next theorem extends the saddle point property of solutions to our multi-period programming in the context of B-properties 12, 13, 14, 19, and 20.

Theorem 2.6 (saddle point). Two sequences of vectors $\{\hat{x}_t\}_1^N$ and $\{\hat{p}_t\}_0^{N-1}$ are parts of optimal paths for I(a) and I(b), that is, there are two vectors $\hat{x}_0 \in C$ and $\hat{p}_N \in D^*$ such that $\{\hat{x}_t\}_1^N$ and $\hat{x}_0$ together is an optimal path for I(a) and $\{\hat{p}_t\}_0^{N-1}$ and $\hat{p}_N$ together is an optimal path for I(b), if and only if $(\{\hat{x}_t\}_1^N, \{\hat{p}_t\}_0^{N-1})$ is a saddle point of the function

$$K(\{x_t\}_1^N, \{p_t\}_0^{N-1}) = f(x_N) + g(p_0) + \sum_{t=1}^{N-1} p_t \cdot x_t$$
$$- \sum_{t=1}^{N} \langle T_{t-1}^{*-1}(p_{t-1}), x_t \rangle, \tag{52}$$

in other words, if and only if,

$$K(\{x_t\}_1^N, \{\hat{p}_t\}_0^{N-1}) \leq K(\{\hat{x}_t\}_1^N, \{\hat{p}_t\}_0^{N-1}) \leq K(\{\hat{x}_t\}_1^N, \{p_t\}_0^{N-1})$$
$$\text{for all } \{x_t\}_1^N \geq 0, \{p_t\}_0^{N-1} \geq 0. \tag{53}$$

For the proof, see Appendix C.6.

2.3.6. Max-min principle

In this subsection, a max-min principle appropriate for our multi-period programming problem is derived. Though this principle is obtained as a simple corollary to Theorem 2.1, it is stated as a theorem because of its importance. The theorem is first demonstrated in our conventional form. Then, with stronger assumptions, it is shown that this representation is exactly equivalent to the more conventional form of the maximum principle in discrete-time optimal control problems. Therefore we can easily interpret the economic meaning of the maximum principle.

First, the next theorem is obtained as a direct corollary to Theorems 2.1 and 2.2.

Theorem 2.7 (max-min principle).   Under either one of the two conditions in Theorem 2.3, a feasible path $\{\hat{x}_t\}_0^N$ for I(a) is an optimal path for that problem if and only if there exists a feasible path $\{\hat{p}_t\}_0^N$ for I(b) such that they together satisfy the next three conditions. In this case, $\{\hat{p}_t\}_0^N$ is an optimal path for I(b). And when $f(x_N) \neq 0$ in I(a), then $\hat{p}_t \geq 0$ for $t = 0, 1, \ldots, N$.

   (i) The hyperplane $\hat{p}_0 \cdot x = \hat{p}_0 \cdot \hat{x}_0$ supports the set C at $\hat{x}_0$, and the hyperplane $p \cdot \hat{x}_N = \hat{p}_N \cdot \hat{x}_N$ supports the set D^* at $\hat{p}_N$.

  (ii) $\hat{p}_t \cdot \hat{x}_t = \max\{\hat{p}_t \cdot x_t | x_t \in T_{t-1}(\hat{x}_{t-1})\}, t = 1, 2, \ldots, N$.

 (iii) $\hat{p}_t \cdot \hat{x}_t = \min\{p_t \cdot \hat{x}_t | p_t \in T_t^*(\hat{p}_{t+1})\}, t = 0, 1, \ldots, N - 1$.

Proof.   Suppose, $\{\hat{x}_t\}_0^N$ is an optimal path for I(a). From Theorem 2.3, under either one of the two conditions in that theorem, there exists an optimal path $\{\hat{p}_t\}_0^N$ for I(b). And, from Theorem 2.1 and Corollary 2.2, they together satisfy the above three conditions, where $\hat{p}_t \geq 0$ ($t = 0, 1, \ldots, N$) when $f(x_N) \neq 0$.

Conversely, suppose a pair of feasible paths $\{\hat{x}_t\}_0^N$ for I(a) and $\{\hat{p}_t\}_0^N$ for I(b) satisfy the above three conditions. (ii) says that $\hat{p}_t \cdot \hat{x}_t = \langle \hat{p}_t, T_{t-1}(\hat{x}_{t-1}) \rangle$, $t = 1, 2, \ldots, N$, and (iii) says that $\hat{p}_{t-1} \cdot \hat{x}_{t-1} = \langle T_{t-1}^*(\hat{p}_t), \hat{x}_{t-1} \rangle$, $t = 1, 2, \ldots, N$. But $\langle \hat{p}_t, T_{t-1}(\hat{x}_{t-1}) \rangle = \langle T_{t-1}^*(\hat{p}_t), \hat{x}_{t-1} \rangle$ for all $\hat{x}_{t-1} \in \mathbf{R}_+^m$ and $\hat{p}_t \in \mathbf{R}_+^{*m}$ by (10). Thus $\hat{p}_{t-1} \cdot \hat{x}_{t-1} = \hat{p}_t \cdot \hat{x}_t$ for $t = 1, 2, \ldots, N$. Then $\hat{p}_0 \cdot \hat{x}_0 = \hat{p}_N \cdot \hat{x}_N$. Therefore, from Corollary 2.2, this fact and condition (i) together imply that $\{\hat{x}_t\}_0^N$ and $\{\hat{p}_t\}_0^N$ are optimal paths for I(a) and I(b), respectively.   Q.E.D.

To compare the above theorem with the conventional form of the maximum principle in the discrete-time optimal control theory, let us consider a special case in which the transformation T_t in I(a) is represented as follows for each $t = 0, 1, \ldots, N - 1$:

$$T_t(x_t) = \{x_{t+1} | x_{t+1} = (x_1^{t+1}, \ldots, x_i^{t+1}, \ldots, x_m^{t+1})$$
$$\text{for some } x_i^{t+1} = f_i^t(x_t, \theta_t), \theta_t \in U_t, \quad i = 1, 2, \ldots, m\}, \quad (54)$$

for each $x_t \in \mathbf{R}_+^m$. Here θ_t is the control vector in period t with an appropriate dimension, U_t is the feasible set of θ_t, and f_i^t is a numerical function defined on $\mathbf{R}_+^m \times U_t$, $i = 1, 2, \ldots, m$.

Note that the transformation formulated by (54) includes the following one as a special case.

$$T_t(x_t) = \{x_{t+1} | x_{t+1} = (x_1^{t+1}, \ldots, x_i^{t+1}, \ldots, x_m^{t+1})$$
$$\text{for some } 0 \leq x_i^{t+1} \leq f_i^t(x_t, \theta_t), \theta_t \in U_t, \quad i = 1, 2, \ldots, m\}.$$

That is, by introducing m new control variables the above transformation can be reformulated in the form of (54).

Corresponding to transformation (54), we introduce the *Hamiltonian function* H_t defined by

$$H_t(p_t, x_{t-1}, \theta_{t-1}) = \sum_{j=1}^m p_j^t f_j^{t-1}(x_{t-1}, \theta_{t-1}). \quad (55)$$

Then the above theorem is restated as follows:

Theorem 2.8 (maximum principle). Suppose T_t in Problem I is represented by (54) where, for each fixed θ_t, each function f_i^t is concave with respect to x_t and continuously differentiable with respect to x_t on $\mathbf{R}_+^m$ ($t = 0, 1, \ldots, N - 1$).[14] Then, under either one of the two conditions in Theorem 2.3, a feasible path $\{\hat{x}_t\}_0^N$ for I(a) and a feasible control sequence $\{\hat{\theta}_t\}_0^{N-1}$ are respectively an optimal path and an optimal control sequence for I(a) if and only if there exists a sequence of vectors $\{\hat{p}_t\}_0^N$ which satisfies the next three conditions. In this case, if $f(x_N) \neq 0$ in I(a), then $\hat{p}_t \geq 0$ for $t = 0, 1, \ldots, N$.

(i) The hyperplane $\hat{p}_0 \cdot x = \hat{p}_0 \cdot \hat{x}_0$ supports the set C at $\hat{x}_0$, and the hyperplane $p \cdot \hat{x}_N = \hat{p}_N \cdot \hat{x}_N$ supports the set D^* at $\hat{p}_N$.

---

[14]When $x_i^t = 0$, the differentiability of f_i^t with respect to x_i^t means it is differentiable from the right. Function $f_j^t(x, \theta)$ is said to be continuously differentiable with respect to x if $\partial f_j^t(x, \theta)/\partial x_i$ is defined by each x_i and continuous on $\mathbf{R}_+^m \times U_t$.

(ii) $H_t(\hat{p}_t, \hat{x}_{t-1}, \hat{\theta}_{t-1}) = \max\{H_t(\hat{p}_t, \hat{x}_{t-1}, \theta_{t-1})|\theta_{t-1} \in U_{t-1}\}$,

(iii) $\hat{p}_i^t = \partial H_{t+1}(\hat{p}_{t+1}, \hat{x}_t, \hat{\theta}_t)/\partial x_i$, $i = 1, 2, \ldots, m$, $t = 0, 1, \ldots, N-1$.

Proof. Since conditions (i) and (ii) are the same in Theorems 2.7 and 2.8, it is sufficient to show that condition (iii) and the feasibility condition, $\hat{p}_t \in T_t^*(\hat{p}_{t+1})$, in Theorem 2.7 can be replaced by condition (iii) in the above. From the definition of adjoint transformation by (25) and from assumption (54) on the form of the transformation T_t, the set $T_t^*(\hat{p}_{t+1})$ is given by

$$T_t^*(\hat{p}_{t+1}) = \{p_t | p_t \cdot x \geq \sum_{j=1}^{m} \hat{p}_j^{t+1} f_j^t(x, \theta) \quad \text{for all } x \geq 0 \text{ and } \theta \in U_t\}.$$

Let us denote $(f_1^t(x, \theta), \ldots, f_m^t(x, \theta))$ by $f_t(x, \theta)$. Then the vector $\hat{p}_t$ associated with the optimal path should satisfy the next condition in order to be in $T_t^*(\hat{p}_{t+1})$.

$$\hat{p}_t \cdot x \geq \hat{p}_{t+1} \cdot f_t(x, \theta) \quad \text{for all } x \geq 0 \text{ and } \theta \in U_t.$$

For each $x \in \mathbf{R}_+^m$, let us define vector $\theta(x)$ in U_t by

$$\hat{p}_{t+1} \cdot f_t(x, \theta(x)) = \max\{\hat{p}_{t+1} \cdot f_t(x, \theta)|\theta \in U_t\}.$$

Such a vector $\theta(x)$ exists for each $x \in \mathbf{R}_+^m$ since $\cup\{f_t(x, \theta)|\theta \in U_t\} = T_t(x)$ and $T_t(x)$ is a monotone set of concave type by assumption. Then, for the optimal $\hat{p}_t$ to be in the set $T_t^*(\hat{p}_{t+1})$, it is necessary and sufficient to satisfy

$$\hat{p}_t \cdot x \geq \hat{p}_{t+1} \cdot f_t(x, \theta(x)) \quad \text{for all } x \geq 0.$$

Recall that condition (iii) in Theorem 2.7 says $\hat{p}_t \cdot \hat{x}_t = \langle T_t^*(\hat{p}_{t+1}), \hat{x}_t \rangle$; $\langle T_t^*(\hat{p}_{t+1}), \hat{x}_t \rangle = \langle \hat{p}_{t+1}, T_t(\hat{x}_t) \rangle = \hat{p}_{t+1} \cdot f_t(\hat{x}_t, \theta(\hat{x}_t))$ from (10) and by definition of $\theta(x)$. Hence, condition (iii) in Theorem 2.7 is equivalent to

$$\hat{p}_t \cdot \hat{x}_t = \hat{p}_{t+1} \cdot f_t(\hat{x}_t, \theta(\hat{x}_t)).$$

Thus, the above inequality is rewritten as follows:

$$\hat{p}_t \cdot (x - \hat{x}_t) \geq \hat{p}_{t+1} \cdot [f_t(x, \theta(x)) - f_t(\hat{x}_t, \theta(\hat{x}_t))] \quad \text{for all } x \geq 0.$$

Next, note by definition of the optimal control vector $\theta(x)$,

$$\sum_{j=1}^{m} \hat{p}_j^{t+1} f_j^t(x, \theta(x)) \geq \sum_{j=1}^{m} \hat{p}_j^{t+1} f_j^t(x, \theta(\hat{x}_t)) \quad \text{for all } x \geq 0,$$

for each $j = 1, 2, \ldots, m$. In addition, since f_j^t is continuously differentiable in x and concave with respect to x, we have

$$f_j^t(x, \theta(\hat{x}_t)) = f_j^t(\hat{x}_t, \theta(\hat{x}_t)) + \sum_{i=1}^{m} ([\partial f_j^t(x, \theta(\hat{x}_t))/\partial x_i]|_{x=\hat{x}_t})(x_i - \hat{x}_i^t)$$

$$+ \delta_j^t(x - \hat{x}_t),$$

where

$$\delta_j^t(x - \hat{x}_t)/\|x - \hat{x}_t\| \to 0 \quad \text{as } \|x - \hat{x}_t\| \to 0, \text{ and } \delta_j^t(x - \hat{x}_t) \leq 0. \tag{56}$$

Therefore, for the optimal price vector $\hat{p}_t$ to be in $T_t^*(\hat{p}_{t+1})$ and to satisfy (iii) in Theorem 2.7, it is necessary that

$$\hat{p}_t \cdot (x - \hat{x}_t) \geq \sum_{i=1}^{m} \left(\sum_{j=1}^{m} \hat{p}_j^{t+1}[\partial f_j^t(x, \theta(\hat{x}_t))/\partial x_i]|_{x=\hat{x}_t} \right)(x_i - \hat{x}_i^t)$$

$$+ \sum_{j=1}^{m} \hat{p}_j^{t+1} \delta_j^t(x - \hat{x}_t)$$

for all $x \geq 0$. That is,

$$\sum_{i=1}^{m} \left(\hat{p}_i^t - \sum_{j=1}^{m} \hat{p}_j^{t+1}[\partial f_j^t(x, \theta(\hat{x}_t))/\partial x_i]|_{x=\hat{x}_t} \right)(x_i - \hat{x}_i^t) \geq \sum_{j=1}^{m} \hat{p}_j^{t+1} \delta_j^t(x - \hat{x}_t)$$

for all $x \geq 0$. From (56), this is true if and only if

$$\hat{p}_i^t = \sum_{j=1}^{m} \hat{p}_j^{t+1}[\partial f_j^t(x, \theta(\hat{x}_t))/\partial x_i]|_{x=\hat{x}_t} \quad \text{when } \hat{x}_i^t > 0,$$

$$\hat{p}_i^t \geq \sum_{j=1}^{m} \hat{p}_j^{t+1}[\partial f_j^t(x, \theta(\hat{x}_t))/\partial x_i]|_{x=\hat{x}_t} \quad \text{when } \hat{x}_i^t = 0. \tag{57}$$

Hence when $\hat{x}_t$ is strictly positive, the vector $\hat{p}_t \in T_t^*(\hat{p}_{t+1})$ satisfying (iii) in Theorem 2.7 can be uniquely determined by (57). Therefore, when vector $\hat{x}^t$ on the optimal path $\{\hat{x}_t\}_0^N$ is strictly positive in every period, the feasibility condition and condition (iii) in Theorem 2.7 on the sequence of vectors $\{\hat{p}_t\}_0^N$ associated with $\{\hat{x}_t\}_0^N$ can be replaced by the condition

$$\hat{p}_i^t = \sum_{j=1}^{m} \hat{p}_j^{t+1}[\partial f_j^t(x, \theta(\hat{x}_t))/\partial x_i]|_{x=\hat{x}_t}, \quad i = 1, 2, \ldots, m,$$

$$t = 0, 1, \ldots, N-1. \tag{58}$$

namely, by (iii) in Theorem 2.8.

Next, take the case in which some vectors, $\hat{x}_t$, on the path $\{\hat{x}_t\}_0^N$ are not strictly positive. In this case take any sequence of vectors $\{\hat{p}_t^t\}$ which satisfies relation (57) and conditions (i) and (ii) in Theorem 2.8.

Then, by using the two facts that when $\hat{x}_t$ is not strictly positive, any vector p_t satisfying (57) has the same value of $p_t \cdot \hat{x}_t$ and that $T_{t-1}^*(p') \subset T_{t-1}^*(p)$ for any $0 \leqq p \leqq p' \in \mathbf{R}_+^{*m}$ since T_{t-1}^* is a m.t.cv. $(t = 1, 2, \ldots, N)$, one easily sees that any sequence of vectors $\{\hat{p}_t\}_0^N$ which satisfies the condition

$$\sum_{j=1}^m \hat{p}_j^{t+1} [\partial f_j^t(x, \theta(\hat{x}_t))/\partial x_i]|_{x=\hat{x}_t} \leqq \hat{p}_i^t \leqq \hat{p}_i', \quad i = 1, 2, \ldots, m,$$
$$t = 0, 1, \ldots, N-1,$$

also satisfies relation (57) and conditions (i) and (ii) in Theorem 2.8. Therefore, the two conditions in Theorem 2.7, condition (iii) and the feasibility condition $\hat{p}_t \in T_t^*(\hat{p}_{t+1})$ for each t, can safely be replaced by (58), that is, by condition (iii) in Theorem 2.8.   Q.E.D.

Conditions (ii) and (iii) in Theorem 2.8 are, respectively, the ordinal forms of the maximum principle and the adjoint equation in discrete-time optimal control theory.[15]

Since Problem I is a special case of Problem II in the next section, as will be shown in Theorem 2.13, the latter half of condition (i) in Theorem 2.8 can be omitted by using the following Hamiltonian function for the last period N.

$$H_N(p_N, x_{N-1}, \theta_{N-1}) = f(f_1^{N-1}(x_{N-1}, \theta_{N-1}), \ldots,$$
$$f_j^{N-1}(x_{N-1}, \theta_{N-1}), \ldots, f_m^{N-1}(x_{N-1}, \theta_{N-1})),$$

where f is the objective function in I(a). The original representation of the maximum principle is more convenient for later use in this book, however.

2.4. Extension of duality theorems to consumption stream problems

2.4.1. Problem formulation

In this section, the results of the preceding section are extended to a more general problem which involves the maximization of the sum of the values of the objective function in each period. The form of the objective function in each period is restricted to a monotone concave gauge.

---

[15]For example, see Fan and Wang (1964), and Canon, Cullum and Polak (1970).

As before, the state of the system at the beginning of period t is denoted by a vector $x_t \in \mathbf{R}_+^m$ $(t = 0, 1, \ldots, N)$. The set of all feasible output vectors at the beginning of the next period is denoted by $T_t(x_t)$. Each vector in the set $T_t(x_t)$ has the form (x_{t+1}, c_{t+1}), where the vector x_{t+1} constitutes the input vector for period $t + 1$ and c_{t+1} may be considered as the part of the outputs consumed in period $t + 1$. Suppose c is a non-negative n-dimensional vector in each period. Then $T_t(x_t)$ is a subset in $\mathbf{R}_+^{m+n}$ and T_t is a multi-valued function from $\mathbf{R}_+^m$ into $\mathbf{R}_+^{m+n}$.[16] The value of the objective function in period $t + 1$ is denoted by $f_{t+1}(x_{t+1}, c_{t+1})$ which may be considered as the welfare or the utility obtained in period $t + 1$.[17] Let us call f_{t+1} the *welfare function* in period $t + 1$ $(t = 0, 1, \ldots, N - 1)$. Durable consumption goods, such as housing, may be included in vector x_{t+1}, or may be in both vectors x_{t+1} and c_{t+1}.

The primal problem in this section is stated as follows:

Problem II. Given a non-singular m.t.cc. T_t $(t = 0, 1, \ldots, N - 1)$ from $\mathbf{R}_+^m$ into $\mathbf{R}_+^{m+n}$, a monotone concave gauge f_t $(t = 1, 2, \ldots, N)$ on $\mathbf{R}_+^{m+n}$, where $f_N(.,.)$ is not identically zero, and the initial vector $0 \le \bar{x}_0 \in \mathbf{R}_+^m$,

II(a) Maximize $\displaystyle\sum_{t=1}^{N} f_t(x_t, c_t)$

subject to $(x_t, c_t) \in T_{t-1}(x_{t-1})$, $\quad t = 1, 2, \ldots, N$,

and $x_0 = \bar{x}_0$.

The dual problem for II(a) cannot be stated at this stage, and will be given later.

Our intention is to obtain duality theorems for the above problem by utilizing the results from the preceding section. For this purpose, the above problem is converted into a final state problem by using the

---

[16]Such a transformation T_t may be obtained as follows. Suppose we are given a transformation T_t' from $\mathbf{R}_+^m$ into $\mathbf{R}_+^m$ which transforms an input commodity vector at the beginning of period t into a set of output commodity vectors at the end of that period. Hence T_t' may be T_t in section 2.3. And we define a transformation T_t from $\mathbf{R}_+^m$ into $\mathbf{R}_+^{m+n}$ by

$$T_t(x_t) = \{(x_{t+1}, c_{t+1}) | x_{t+1} + (c_{t+1}, 0) \in T_t'(x_t), x_{t+1} \ge 0, c_{t+1} \ge 0\},$$

where $x_{t+1} \in \mathbf{R}_+^m$, $c_{t+1} \in \mathbf{R}_+^n$, $m \ge n \ge 0$ and $(m - n)$ represents the number of commodities prohibited to be consumed, for example, man. Then one easily sees that when T_t' is a m.t.cc. from $\mathbf{R}_+^m$ into $\mathbf{R}_+^m$, T_t defined above is a m.t.cc. from $\mathbf{R}_+^m$ into $\mathbf{R}_+^{m+n}$.

[17]For example, $f_t(x_t, c_t) = (1 + \rho)^{-t} L_t f(c_t/L_t)$, where L_t is the total population in period t which is given exogenously and ρ is the social discount rate.

augmented transformation $\mathcal{T}_t$ $(t = 0, 1, \ldots, N - 1)$ which is defined by

$$\mathcal{T}_t(x_t, x^t_{m+1}) = \{(x_{t+1}, x^{t+1}_{m+1}) | 0 \leq x^{t+1}_{m+1} \leq x^t_{m+1} + f_{t+1}(x_{t+1}, c_{t+1}),$$
$$(x_{t+1}, c_{t+1}) \in T_t(x_t)\}, \tag{59}$$

for each $(x_t, x^t_{m+1}) \in \mathbf{R}^{m+1}_+$. The new scalar variable x^t_{m+1} represents the amount of "welfare stock" in period t. In the context of Problem II, one easily sees that $\mathcal{T}_t$ defined by (59) is a non-singular m.t.cc. from $\mathbf{R}^{m+1}_+$ into $\mathbf{R}^{m+1}_+$. Hence, from definition (5) the adjoint transformation of $\mathcal{T}_t$ is defined by

$$\mathcal{T}^*_t(p_{t+1}, p^{t+1}_{m+1}) = \{(p_t, p^t_{m+1}) \in \mathbf{R}^{*m+1}_+ | p_t \cdot x_t + p^t_{m+1}x^t_{m+1} \geq p_{t+1} \cdot x_{t+1}$$
$$+ p^{t+1}_{m+1}(x^t_{m+1} + f_{t+1}(x_{t+1}, c_{t+1}))$$
$$\text{for all } (x_{t+1}, c_{t+1}) \in T_t(x_t), x_t \geq 0 \text{ and } x^t_{m+1} \geq 0\}, \tag{60}$$

for each $(p_{t+1}, p^{t+1}_{m+1}) \in \mathbf{R}^{m+1}_+$. Here, p_t is the price vector corresponding to the commodity vector x_t, and p^t_{m+1} corresponds to the welfare stock x^t_{m+1}. By using the above two transformations, the next pair of problems are defined.

Problem II′. Let T_t $(t = 0, 1, \ldots, N - 1)$, f_t $(t = 1, 2, \ldots, N)$ and $\bar{x}_0$ be those given in Problem II. And let us define $\mathcal{T}_t$ and $\mathcal{T}^*_t$ $(t = 0, 1, \ldots, N - 1)$ by (59) and (60), respectively. Then a pair of problems are

II′(a)   Maximize x^N_{m+1},

subject to $(x_{t+1}, x^{t+1}_{m+1}) \in \mathcal{T}_t(x_t, x^t_{m+1})$, $t = 0, 1, \ldots, N - 1$,

and $(x_0, x^0_{m+1}) = (\bar{x}_0, 0)$,

II′(b)   Minimize $(p_0, p^0_{m+1}) \cdot (\bar{x}_0, 0)$

subject to $(p_t, p^t_{m+1}) \in \mathcal{T}^*_t(p_{t+1}, p^{t+1}_{m+1})$, $t = 0, 1, \ldots, N - 1$,

and $(p_N, p^N_{m+1}) = (0, 1)$.

Since the relation $x^{t+1}_{m+1} \leq x^t_{m+1} + f_{t+1}(x_{t+1}, c_{t+1})$ should be satisfied with an equality on the optimal path, it is clear that problems II(a) and II′(a) are equivalent to each other. Therefore, by examining the solutions of II′(a) and II′(b), the property of the optimal path for II(a) can be examined. Since Problem II′ is a special case of Problem I′, all the results in the preceding section are also valid for the optimal paths for Problem II′. To apply theorems from the preceding section to our present problem, let us begin with:

Lemma 2.5. Suppose a sequence of vectors $\{(\hat{p}_t, \hat{p}^t_{m+1})\}_0^N$ is an optimal path for II'(b). Then $\{(\hat{p}_t, 1)\}_0^N$ is also an optimal path for that problem.

For the proof, see Appendix C.7. Thus, we can always assume that $\hat{p}^t_{m+1} = 1$ ($t = 0, 1, \ldots, N$) on the optimal path. Therefore, if we define a transformation T^*_t by

$$T^*_t(p_{t+1}) = \{p_t \in \mathbf{R}^{*m}_+ | p_t \cdot x_t \geqq p_{t+1} \cdot x_{t+1} + f_{t+1}(x_{t+1}, c_{t+1})$$
$$\text{for all } (x_{t+1}, c_{t+1}) \in T_t(x_t), x_t \geqq 0\}, \tag{61}$$

for each $p_{t+1} \in \mathbf{R}^{*m}_+$, the above lemma together with definition (60) says that a sequence of vectors $\{(\hat{p}_t, 1)\}_0^N$ is an optimal path for II'(b) if and only if $\{\hat{p}_t\}_0^N$ is an optimal path for the following problem.

II(b) Minimize $p_0 \cdot \bar{x}_0$,
  subject to $p_t \in T^*_t(p_{t+1})$, $\quad t = 0, 1, \ldots, N - 1$,
  and $p_N = 0$.

Consequently, the above problem II(b) can be considered as the dual problem of II(a). Moreover, the economic meaning of the pair of problems II(a) and II(b) is very clear. Thus, in the remainder of this section, theorems from the preceding section are restated in terms of II(a) and II(b).

2.4.2. *Duality theorem*

For problems II(a) and II(b), the duality theorem is stated as follows:

Theorem 2.9 (duality theorem for Problem II).   Suppose $[\{\hat{x}_t\}_0^N, \{\hat{c}_t\}_0^N]$ is an optimal path for II(a) and $\{\hat{p}_t\}_0^N$ is an optimal path for II(b). Then, we have

(i) $\mu = \displaystyle\sum_{t=1}^N f_t(\hat{x}_t, \hat{c}_t) = \hat{p}_{N-1} \cdot \hat{x}_{N-1} + \displaystyle\sum_{\tau=1}^{N-1} f_\tau(\hat{x}_\tau, \hat{c}_\tau) = \cdots$

$\qquad = \hat{p}_t \cdot \hat{x}_t + \displaystyle\sum_{\tau=1}^t f_\tau(\hat{x}_\tau, \hat{c}_\tau) = \cdots = \hat{p}_0 \cdot \hat{x}_0,$

where μ is a non-negative real number.

(ii) $f_t(\hat{x}_t, \hat{c}_t) + \hat{p}_t \cdot \hat{x}_t = \max\{f_t(x_t, c_t) + \hat{p}_t \cdot x_t | (x_t, c_t) \in T_t(\hat{x}_{t-1})\}$,
$$t = 1, 2, \ldots, N,$$

$f_t(\hat{x}_t, \hat{c}_t) + \hat{p}_t \cdot \hat{x}_t = f_t(\hat{x}_t, \hat{c}_t) + \min\{p_t \cdot \hat{x}_t | p_t \in T^*_t(\hat{p}_{t+1})\}$,
$$t = 0, 1, \ldots, N - 1.$$

(iii) $\hat{p}_t \geq 0$ for $t = 0, 1, \ldots, N-1$ and $\hat{p}_N = 0$.

Conversely, if a pair of feasible paths, $[\{\hat{x}_t\}_0^N, \{\hat{c}_t\}_1^N]$ for II(a) and $\{\hat{p}_t\}_0^N$ for II(b), satisfy the condition $\Sigma_{t=1}^N f_t(\hat{x}_t, \hat{c}_t) = \hat{p}_0 \cdot \bar{x}_0$, then they are optimal paths for II(a) and II(b), respectively.

Proof.   (i), (ii) and (iv) are obtained immediately by applying Theorem 2.1 to Problem II′ with the consideration that the pair of Problems II′(a) and II′(b) are equivalent to the pair of problems II(a) and II(b). Therefore, let us prove (iii). By definition of the optimal path for II(b), $\hat{p}_N = 0$. Then, for $\hat{p}_{N-1} \in T^*_{N-1}(\hat{p}_N = 0)$, it should be true that

$$\hat{p}_{N-1} \cdot x_{N-1} \geq f_N(x_N, c_N) \quad \text{for all } (x_N, c_N) \in T_{N-1}(x_{N-1}), x_{N-1} \geq 0.$$

Then, from the assumptions that $f_N(x_N, c_N) > 0$ for some $(x_N, c_N) \geq 0$ and that T_{N-1} is non-singular, we must have $\hat{p}_{N-1} \geq 0$. Next, for $\hat{p}_{N-2} \in T^*_{N-2}(\hat{p}_{N-1})$, it should be true that

$$\hat{p}_{N-2} \cdot x_{N-2} \geq \hat{p}_{N-1} \cdot x_{N-1} + f_{N-1}(x_{N-1}, c_{N-1})$$
$$\text{for all } (x_{N-1}, c_{N-1}) \in T_{N-2}(x_{N-2}), x_{N-2} \geq 0.$$

Then, since $\hat{p}_{N-1} \geq 0$ and T_{N-2} is non-singular by assumption, we must have $\hat{p}_{N-2} \geq 0$. Repeating the same procedure, we conclude $p_t \geq 0$ for $t = 0, 1, \ldots, N-1$.   Q.E.D.

2.4.3.   Existence theorem

To obtain the existence theorem for II(b), first we observe

Lemma 2.6.   In the context of Problem II, if T_t and f_{t+1} satisfy the Lipschitz condition,[18] respectively, then $\mathcal{T}_t$ defined by (59) also satisfies this condition ($t = 0, 1, \ldots, N-1$).

For the proof, see Appendix C.8. Then, by applying Theorem 2.2(i) and Theorem 2.3(a) and (b), we obtain

Theorem 2.10 (existence theorem for Problem II).

(i) There always exists an optimal path for II(a).

(ii) An optimal path exists for II(b) under either one of two conditions:

---

[18]It is said that f_t satisfies the Lipschitz condition when there exists a scalar α_t such that $\|f_t(y, c) - f_t(y', c')\| \leq \alpha_t \|(y, c) - (y', c')\|$ for all $(y, c), (y', c') \in \mathbf{R}_+^{m+n}$.

(a) T_t and f_{t+1} satisfy the Lipschitz condition, respectively ($t = 0, 1, \ldots, N - 1$).

(b) $\bar{x}_0 > 0$.

Proof. (i): By applying Theorem 2.2(i) to problem II'(a), the existence of an optimal path for this problem is guaranteed. Then, since II(a) and II'(a) are equivalent to each other, an optimal path also exists for II(a).

(ii)-(a): Let $f(x_N) = x_{m+1}^N$. Then, the steepness of this function has an upper bound on $\mathbf{R}_+^m$. Hence, from Lemma 2.6 and Theorem 2.3(a), II'(b) has an optimal solution. Then, since II(b) and II'(b) are equivalent to each other, II(b) also has an optimal solution.

(ii)-(b): Let us consider the next pair of problems.

II''(a) Maximize x_{m+1}^N,
    subject to $(x_{t+1}, x_{m+1}^{t+1}) \in \mathcal{T}_t(x_t, x_{m+1}^t)$, $t = 0, 1, \ldots, N - 1$,
    and $(x_0, x_{m+1}^0) = (\bar{x}_0, \bar{x}_{m+1}^0) > 0$.

II''(b) Minimize $(p_0, p_{m+1}^0) \cdot (\bar{x}_0, \bar{x}_{m+1}^0)$,
    subject to $(p_t, p_{m+1}^t) \in \mathcal{T}_t^*(p_{t+1}, p_{m+1}^{t+1})$, $t = 0, 1, \ldots, N - 1$,
    and $(p_N, p_{m+1}^N) = (0, 1)$.

Then, from Theorem 2.3(b), both problems have optimal paths. Let us denote them by $\{(\hat{x}_t, \hat{x}_{m+1}^t)\}_0^N$ and $\{(\hat{p}_t, \hat{p}_{m+1}^t)\}_0^N$, respectively. Then, similarly to the proof of Lemma 2.5, we can prove that $\{(\hat{p}_t, 1)\}_0^N$ is also an optimal path for II''(b). Hence, from (i) in Theorem 2.1', $\hat{x}_{m+1}^N = \hat{p}_0 \cdot \bar{x}_0 + \bar{x}_{m+1}^0$.

Next, comparing II'(a) and II''(a), when $\{(\hat{x}_t, \hat{x}_{m+1}^t)\}_0^N$ is an optimal path for II''(a), $\{(\hat{x}_t, \hat{x}_{m+1}^t - \bar{x}_{m+1}^0)\}_0^N$ is a feasible path for II'(a). Likewise, when $\{(\hat{p}_t, 1)\}_0^N$ is an optimal path for II''(b), it is a feasible path for II'(b). And this pair of feasible paths for II'(a) and II'(b) satisfies the condition, $\hat{x}_{m+1}^N - x_{m+1}^0 = \hat{p}_0 \cdot \bar{x}_0$. Therefore, from the latter half of Theorem 2.1', $\{(\hat{p}_t, 1)\}_0^N$ is an optimal path for II'(b). Then, since $\{(\hat{p}_t, 1)\}_0^N$ is an optimal path for II'(b) if and only if $\{\hat{p}_t\}_0^N$ is an optimal path for II(b), $\{\hat{p}_t\}_0^N$ is an optimal path for II(b).   Q.E.D.

2.4.4. Reciprocity principle

In the context of Problem II, we define a pair of problems as follows:

Problem II'(reciprocal). Given two vectors $(\bar{x}_N, \bar{x}_{m+1}^N) \in \mathbf{R}_+^{m+1}$ and

$(\bar{p}_0, \bar{p}^0_{m+1}) \in \mathbf{R}^{*m+1}_+$;

II'(c)   Minimize $(\bar{p}_0, \bar{p}^0_{m+1}) \cdot (x_0, x^0_{m+1})$,
         subject to $(x_t, x^t_{m+1}) \in \mathcal{T}^{-1}_t(x_{t+1}, x^{t+1}_{m+1})$, $t = 0, 1, \ldots, N - 1$,
         and $(x_N, x^N_{m+1}) = (\bar{x}_N, \bar{x}^N_{m+1})$.

II'(d)   Maximize (p_N, p^N_{m+1}),
         subject to $(p_{t+1}, p^{t+1}_{m+1}) \in \mathcal{T}^{*-1}_t(p_t, p^t_{m+1})$, $t = 0, 1, \ldots, N - 1$,
         and $(p_0, p^0_{m+1}) = (\bar{p}_0, \bar{p}^0_{m+1})$.

Here $\mathcal{T}^{-1}_t$ and $\mathcal{T}^{*-1}_t$ ($t = 0, 1, \ldots, N - 1$) are, respectively, the inverse transformations of $\mathcal{T}_t$ and $\mathcal{T}^*_t$ in Problem II'.

We can now state the reciprocity principle in terms of Problem II' and Problem II' (reciprocal) as follows:

Theorem 2.11 (reciprocity principle for Problem II').   A pair of paths $\{(\hat{x}_t, \hat{x}^t_{m+1})\}^N_0$ and $\{(\hat{p}_t, \hat{p}^t_{m+1})\}^N_0$ are optimal paths for II'(a) and II'(b) with $(\bar{x}_0, \bar{x}^0_{m+1}) = (\hat{x}_0, \hat{x}^0_{m+1})$ and $(\bar{p}_N, \bar{p}^N_{m+1}) = (\hat{p}_N, \hat{p}^N_{m+1})$ if and only if they are optimal paths for II'(c) and II'(d) with $(\bar{x}_N, \bar{x}^N_{m+1}) = (\hat{x}_N, \hat{x}^N_{m+1})$ and $(\bar{p}_0, \bar{p}^0_{m+1}) = (\hat{p}_0, \hat{p}^0_{m+1})$.

This can be obtained immediately by applying Theorem 2.4 to Problem II' and Problem II' (reciprocal). Then, since Problem II' is equivalent to Problem II, we have in terms of Problem II:

Corollary 2.6.   Suppose $[\{\hat{x}_t\}^N_0, \{\hat{c}_t\}^N_1]$ and $\{\hat{p}_t\}^N_0$ are optimal paths for II(a) and II(b), respectively. Then they satisfy not only (i), (ii) and (iii) in Theorem 2.9 but also the following:

(iv) $\hat{p}_t \cdot \hat{x}_t + \sum_{w=\tau+1}^{t} f_w(\hat{x}_w, \hat{c}_w) = \max\{\hat{p}_t \cdot x_t + \sum_{w=\tau+1}^{t} f_w(x_w, c_w) | (x_{w+1},$

$\qquad c_{w+1}) \in T_w(x_w), w = \tau, \tau + 1, \ldots, t - 1, \text{ and } x_\tau = \hat{x}_\tau\}$,

$\qquad N \geq t > \tau \geq 0$,

(v) $\hat{p}_t \cdot \hat{x}_t = \min\{\hat{p}_t \cdot x_t | (x_{w+1}, c_{w+1}) \in T_w(x_w), w = t, t + 1, \ldots, \tau' - 1,$

$\qquad \text{and } x_{\tau'} = \hat{x}_{\tau'}, \sum_{w=t+1}^{\tau'} f_w(x_w, c_w) \geq \sum_{w=t+1}^{\tau'} f_w(\hat{x}_w, \hat{c}_w)\}$,

$\qquad N \geq \tau' > t \geq 0$,

(vi) $\hat{p}_t \cdot \hat{x}_t = \min\{p_t \cdot \hat{x}_t | p_w \in T^*_w(p_{w+1}), w = t, t + 1, \ldots, \tau' - 1,$

$\qquad \text{and } p_{\tau'} = \hat{p}_{\tau'}\}$, $N \geq \tau' > t \geq 0$,

(vii) $\hat{p}_t \cdot \hat{x}_t = \max\{p_t \cdot \hat{x}_t | p_w \in T^*_w(p_{w+1}), w = \tau, \tau + 1, \ldots, t - 1,$

$\qquad \text{and } p_\tau = \hat{p}_\tau\}$, $N \geq t > \tau \geq 0$.

This corollary is directly obtained by applying Corollary 2.3 to Problem II′ and Problem II′ (reciprocal) and restating the results in terms of Problem II. Each result in the above corollary is simply a restatement of the general principle of optimality in dynamic programming that "any portion of an optimal path is also an optimal path." In particular, considering the facts that $\hat{p}_N = \bar{p}_N = 0$ and that T_{t-1} has the property of free disposability of outputs, we get from (v) with $\tau' = N$:

Corollary 2.6′. Suppose $[\{\hat{x}_t\}_0^N, \{\hat{c}_t\}_1^N]$ and $\{\hat{p}_t\}_0^N$ are optimal paths for II(a) and II(b), respectively. Then, for each t $(t = 0, 1, \ldots, N-1)$, $\hat{x}_t$ has the minimum value of the input cost $x_t \cdot \hat{p}_t$ among all the input vectors in period t from which the welfare stock $\Sigma_{w=t+1}^N f_w(\hat{x}_w, \hat{c}_w)$ can be obtained thereafter.

2.4.5. Max-min principle

Let us define the Hamiltonian function for Problem II by

$$H_t(x_t, c_t, p_t) = f_t(x_t, c_t) + p_t \cdot x_t. \tag{62}$$

Then we have

Theorem 2.12 (max-min principle for Problem II).   Under either one of the two conditions in Theorem 2.10, a feasible path $[\{\hat{x}_t\}_0^N, \{\hat{c}_t\}_1^N]$ for II(a) is an optimal path for that problem if and only if there exists a feasible path $\{\hat{p}_t\}_0^N$ for II(b) with $\hat{p}_t \geq 0$ $(t = 0, 1, \ldots, N-1)$ such that they together satisfy the next two conditions. In this case, $\{\hat{p}_t\}_0^N$ is an optimal path for II(b).

(i) $H_t(\hat{x}_t, \hat{c}_t, \hat{p}_t) = \text{Max}\{H_t(x_t, c_t, \hat{p}_t) | (x_t, c_t) \in T_t(\hat{x}_{t-1})\}, \quad t = 1, 2, \ldots, N.$

(ii) $H_t(\hat{x}_t, \hat{c}_t, \hat{p}_t) = \text{Min}\{H_t(\hat{x}_t, \hat{c}_t, p_t) | p_t \in T_t^*(\hat{p}_{t+1})\}, \quad t = 0, 1, \ldots, N-1.$

This theorem is directly obtained by applying Theorem 2.7 to Problem II′ and by using Theorem 2.10.

To obtain the conventional representation of the maximum principle for Problem II, let us assume, for example, that T_t $(t = 0, 1, \ldots, N-1)$ in II(a) is represented by

$$T_t(x_t) = \{(x_{t+1}, c_{t+1}) | (x_{t+1}, c_{t+1}) = (x_1^{t+1}, \ldots, x_i^{t+1}, \ldots,$$
$$x_m^{t+1}, c_1^{t+1}, \ldots, c_j^{t+1}, \ldots, c_n^{t+1}) \quad \text{for some } x_i^{t+1} = f_i^t(x_t, \theta_t),$$
$$i = 1, \ldots, m, c_j^{t+1} = g_j^t(x_t, \theta_t), \quad j = 1, \ldots, n, \theta_t \in U_t\}, \tag{63}$$

for each $x_t \in \mathbf{R}_+^m$, $t = 0, 1, \ldots, N - 1$. Note that the above formulation includes the next one as a special case.

$$T_t(x_t) = \{(x_{t+1}, c_{t+1})|0 \le x_{t+1} + (c_{t+1}, 0) \le f_t(x_t, \theta_t)$$
$$\theta_t \in U_t, \text{ where } x_{t+1} \in \mathbf{R}_+^m, c_{t+1} \in \mathbf{R}_+^n\}.$$

Here, $m \ge n \ge 0$, and f_t is a m-dimensional vector function.

Then the Hamiltonian function (62) is rewritten as follows:

$$H_t(p_t, x_{t-1}, \theta_{t-1}) = f_t(f_1^{t-1}(x_{t-1}, \theta_{t-1}), \ldots, g_n^{t-1}(x_{t-1}, \theta_{t-1}))$$
$$+ \sum_{i=1}^m p_i^t f_i^{t-1}(x_{t-1}, \theta_{t-1}). \tag{64}$$

Hence, by applying Theorem 2.8 to Problem II′, we immediately have

Theorem 2.13 (maximum principle for Problem II). Suppose T_t in Problem II is represented by (63) in which each f_i^t and each g_j^t are concave and continuously differentiable in x_t on $\mathbf{R}_+^m$. Then under either one of two conditions in Theorem 2.10, a feasible path $[\{\hat{x}_t\}_0^N, \{\hat{c}_t\}_1^N]$ for II(a) and a feasible control sequence $\{\hat{\theta}_t\}_0^{N-1}$ are an optimal path and an optimal control sequence for II(a), respectively, if and only if there exists a sequence of vectors $\{\hat{p}_t\}_0^N$ which satisfies the next two conditions.

(i) $H_t(\hat{p}_t, \hat{x}_{t-1}, \hat{\theta}_{t-1}) = \max \{H_t(\hat{p}_t, \hat{x}_{t-1}, \theta_{t-1})|\theta_{t-1} \in U_{t-1}\}$,
$$t = 1, 2, \ldots, N.$$

(ii) $\hat{p}_i^t = \partial H_{t+1}(\hat{p}_{t+1}, \hat{x}_t, \hat{\theta}_t)/\partial x_i$, $i = 1, 2, \ldots, m$,
$$t = 0, 1, \ldots, N - 1, \text{ and } \hat{p}_N = 0.$$

2.5. Extension of duality theorems to generalized monotone programming

Compared to general convex programming, the biggest restriction on Problem II is the assumption of the first-degree homogeneity of transformations and objective functions. This assumption is removed in this section, and the results from the preceding sections are extended to this new class of problems.

A transformation T from $\mathbf{R}_+^m$ into $\mathbf{R}_+^n$ may be called a *generalized monotone transformation of concave type* (abbreviated to generalized m.t.cc.) from $\mathbf{R}_+^m$ into $\mathbf{R}_+^n$ if it satisfies the following five conditions.

(i)　$T(\lambda x^1 + (1 - \lambda)x^2) \supset \lambda T(x^1) + (1 - \lambda)T(x^2)$
$$\text{for all } x^1, x^2 \in \mathbf{R}_+^m \text{ and } 1 > \lambda > 0,$$

(ii)　$T(0)$ is a monotone set of concave type in $\mathbf{R}_+^n$,

(iii)　it is a closed mapping,

(iv)　$0 \leqq x^1 \leqq x^2$ implies $T(x^1) \subset T(x^2)$　for all $x^1, x^2 \in \mathbf{R}_+^m$,

(v)　$0 \leqq y^1 \leqq y^2 \in T(x)$ implies $y^1 \in T(x)$　for all $y^1, y^2 \in \mathbf{R}_+^n$.

A generalized m.t.cc. is called *non-singular* when $T(x)$ has a non-empty interior for $x > 0$. This implies, as before, that if x is strictly positive, then there is a strictly positive output vector in $T(x)$.

A continuous real valued function f on $\mathbf{R}_+^m$ is called a *generalized monotone concave gauge* on $\mathbf{R}_+^m$ if it satisfies the next three conditions:

(a)　$f(\lambda x^1 + (1 - \lambda)x^2) \geqq \lambda f(x^1) + (1 - \lambda)f(x^2)$
$$\text{for all } x^1, x^2 \in \mathbf{R}_+^m \text{ and } 1 > \lambda > 0.$$

(b)　$f(0) \geqq 0.$

(c)　$f(x^2) \geqq f(x^1)$　for $x^2 \geqq x^1 \geqq 0.$

Note, a m.t.cc. and a monotone concave gauge are also, respectively, a generalized m.t.cc. and a generalized monotone concave gauge. A problem in which each T_t and f_t in Problem II is replaced by a generalized m.t.cc. and a generalized monotone concave gauge, respectively, may be called *generalized monotone programming*.

Using these terms, the primal problem in this section is stated as follows:

Problem III.　Given a non-singular generalized m.t.cc. T_t from $\mathbf{R}_+^m$ into $\mathbf{R}_+^{m+n}$ $(t = 0, 1, \ldots, N - 1)$, a generalized monotone concave gauge f_t on $\mathbf{R}_+^{m+n}$ $(t = 1, 2, \ldots, N)$ where $f_N(x_N, c_N) \neq f(0, 0)$, and an initial vector $0 \leq \bar{x}_0 \in \mathbf{R}_+^m$:

III(a)　Maximize $\displaystyle\sum_{t=1}^{N} f_t(x_t, c_t),$
　　　　subject to $(x_t, c_t) \in T_{t-1}(x_{t-1}), t = 1, 2, \ldots, N,$
　　　　and $x_0 = \bar{x}_0.$

The intention in this section is to obtain duality theorems for the above problem by utilizing the results from the preceding two sections. Thus, since it was required in the preceding two sections that every function and transformation in a problem have the property of first-degree homogeneity, the above problem should be changed into an equivalent problem in which each welfare function and transformation is linearly homogeneous.

Given a numerical function $f(x)$ on $\mathbf{R}_+^m$ which is not linearly homogeneous, a simple way to change it into a linearly homogeneous function is to introduce a fictitious scalar variable, say η, and define a new function $h(x, \eta)$ by $h(x, \eta) = \eta f(x/\eta)$ for $x \geqq 0$ and $\eta > 0$. Then $h(x, \eta)$ is linearly homogeneous with respect to (x, η). Of course, the value of function h at $\eta = 0$ should be defined appropriately so as to ensure the continuity of h. Similarly, a transformation $T(x)$ which is not linearly homogeneous can be changed into a linearly homogeneous function, $\zeta T(x/\xi)$, with the help of a fictitious scalar variable ξ. More precisely, we have

Lemma 2.7. In the context of Problem III, a new function $h_t(x, c, \eta)$ of which the graph given by the closure of the set

$$\left\{((x, c, \eta), y) | y = \eta f_t\left(\frac{x}{\eta}, \frac{c}{\eta}\right), (x, c) \in \mathbf{R}_+^{m+n} \quad \text{and } \eta > 0\right\}, \qquad (65)$$

is a monotone concave gauge on $\mathbf{R}_+^{m+n+1}$ $(t = 1, 2, \ldots, N)$, and $h_t(x, c, \eta) = \eta f_t(x/\eta, c/\eta)$ for all $(x, c) \in \mathbf{R}_+^{m+n}$ and $\eta > 0$. Similarly, in the context of Problem III, a new function S_t of which the graph given by the closure of the set

$$\left\{((x, \xi), z) | z \in \xi T_t\left(\frac{x}{\xi}\right), x \in \mathbf{R}_+^m \quad \text{and } \xi > 0\right\}, \qquad (66)$$

is a m.t.cc. from $\mathbf{R}_+^{m+1}$ into $\mathbf{R}_+^{m+n}$ $(t = 0, 1, \ldots, N-1)$, and $S_t(x, \xi) = \xi T_t(x/\xi)$ for all $x \in \mathbf{R}_+^m$ and $\xi > 0$.

For the proof, see Appendix C.9. However, this concept is not difficult to demonstrate graphically. For example, take a generalized monotone concave gauge $f(x)$ from $\mathbf{R}_+$ into $\mathbf{R}_+$. Then, Figure 2.9 depicts how to obtain a monotone concave gauge $h(x, \eta)$ from $\mathbf{R}_+^2$ into $\mathbf{R}_+$ by the closure operation represented by (65). Similarly, if we consider in Figure 2 that the curve $f(x)$ in that figure represents the upper boundary of the generalized m.t.cc., $T(x) = \{y | 0 \leqq y \leqq f(x)\}$, from $\mathbf{R}_+$ into $\mathbf{R}_+$, then that figure also depicts how to obtain a m.t.cc. from $\mathbf{R}_+^2$ into $\mathbf{R}_+$ by closure operation (66).

Next, considering that these new variables η and ξ represent just two new (fictitious) commodities, we define a new transformation T_t'

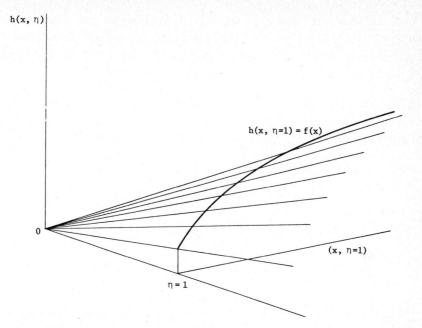

Figure 2.9. Generation of a monotone concave gauge from a generalized monotone concave gauge.

$(t = 0, 1, \ldots, N - 1)$ by

$$T'_t(x_t, \eta_t, \xi_t) = \{(x_{t+1}, c_{t+1}, \eta_{t+1}, \xi_{t+1})|(x_{t+1}, c_{t+1}) \in S_t(x_t, \xi_t),$$
$$0 \leq \eta_{t+1} \leq \eta_t \quad \text{and} \quad 0 \leq \xi_{t+1} \leq \xi_t\}, \tag{67}$$

for each $(x_t, \eta_t, \xi_t) \in \mathbf{R}_+^{m+2}$. In the context of Problem III, T'_t is clearly a non-singular m.t.cc. from $\mathbf{R}_+^{m+2}$ into $\mathbf{R}_+^{m+n+2}$ $(t = 0, 1, \ldots, N - 1)$, and the new problem is:

III'(a)   Maximize $\displaystyle\sum_{t=1}^{N} h_t(x_t, c_t, \eta_t)$,

subject to $(x_t, c_t, \eta_t, \xi_t) \in T'_{t-1}(x_{t-1}, \eta_{t-1}, \xi_{t-1})$, $t = 1, 2, \ldots, N$,
and $(x_0, \eta_0, \xi_0) = (\bar{x}_0, 1, 1)$.

Since all the conditions on f_t and T_t in Problem II are satisfied, respectively, by h_t and T'_t in this problem, all the results in the preceding sections can be directly applied to problem III'(a).

Recall that h_t is a non-decreasing function of η, and S_t also is a "non-decreasing function" of ξ from condition (v) of m.t.cc. Therefore if $[\{(\hat{x}_t, \hat{\eta}_t, \hat{\xi}_t)\}_0^N, \{\hat{c}_t\}_1^N]$ is an optimal path for III'(a), then $[\{(\hat{x}_t, 1, 1)\}_0^N,$

$\{\hat{c}_t\}_1^N]$ is also an optimal path for III'(a). Hence, we can assume $\hat{\eta}_t = 1$ and $\hat{\xi} = 1$ $(t = 0, 1, \ldots, N)$ on the optimal path for III'(a) without loss of generality. Then, since III'(a) is exactly the same as III(a) when $\eta_t = 1$ and $\xi_t = 1$ $(t = 0, 1, \ldots, N)$ in III'(a), we have:

Lemma 2.8. $[\{(\hat{x}_t, \hat{\eta}_t, \hat{\xi}_t)\}_0^N, \{\hat{c}_t\}_1^N]$ is a solution for III'(a) only if $[\{\hat{x}_t\}_0^N, \{\hat{c}_t\}_1^N]$ is a solution for III(a). Conversely, if $[\{\hat{x}_t\}_0^N, \{\hat{c}_t\}_1^N]$ is a solution for III(a), then $[\{(\hat{x}_t, 1, 1)\}_0^N, \{\hat{c}_t\}_1^N]$ is a solution for III'(a).

Next, applying adjoint operation (61), and taking into account the fact that h_t is continuous on $\mathbf{R}_+^{m+n+1}$ and T_t' is closed mapping from $\mathbf{R}_+^{m+2}$ into $\mathbf{R}_+^{m+n+2}$, the adjoint transformation of T_t' is given by

$$T_t'^*(p_{t+1}, p_\eta^{t+1}, p_\xi^{t+1}) = \Big\{(p_t, p_\eta^t, p_\xi^t) \in \mathbf{R}_+^{*m+2} \big| p_t \cdot x_t + p_\eta^t \eta + p_\xi^t \xi$$

$$\geq p_{t+1} \cdot x_{t+1} + p_\eta^{t+1} \eta + p_\xi^{t+1} \xi + \eta f_{t+1}\left(\frac{x_{t+1}}{\eta}, \frac{c_{t+1}}{\eta}\right)$$

$$\text{for all } (x_{t+1}, c_{t+1}) \in S_t(x_t, \xi), \ x_t \geq 0,$$

$$\eta > 0 \text{ and } \xi > 0\Big\}, \tag{68}$$

for each $(p_{t+1}, p_\eta^{t+1}, p_\xi^{t+1}) \in \mathbf{R}^{m+2}$. Now the dual problem of III(a) is formulated as follows:

III(b)   Minimize $p_0 \cdot \bar{x}_0 + p_\eta^0 + p_\xi^0$,
   subject to $(p_t, p_\eta^t, p_\xi^t) \in T_t'^*(p_{t+1}, p_\eta^{t+1}, p_\xi^{t+1})$, $t = 0, 1, \ldots, N-1$,
   and $p_N = 0, p_\eta^N = 0, p_\xi^N = 0$.

Then, by applying Theorem 2.9 to III'(a) and III(b), and recalling Lemma 2.8, the next theorem is immediately obtained.

Theorem 2.14 (duality theorem for Problem III). Suppose $[\{\hat{x}_t\}_0^N, \{\hat{c}_t\}_1^N]$ is an optimal path for III(a), and $\{(\hat{p}_t, \hat{p}_\eta^t, \hat{p}_\xi^t)\}_0^N$ is an optimal path for III(b). Then, we have:

(i) $\displaystyle\sum_{t=1}^N f_t(\hat{x}_t, \hat{c}_t) = p_{N-1} \cdot \hat{x}_{N-1} + \hat{p}_\eta^{N-1} + \hat{p}_\xi^{N-1} + \sum_{\tau=1}^{N-1} f_\tau(\hat{x}_\tau, \hat{c}_\tau) = \cdots$

$$= \hat{p}_t \cdot \hat{x}_t + \hat{p}_\eta^t + \hat{p}_\xi^{N-1} + \sum_{\tau=1}^t f_\tau(\hat{x}_\tau, \hat{c}_\tau) = \cdots = \hat{p}_0 \cdot \bar{x}_0 + \hat{p}_\eta^0 + \hat{p}_\xi^0.$$

(ii) $f_t(\hat{x}_t, \hat{c}_t) + \hat{p}_t \cdot \hat{x}_t + \hat{p}_\eta^t + \hat{p}_\xi^t = \max\{f_t(x_t, c_t) + \hat{p}_t \cdot x_t | (x_t, c_t)$

$\in T_t(\hat{x}_{t-1})\} + \hat{p}_\eta^t + \hat{p}_\xi^t$, $t = 1, 2, \ldots, N$.

$f_t(\hat{x}_t, \hat{c}_t) + \hat{p}_t \cdot \hat{x}_t + \hat{p}_\eta^t + \hat{p}_\xi^t = f_t(\hat{x}_t, \hat{c}_t) + \min\{(p_t \cdot \hat{x}_t + p_\eta^t + p_\xi^t) |$

$(p_t, p_\eta^t, p_\xi^t) \in T_t'^*(\hat{p}_{t+1}, \hat{p}_\eta^{t+1}, \hat{p}_\xi^{t+1})\}$, $t = 0, 1, \ldots, N-1$.

(iii) $(\hat{p}_t, \hat{p}^t_\eta, \hat{p}^t_\xi) \geq 0$ for $t = 0, 1, \ldots, N-1$,

and $\hat{p}_N = 0, \hat{p}^N_\eta = 0, \hat{p}^N_\xi = 0$.

(iv) $\hat{p}^t_\eta \geq \hat{p}^{t+1}_\eta, \hat{p}^t_\xi \geq \hat{p}^{t+1}_\xi$ for $t = 0, 1, \ldots, N$.

(v) Conversely, if a pair of feasible paths, $[\{\hat{x}_t\}^N_0, \{\hat{c}_t\}^N_1]$ for III(a) and $\{(\hat{p}_t, \hat{p}^t_\eta, \hat{p}^t_\xi)\}^N_0$ for III(b), satisfy the condition, $\Sigma^N_{t=1} f_t(\hat{x}_t, \hat{c}_t) = \hat{p}_0 \cdot \bar{x}_0 + \hat{p}^0_\eta + \hat{p}^0_\xi$, then they are optimal paths for III(a) and III(b), respectively.

Note that (iv) in the above theorem immediately follows from the definition of adjoint transformation T'^*_t by (68).

From (i) and (iv), we see that

$$\left(\hat{p}_{t+1} \cdot \hat{x}_{t+1} + \sum_{\tau=1}^{t+1} f_\tau(\hat{x}_\tau, \hat{c}_\tau)\right) - \left(\hat{p}_t \cdot \hat{x}_t + \sum_{\tau=1}^{t} f_\tau(\hat{x}_\tau, \hat{c}_\tau)\right)$$

$$= (\hat{p}^t_\eta - \hat{p}^{t+1}_\eta) + (\hat{p}^t_\xi - \hat{p}^{t+1}_\xi) \geq 0, \quad t = 0, 1, \ldots, N-1.$$

Hence, if the values of the fictitious variables (i.e., the payments $\hat{p}^t_\eta$ and $\hat{p}^t_\xi$ to the fictitious variables $\hat{\eta}_t = 1$ and $\hat{\xi}_t = 1$, and the revenues $\hat{p}^{t+1}_\eta$ and $\hat{p}^{t+1}_\xi$ from fictitious variables $\hat{\eta}_{t+1} = 1$ and $\hat{\xi}_{t+1} = 1$) are omitted, the optimal activity in each period may realize positive profits under the price sequence $\{\hat{p}_t\}^N_0$. The differences $\hat{p}^t_\eta - \hat{p}^{t+1}_\eta$ and $\hat{p}^t_\xi - \hat{p}^{t+1}_\xi$ may be considered to represent the marginal productivity of the fictitious variables η and ξ during the period t on the optimal path. Finally, comparing Theorem 2.14 with Theorem 2.9, we see that we can choose $\hat{p}^t_\eta = 0$ $(t = 0, 1, \ldots, N)$ if every f_t $(t = 1, 2, \ldots, N)$ is a monotone concave guage in Problem III and that we can choose $\hat{p}^t_\xi = 0$ $(t = 0, 1, \ldots, N)$ if every T_t $(t = 1, 2, \ldots, N)$ is a m.t.cc. in Problem III.

Next, for the existence theorem for Problem III, we need the next lemma of which proof is given in Appendix C.10.

Lemma 2.9. In the context of Problem III, if f_{t+1} satisfies the Lipshitz condition, and if f_{t+1} is a monotone concave gauge or there exists a real number K_{t+1} such that $f_{t+1}(x, c) \leq K_{t+1}$ on $\mathbf{R}^{m+n}_+$, then h_{t+1} satisfies the Lipschitz condition where h_{t+1} is the function defined by the closure operation in Lemma 2.7 and $t = 0, 1, 2, \ldots, N-1$. Likewise, if T_t satisfies the Lipshitz condition, and if T_t is a m.t.cc. or there exists a bounded set D_t in $\mathbf{R}^m_+$ such that $T_t(x) \subset D_t$ for any $x \in \mathbf{R}^m_+$, then T'_t satisfies the Lipschitz condition where T'_t is the function defined by (67) and $t = 0, 1, \ldots, N-1$.

Hence, by applying Theorem 2.10 to a pair of problems III'(a) and III(b), and recalling Lemma 2.8, we have

Theorem 2.15 (existence theorem for Problem III)
 (i) There always exists an optimal path for III(a).
 (ii) An optimal path exists for III(b) under either one of the next two
      conditions.
   (a) $\bar{x}_0 > 0$.
   (b) T_t and f_{t+1} satisfy the Lipschitz condition, f_{t+1} is a monotone
       concave gauge or there exists a real number K_{t+1} such that
       $f_{t+1}(x, c) \leq K_{t+1}$ on $\mathbf{R}_+^{m+n}$, and T_t is a m.t.cc. or there exists a
       bounded set D_t in $\mathbf{R}_+^m$ such that $T_t(x) \subset D_t$ for any $x \in \mathbf{R}_+^m$. Here,
       $t = 0, 1, \ldots, N - 1$.

In (b) of above theorem, both f_{t+1} and T_t are assumed to be bounded on
their domain in order to directly apply Theorem 2.10. But we know that
both η_t and ξ_t can be fixed on the optimal path for III'(a). And
$h_t(x_t, c_t, \eta_t) = f_t(x_t, c_t)$ when $\eta_t = 1$, and $S_t(x_t, \xi_t) = T_t(x_t)$ when $\xi_t = 1$.
Hence, it will be possible to prove that the boundedness of both
functions, f_{t+1} and T_t, is not actually needed in the above theorem.

Next, the Hamiltonian function for the pair of problems III'(a) and
III(b) is defined by

$$H_t''(x_t, c_t, \eta_t, \xi_t, p_t, p_\eta^t, p_\xi^t) = h_t(x_t, c_t, \eta_t) + p_t \cdot x_t + p_\eta^t \eta_t + p_\xi^t \xi_t. \quad (69)$$

But, since the values of η_t and ξ_t can each be fixed to 1, it is
convenient to abbreviate $H_t''(x_t, c_t, 1, 1, p_t, p_\eta^t, p_\xi^t)$ to $H_t'(x_t, c_t, p_t, p_\eta^t, p_\xi^t)$.
That is, we define H_t' by

$$H_t'(x_t, c_t, p_t, p_\eta^t, p_\xi^t) = f_t(x_t, c_t) + p_t \cdot x_t + p_\eta^t + p_\xi^t.$$

Then, by applying Theorem 2.12 to the pair of Problems III'(a) and
III(b) and by using Lemma 2.8, the max-min principle for Problem III
is immediately obtained as follows.

Theorem 2.16 (max-min principle for Problem III).   Under either one
of the two conditions in Theorem 2.15, a feasible path $[\{\hat{x}_t\}_0^N, \{\hat{c}_t\}_1^N]$ for
III(a) is an optimal path for that problem if and only if there exists a
feasible path $\{(\hat{p}_t, \hat{p}_\eta^t, \hat{p}_\xi^t)\}_0^N$ for III(b) with $(\hat{p}_t, \hat{p}_\eta^t, \hat{p}_\xi^t) \geq 0$ $(t = 0,$
$1, \ldots, N - 1)$ such that they together satisfy the next two conditions.
In this case, $\{(\hat{p}_t, \hat{p}_\eta^t, \hat{p}_\xi^t)\}_0^N$ is an optimal path for III(b).

 (i) $H_t'(\hat{x}_t, \hat{c}_t, \hat{p}_t, \hat{p}_\eta^t, \hat{p}_\xi^t) = \max \{H_t'(x_t, c_t, \hat{p}_t, \hat{p}_\eta^t, \hat{p}_\xi^t) | (x_t, c_t) \in T_{t-1}(\hat{x}_{t-1})\},$
$$t = 1, 2, \ldots, N.$$

(ii) $\quad H'_t(\hat{x}_t, \hat{c}_t, \hat{p}_t, \hat{p}^t_\eta, \hat{p}^t_\xi) = \min \{H'_t(\hat{x}_t, \hat{c}_t, p_t, p^t_\eta, p^t_\xi)|(p_t, p^t_\eta, p^t_\xi)$

$$\in T'^*_t(\hat{p}_{t+1}, \hat{p}^{t+1}_\eta, \hat{p}^{t+1}_\xi)\}, \quad t = 0, 1, \ldots, N-1.$$

A more interesting result is the following. Suppose T_t ($t = 0, 1, \ldots, N-1$) in Problem III is formulated in the form of (63). Therefore suppose that the transformation S_t, of which the graph given by the closure of the set (66), is represented by

$$S_t(x_t, \xi_t) = \{(x_{t+1}, c_{t+1})|(x_{t+1}, c_{t+1}) = (x^{t+1}_1, \ldots, x^{t+1}_i, \ldots, x^{t+1}_m,$$
$$c^{t+1}_1, \ldots, c^{t+1}_j, \ldots, c^{t+1}_n) \quad \text{for some } x^{t+1}_i = \tilde{f}^t_i(x_t, \theta_t, \xi_t),$$
$$i = 1, 2, \ldots, m, c^{t+1}_j = \tilde{g}^t_j(x_t, \theta_t, \xi_t), j = 1, 2, \ldots, n, \theta_t \in U_t\}$$

where

$$\tilde{f}^t_i(x, \theta, \xi) = \xi f^t_i(x/\xi, \theta) \quad \text{for } \xi > 0, i = 1, 2, \ldots, m,$$
$$\tilde{g}^t_i(x, \theta, \xi) = \xi g^t_j(x/\xi, \theta) \quad \text{for } \xi > 0, j = 1, 2, \ldots, n.$$

Here, f^t_i and g^t_j are those functions specified in the transformation (63). Then, the Hamiltonian function (69) can be rewritten as follows:

$$H''_t(x_{t-1}, \eta_t, \xi_{t-1}, \theta_{t-1}, p_t, p^t_\eta, p^t_\xi) = h_t(\tilde{f}^{t-1}_1(x_{t-1}, \theta_{t-1}, \xi_{t-1}), \ldots,$$
$$\tilde{g}^{t-1}_n(x_{t-1}, \theta_{t-1}, \xi_{t-1}), \eta_t)$$
$$+ \sum_{i=1}^m p^t_i \tilde{f}^{t-1}_i(x_{t-1}, \theta_{t-1}, \xi_{t-1})$$
$$+ p^t_\eta \eta_t + p^t_\xi \xi_t.$$

For convenience, let us introduce one more function H_t:

$$H_t(x_{t-1}, \theta_{t-1}, p_t) = f_t(f^{t-1}_1(x_{t-1}, \theta_{t-1}), \ldots, g^{t-1}_n(x_{t-1}, \theta_{t-1}))$$
$$+ \sum_{i=1}^m p^t_i f^{t-1}_i(x_{t-1}, \theta_{t-1}).$$

Then, since on the optimal path we can set $\hat{\eta}_t = 1$ ($t = 1, 2, \ldots, N$) and $\hat{\xi}_t = 1$ ($t = 0, 1, \ldots, N-1$), we have

$$\partial H''_t(\hat{x}_{t-1}, \hat{\eta}_t, \hat{\xi}_{t-1}, \hat{\theta}_{t-1}, \hat{p}_t, \hat{p}^t_\eta, \hat{p}^t_\xi)/\partial x_i = \partial H_t(\hat{x}_{t-1}, \hat{\theta}_{t-1}, \hat{p}_t)/\partial x_i,$$
$$i = 1, 2, \ldots, m,$$

and

$$H''_t(\hat{x}_{t-1}, \hat{\eta}_t, \hat{\xi}_{t-1}, \hat{\theta}_{t-1}, \hat{p}_t, \hat{p}^t_\eta, \hat{p}^t_\xi) = H_t(\hat{x}_{t-1}, \hat{p}_t, \hat{\theta}_{t-1}) + \hat{p}^t_\eta + \hat{p}^t_\xi.$$

Therefore, by applying Theorem 2.13 to the pair of problems III′(a) and III(b) and by neglecting unnecessary information about $\hat{p}^t_\eta$ and $\hat{p}^t_\xi$, we obtain the following theorem.

Theorem 2.17 (maximum principle for Problem III).   Suppose T_t in problem III(a) is represented by (63) in which each $f_i^t(x_t, \theta_t)$ and each $g_j^t(x_t, \theta_t)$ are concave and continuously differentiable with respect to x_t on $\mathbf{R}_+^m$. Then under either one of the two conditions in Theorem 2.15, a feasible path $[\{\hat{x}_t\}_0^N, \{\hat{c}_t\}_1^N]$ for III(a) and a feasible control sequence $\{\hat{\theta}_t\}_0^{N-1}$ are, respectively, an optimal path and an optimal control sequence for III(a) if and only if there exists a sequence of vectors $\{\hat{p}_t\}_0^N$ which satisfies the next two conditions.

(i) $H_t(\hat{x}_{t-1}, \hat{\theta}_{t-1}, \hat{p}_t) = \max \{H_t(\hat{x}_{t-1}, \theta_{t-1}, \hat{p}_t) | \theta_{t-1} \in \bar{U}_{t-1}\}$,

$$t = 1, 2, \ldots, N.$$

(ii) $\hat{p}_i^t = \partial H_{t+1}(\hat{x}_t, \hat{\theta}_t, \hat{p}_{t+1})/\partial x_i \quad i = 1, 2, \ldots, m, \quad t = 0, 1, \ldots, N-1$,

and

$$\hat{p}_i^N = 0, \qquad\qquad\qquad i = 1, 2, \ldots, m.$$

Notice that the fictitious variables η and ξ and the corresponding price variables p_η and p_ξ have completely disappeared from the above theorem despite the fact that it was obtained by using these fictitious variables. Moreover, the above theorem is the same as the ordinary form of the maximum principle for Problem III(a) when T_t in the problem is represented by (63).

Though the above two conditions (i) and (ii) provide sufficient information to solve Problem III(a), the following information is also obtained when Theorem 2.13 is applied to the pair of Problems III'(a) and III(b).

$$\hat{p}_\eta^t = \partial H_{t+1}''(\hat{x}_t, \hat{\eta}_{t+1}, \hat{\xi}_t, \hat{\theta}_t, \hat{p}_{t+1}, \hat{p}_\eta^t, \hat{p}_\xi^t)/\partial \eta$$

$$= \left. \frac{\partial \eta f_{t+1}(\hat{x}_{t+1}/\eta, \hat{c}_{t+1}/\eta)}{\partial \eta} \right|_{\eta=1} + \hat{p}_\eta^{t+1},$$

$$\hat{p}_\xi^t = \partial H_{t+1}''(\hat{x}_t, \hat{\eta}_{t+1}, \hat{\xi}_t, \hat{\theta}_t, \hat{p}_{t+1}, \hat{p}_\eta^t, \hat{p}_\xi^t)/\partial \xi$$

$$= \left. \frac{\partial f_{t+1}(\xi f_1^t(\hat{x}_t/\xi, \hat{\theta}_t), \ldots, \xi g_n^t(\hat{x}_t/\xi, \hat{\theta}_t))}{\partial \xi} \right|_{\xi=1}$$

$$+ \sum_{i=1}^m \hat{p}_i^{t+1} \left. \frac{\partial \xi f_i^t(\hat{x}_t/\xi, \hat{\theta}_t)}{\partial \xi} \right|_{\xi=1} + \hat{p}_\xi^{t+1}.$$

Hence we have

$$\hat{p}_\eta^t - \hat{p}_\eta^{t+1} = f_{t+1}(\hat{x}_{t+1}, \hat{c}_{t+1}) - (\hat{x}_{t+1}, \hat{c}_{t+1}) \cdot \nabla f_{t+1}(\hat{x}_{t+1}, \hat{c}_{t+1}), \qquad (70a)$$

$$\hat{p}_{\xi}^{t} - \hat{p}_{\xi}^{t+1} = \sum_{i=1}^{m} \frac{\partial f_{t+1}(\hat{x}_{t+1}, \hat{c}_{t+1})}{\partial x_{i}} (f_{i}^{t}(\hat{x}_{t}, \hat{\theta}_{t}) - \hat{x}_{t} \cdot \nabla f_{i}^{t}(\hat{x}_{t}, \hat{\theta}_{t}))$$

$$+ \sum_{j=1}^{n} \frac{\partial f_{t+1}(\hat{x}_{t+1}, \hat{c}_{t+1})}{\partial c_{j}} (g_{j}^{t}(\hat{x}_{t}, \hat{\theta}_{t}) - \hat{x}_{t} \cdot \nabla g_{j}^{t}(\hat{x}_{t}, \hat{\theta}_{t}))$$

$$+ \sum_{i=1}^{m} \hat{p}_{i}^{t+1}(f_{i}^{t}(\hat{x}_{t}, \hat{\theta}_{t}) - \hat{x}_{t} \cdot \nabla f_{i}^{t}(\hat{x}_{t}, \hat{\theta}_{t})). \tag{70b}$$

Here, each of ∇f_{t+1}, ∇f_{i}^{t} and ∇g_{j}^{t} represents the gradient of the corresponding function. Therefore, when function f_{t+1} is linearly homogeneous, $\hat{p}_{\eta}^{t} = \hat{p}_{\eta}^{t+1}$. Similarly, when all of f_{i}^{t} ($i = 1, 2, \ldots, m$) and g_{j}^{t} ($j = 1, 2, \ldots, n$) are linearly homogeneous with respect to x, then $\hat{p}_{\xi}^{t} = \hat{p}_{\xi}^{t+1}$, as was expected.

Note that (i) and (ii) in Theorem 2.17 can be solved independently of (70). Then (70) is solved by using those results.

2.6. Monotone programming with infinite horizons

So far it has been assumed that the length of the plan period was finite. But in many dynamic optimization problems in economics, especially in theoretical analyses, one is led to consider programming problems with an infinite time horizon. For example, consider consumption-investment planning with a finite time horizon. In this case, one must set certain targets for capital accumulation for the end of the plan period; otherwise no capital would be left at the end of the plan period. The question then arises as to how to determine these target levels of capital. This difficulty exists whether the plan period is short or long. One rational way to avoid this difficulty is to consider development programs over an infinite time horizon.[19]

In this section, results from the previous sections are extended to programming problems involving infinite horizons. But this extension is not straightforward since there is no "end period" for a programming problem with an infinite horizon. Recall that all the previous results were obtained by examining the relations between the primal optimal path and the dual optimal path, and that the programming of the dual path (i.e., price path) started from the end period. But, since there is no end period for a program with infinite horizons, the dual

---

[19]For the necessity of development programs with infinite horizon, for example, see Gale (1967a).

problem with an infinite time horizon cannot be constructed. Hence, instead of examining the relation between the two independent programs (i.e., the primal and the dual) with infinite horizons, we must devise some "limiting procedures" which utilize the dual properties of a sequence of programs with finite horizons.

In the first subsection, properties of frontier paths with infinite horizons are examined. Then, in the second subsection, we exploit the character of the optimal paths for consumption stream problems with an infinite time horizon. Finally, these results are extended to generalized monotone programming with infinite horizons.

2.6.1. *Frontier paths with infinite horizons*

In this section, we extend the results in Section 2.3 (duality and related theorems in final state problems) to monotone programs with infinite horizons. The base for the theory of monotone programs with finite horizons has been the price-character of the final state optimal paths (i.e., optimal paths for Problem I(a)). On the other hand, the base for the theory of monotone programming with infinite horizons is the price-character of a frontier path with an infinite time horizon.

To define a frontier path of infinite horizons, let us introduce some notation. As in Section 2.3, the state of the system at the beginning of period t is denoted by x_t which is a vector in $\mathbf{R}_+^m$, and the set of all feasible states (i.e., all feasible output vectors) of the system at the beginning of the next period $t + 1$ is denoted by $T_t(x_t)$ which is a subset of $\mathbf{R}_+^m$. Throughout this section, it is assumed that T_t is a non-singular m.t.cc. from $\mathbf{R}_+^m$ into $\mathbf{R}_+^m$ ($t = 0, 1, 2, \ldots$). Then, by using definition (28a), we set

$$A_t(\bar{x}_0) = (T_{t-1} \cdot T_{t-2} \cdot \ldots \cdot T_1 \cdot T_0)(\bar{x}_0),$$

where $\bar{x}_0$ ($\in \mathbf{R}_+^m$) is the initial vector specified in the problem. Hence $A_t(\bar{x}_0)$ gives the set of all output vectors at the beginning of period t which are feasible in t periods starting from the initial vector $\bar{x}_0$. By using the notation in Section 2.3.5, we denote the frontier set of $A_t(\bar{x}_0)$ by $F(A_t(\bar{x}_0))$. But, since we specify $\bar{x}_0$ to be a fixed vector in each problem, we simply represent $F(A_t(\bar{x}_0))$ by F_t whenever there would be no confusion. That is,

$$F_t = F(A_t(\bar{x}_0)), \quad t = 1, 2, \ldots$$

Then, a commodity path $\{\hat{x}_t\}_0^\infty$ is called a *frontier path with infinite horizons* if it satisfies the following conditions.

$$\hat{x}_t \in T_{t-1}(\hat{x}_{t-1}), \hat{x}_t \in F_t \quad \text{for } t = 1, 2, \ldots, \text{ and } \hat{x}_0 = \bar{x}_0.$$

Frontier paths with infinite horizons are simply called *frontier paths* whenever there would be no confusion.

By using this notation, the problem in this section is stated as follows.

Problem IV. Given non-singular m.t.cc. T_t from $\mathbf{R}_+^m$ into $\mathbf{R}_+^m$ ($t = 0, 1, \ldots$) and an initial input vector $\bar{x}_0 \in \mathbf{R}_+^m$, find a frontier path $\{\hat{x}_t\}_0^\infty$ with infinite horizons, that is, a path $\{\hat{x}_t\}_0^\infty$ which satisfies the following condition.

$$\hat{x}_t \in T_{t-1}(\hat{x}_{t-1}), \hat{x}_t \in F_t \quad \text{for } t = 1, 2, \ldots, \text{ and } \hat{x}_0 = \bar{x}_0. \tag{71}$$

Here, $F_t = F(A_t(\bar{x}_0))$, $A_t = T_{t-1} \cdot T_{t-2} \cdot \ldots \cdot T_1 \cdot T_0$, and it is assumed that

$$A_t(\bar{x}_0) \neq \{0\} \quad \text{for all } t. \tag{72}$$

Note that if $A_t(\bar{x}_0) = \{0\}$ for some $t = t'$, then $A_t(\bar{x}_0) = \{0\}$ for all $t \geq t'$. In this case the above problem degenerates to Problem I'(a), i.e., to a final state problem. Hence, throughout this section we assume that condition (72) is satisfied. This condition is met, for example, if $\bar{x}_0 > 0$.

The purpose of this section is to characterize the properties of frontier paths for Problem IV in terms of (efficiency) prices. For this purpose, we must first confirm that Problem IV has frontier paths with infinite horizons.

We denote by $X(\bar{x}_0)$ the set of all feasible paths (with an infinite time horizon) for Problem IV starting from the initial point $\bar{x}_0$. Each path $\{x_t\}_0^\infty$ in $X(\bar{x}_0)$ can be identified by a point, having x_t as the tth coordinate, in the Cartesian product Ω of the infinite number of Euclidian spaces $\mathbf{R}^m$ (i.e., $\Omega = \mathbf{R}^m \times \mathbf{R}^m \times \cdots$). Hence, $X(\bar{x}_0)$ can be considered as a subset of Ω. Similarly, denote by $\hat{X}_t(\bar{x}_0)$ the set of all feasible paths (with an infinite time horizon) starting from $\bar{x}_0$ and passing through some point on the frontier set F_t. $\hat{X}_t(\bar{x}_0)$ also can be regarded as a subset of Ω.

Lemma 2.10. In the context of Problem IV,
(i) F_t is a nonempty compact subset of $\mathbf{R}_+^m$ ($t = 1, 2, \ldots$).
(ii) If $x_{t-1} \in A_{t-1}(\bar{x}_0)$ and $x_{t-1} \notin F_{t-1}$, then $T_{t-1}(x_{t-1}) \cap F_t = \emptyset$ ($t = 2, 3, \ldots$).

(iii) $X(\bar{x}_0)$ is a nonempty compact subset of Ω.

(iv) $\hat{X}_t(\bar{x}_0)$ is a nonempty compact subset of Ω $(t = 1, 2, \ldots)$.

For the proof, see Appendix C.11. Statement (ii) implies that any feasible path passing through the frontier set F_t in period t must pass through some point on the frontier set F_{t-1} in period $t - 1$.

By using the above lemma, we have

Theorem 2.18 (existence of a frontier path).   There always exists a frontier path (with an infinite time horizon) for Problem IV. The set of all frontier paths constitutes a compact subset of Ω.

Proof.   We denote by $\hat{X}(\bar{x}_0)$ the set of all frontier paths (with infinite horizons) starting from $\bar{x}_0$. Then, by the definition of sets $\hat{X}_t(\bar{x}_0)$ $(t = 1, 2, \ldots)$,

$$\hat{X}(\bar{x}_0) = \bigcap_{t=1}^{\infty} \hat{X}_t(\bar{x}_0),$$

and from (ii) of Lemma 2.10,

$$\hat{X}_1(\bar{x}_0) \supset \hat{X}_2(\bar{x}_0) \supset \cdots \supset \hat{X}_t(\bar{x}_0) \supset \cdots.$$

Recall from (iv) of Lemma 2.10 that each $\hat{X}_t(\bar{x}_0)$ is a nonempty compact subset of Ω $(t = 1, 2, \ldots)$. Then, since the intersection of any number of compact sets is compact, $\hat{X}(\bar{x}_0)$ is compact. In addition, since the sequence of $\hat{X}_t(\bar{x}_0)$ is "decreasing," by the finite intersection axiom, we have $\hat{X}(\bar{x}_0) \neq \emptyset$. Consequently, $\hat{X}(\bar{x})$ is a nonempty compact subset of Ω.   Q.E.D.

We now proceed to characterize frontier paths in terms of efficient price vectors. This can be achieved by utilizing the results from Section 2.3. One sees that Problems I(c) and I(d) are more useful for this purpose than Problems I(a) and I(b) since there is no end period for Problem IV.

Suppose $\{\hat{x}_t\}_0^{\infty}$ is a frontier path for Problem IV. Take the optimal vector $\hat{x}_t$ in period t, and consider the supporting hyperplane of the "inverse set"

$$A_t^{-1}(\hat{x}_t) \equiv (T_0^{-1} \cdot T_1^{-1} \cdot \ldots \cdot T_{t-2}^{-1} \cdot T_{t-1}^{-1})(\hat{x}_t),$$

at the initial point $\bar{x}_0$. That is, consider a hyperplane $\hat{p}_0 \cdot x = \hat{p}_0 \cdot \bar{x}_0$ such that

$$0 < \hat{p}_0 \cdot \bar{x}_0 = \min\{\hat{p}_0 \cdot x_0 | x_0 \in A_t^{-1}(\hat{x}_t)\}, \quad \hat{p}_0 \in \mathbf{R}_+^{*m}.$$

We denote by $N_t(\bar{x}_0)$ the normal vectors of all hyperplanes supporting the set $A_t^{-1}(\hat{x}_t)$ at $\bar{x}_0$. That is,

$$N_t(\bar{x}_0) = \{\hat{p}_0 \in \mathbf{R}_+^{*m} | 0 < \hat{p}_0 \cdot \bar{x}_0 = \min\{\hat{p}_0 \cdot x_0 | x_0 \in A_t^{-1}(\hat{x}_t)\}\}. \tag{73}$$

We can construct the set $N_t(\bar{x}_0)$ for each period t ($t = 1, 2, \ldots$), and we have

Lemma 2.11. Suppose $\bar{x}_0 > 0$ in Problem IV. Then, if $\{\hat{x}_t\}_0^\infty$ is a frontier path for Problem IV, the set

$$N(\bar{x}_0) \equiv \bigcap_{t=1}^\infty N_t(\bar{x}_0) \tag{74}$$

is nonempty. Here each set $N_t(\bar{x}_0)$ is defined by (73).

For the proof, see Appendix C.12. By using the above lemma, we obtain

Theorem 2.19 (price sustainability of the frontier path). Suppose $\bar{x}_0 > 0$ in Problem IV. Then if $\{\hat{x}_t\}_0^\infty$ is a frontier path, there exists a sequence of price vectors $\{\hat{p}_t\}_0^\infty$ which satisfies the following two conditions.

(i) $0 < \hat{p}_0 \cdot \bar{x}_0 = \hat{p}_1 \cdot \hat{x}_1 = \cdots = \hat{p}_t \cdot \hat{x}_t = \cdots.$

(ii) $\hat{p}_t \in T_t^*(\hat{p}_{t+1}), \quad t = 0, 1, \ldots.$

Conversely, a feasible path $\{\hat{x}_t\}_0^\infty$ for Problem IV (here the condition, $\bar{x}_0 > 0$, is not required) is a frontier path if there exists a sequence of price vectors $\{\hat{p}_t\}_0^\infty$ which satisfies the above two conditions.

Proof. Let us first prove the latter half (i.e., sufficiency). Consider a pair of problems ($t = 1, 2, \ldots$):

I(a) Maximize $\hat{p}_t \cdot x_t$,

  subject to $x_\tau \in T_{\tau-1}^*(x_{\tau-1}), \quad \tau = 1, 2, \ldots, t,$

  and $x_0 = \bar{x}_0$.

I(b) Minimize $p_0 \cdot \bar{x}_0$,

  subject to $p_\tau \in T_\tau^*(p_{\tau+1}), \quad \tau = 0, 1, \ldots, t - 1,$

  and $p_t = \hat{p}_t$.

From the latter half of Theorem 2.1', conditions (i) and (ii) in the above theorem guarantee that the portion $\{\hat{x}_\tau\}_0^t$ of the path $\{\hat{x}_t\}_0^\infty$ is an optimal path for I(a) and the portion $\{\hat{p}_\tau\}_0^t$ of the path $\{\hat{p}_t\}_0^\infty$ is an optimal path for I(b). Hence, we have

$$0 < \hat{p}_t \cdot \hat{x}_t = \max \{\hat{p}_t \cdot x_t | x_t \in A_t(\bar{x}_0)\}, \quad t = 1, 2, \ldots.$$

Consequently, from Lemma 2.4, $\{\hat{x}_t\}_0^\infty$ is a frontier path for Problem IV.

Next, let us prove the first half of the above theorem. Take any price vector $\hat{p}_0 \in N(\bar{x}_0)$, where $N(\bar{x}_0)$ is the set defined by (74). Since $N(\bar{x}_0)$ is nonempty from Lemma 2.10, such a price vector exists. From the construction of set $N(\bar{x}_0)$ by (73) and (74),

$$0 < \hat{p}_0 \cdot \bar{x}_0 = \min \{\hat{p}_0 \cdot x_0 | x_0 \in A_t^{-1}(\hat{x}_t)\}, \quad t = 1, 2, \ldots.$$

Recall from (10) that $\langle \hat{p}_0, A_t^{-1}(\hat{x}_t) \rangle = \langle A_t^{*-1}(\hat{p}_0), \hat{x}_t \rangle$ since A_t^{*-1} is a m.t.cc. and A_t^{-1} is its adjoint transformation. In addition, $\langle A_t^{*-1}(\hat{p}_0), \hat{x}_t \rangle = \max \{p_t \cdot \hat{x}_t | p_t \in A_t^{*-1}(\hat{p}_0)\}$ since $A_t^{*-1}(\hat{p}_0)$ is a compact set. Therefore,

$$0 < \hat{p}_0 \cdot \bar{x}_0 = \min \{\hat{p}_0 \cdot x_0 | x_0 \in A_t^{-1}(\hat{x}_t)\}$$
$$= \max \{p_t \cdot \hat{x}_t | p_t \in A_t^{*-1}(\hat{p}_0)\}, \quad t = 1, 2, \ldots. \tag{75}$$

Next, denote by $P(\hat{p}_0)$ the set of all price paths (with an infinite time horizon) which start from $\hat{p}_0$ and satisfy the following feasibility condition:

$$p_{t+1} \in T_t^{*-1}(p_t), \quad t = 0, 1, 2, \ldots,$$

that is

$$p_t \in T_t^*(p_{t+1}), \quad t = 0, 1, 2, \ldots.$$

The set $P(\hat{p}_0)$ can be considered as a subset of the Cartesian product Ω^* of the infinite number of Euclidian spaces $\mathbf{R}^{*m}$ (i.e., $\Omega^* = \mathbf{R}^{*m} \times \mathbf{R}^{*m} \times \cdots$). By the same method as was used in the proof of (iii) in Lemma 2.10, it can be proved that $P(\hat{p}_0)$ is a nonempty compact subset of Ω. Furthermore, define the set Q_t ($t = 1, 2, \ldots$) by

$$Q_t = \{p_t | p_t \cdot \hat{x}_t = \max \{p \cdot \hat{x}_t | p \in A_t^{*-1}(\hat{p}_0)\}, \quad p_t \in A_t^{*-1}(\hat{p}_0)\}.$$

Then, since $A_t^{*-1}(\hat{p}_0)$ is a monotone set of concave type, one easily sees that Q_t is a nonempty compact subset of $\mathbf{R}^{*m}$. Therefore, if we denote by $\hat{P}_t(\hat{p}_0)$ the set of all feasible price paths which start from $\hat{p}_0$ and pass through some point on Q_t, then similarly to the proof of (iv) in Lemma 2.10, it can be proved that $\hat{P}_t(\hat{p}_0)$ is a nonempty compact subset of Ω^* ($t = 1, 2, \ldots$).

Moreover, from Theorem 2.4 (reciprocity principle) and Corollary 2.4, one observes that, for any given t (≥ 1), any path $\{p'_\tau\}_0^\infty \in \hat{P}_t(\hat{p}_0)$ satisfies the following conditions:

$$p'_\tau \in A_\tau^{*^{-1}}(\hat{p}_0), \quad p'_\tau \cdot \hat{x}_\tau = \max\{p_\tau \cdot \hat{x}_\tau | p_\tau \in A_\tau^{*^{-1}}(\hat{p}_0)\}, \quad \tau = 1, 2, \ldots t, \tag{76}$$

$$0 < \hat{p}_0 \cdot \bar{x}_0 = p'_1 \cdot \hat{x}_1 = \cdots = p'_\tau \cdot \hat{x}_\tau = \cdots = p'_t \cdot \hat{x}_t. \tag{77}$$

Condition (76) implies $p'_\tau \in Q_\tau$ for $\tau = 1, 2, \ldots, t$, and hence $\{p'_\tau\}_0^\infty \in \hat{P}_\tau(\hat{p}_0)$, $\tau = 1, 2, \ldots, t$. Therefore,

$$\hat{P}_1(\hat{p}_0) \supset \hat{P}_2(\hat{p}_0) \supset \cdots \supset \hat{P}_t(\hat{p}_0) \supset \cdots.$$

Consequently, recalling that each $\hat{P}_t(\hat{p}_0)$ is nonempty and compact, we conclude that the intersection

$$\hat{P}(\hat{p}_0) \equiv \bigcap_{t=1}^\infty \hat{P}_t(\hat{p}_0)$$

is a nonempty compact subset of Ω^*. Then, recalling (77) and that each path in $\hat{P}_t(\hat{p}_0)$ is a feasible path ($t = 1, 2, \ldots$), any path in $\hat{P}(\hat{p}_0)$ satisfies conditions (i) and (ii) of Theorem 2.19.   Q.E.D.

Of course, from (i) and (ii) of Theorem 2.19, any sequence of price vectors $\{\hat{p}_t\}_0^\infty$ in Theorem 2.19 satisfies the following condition.

$$\hat{p}_t \geq 0, \quad t = 1, 2, \ldots. \tag{78}$$

We may call a feasible path $\{\hat{x}_t\}_0^\infty$ for Problem IV an *efficient path* (with an infinite time horizon) if each $\hat{x}_t$ is on the efficient set of $A_t(\bar{x}_0)$ ($t = 1, 2, \ldots$). Then, since an efficient path is a frontier path, we have

Corollary 2.7. Suppose $\bar{x}_0 > 0$ in Problem IV. Then if $\{\hat{x}_t\}_0^\infty$ is an efficient path for Problem IV, then there exists a sequence of price vectors $\{\hat{p}_t\}_0^\infty$ which satisfies the two conditions in Theorem 2.19.

Note that the latter half of Theorem 2.19 does not always hold for efficient paths since a frontier path is not necessarily an efficient path.

The next results immediately follow from Theorem 2.1' and Corollary 2.3.

Corollary 2.8.   In the context of Theorem 2.19, if a pair of paths $\{\hat{x}_t\}_0^\infty$ and $\{\hat{p}_t\}_0^\infty$ satisfy conditions (i) and (ii) of Theorem 2.19, then they also satisfy condition (ii) in Theorem 2.1' for any $N \geq 1$ and condition (iv) in Corollary 2.3 for any $N \geq 1$.

The next theorem gives the max-min principle for Problem IV.

Theorem 2.20 (max-min principle for Problem IV).   Suppose $\bar{x}_0 > 0$ in Problem IV. Then, a feasible path $\{\hat{x}_t\}_0^\infty$ for Problem IV is a frontier path for that problem if and only if there exists a feasible price path $\{\hat{p}_t\}_0^\infty$ (i.e., a price path $\{\hat{p}_t\}_0^\infty$ which satisfies the feasibility condition (ii) of Theorem 2.19) such that they together satisfy the next two conditions. In this case, $\hat{p}_t \geq 0$ for $t = 0, 1, \ldots$.

 (i) $0 < \hat{p}_t \cdot \hat{x}_t = \max \{\hat{p}_t \cdot x_t | x_t \in T_{t-1}(\hat{x}_{t-1})\}, \quad t = 1, 2, \ldots$.
 (ii) $0 < \hat{p}_t \cdot \hat{x}_t = \min \{p_t \cdot \hat{x}_t | p_t \in T_t^*(\hat{p}_{t+1})\}, \quad t = 0, 1, \ldots$.

Note that if we do not assume $\bar{x}_0 > 0$ in the above theorem, the "if" part is always true but the "only if" part may not be always true. Similarly to the case of Theorem 2.7, the above theorem immediately derives from Theorem 2.19. Likewise, as in the case of Theorem 2.8, the next maximum principle directly follows from Theorem 2.20.

Theorem 2.21 (maximum principle for Problem IV).   Suppose T_t in Problem IV is represented by (54) where, for each fixed θ_t, each function f_j^t is concave and continuously differentiable with respect to x_t on $\mathbf{R}_+^m$. Then, if $\bar{x} > 0$, a feasible path $\{\hat{x}_t\}_0^\infty$ for Problem IV is a frontier path if and only if there exists a sequence of price vectors $\{\hat{p}_t\}_0^\infty$ and a control sequence $\{\hat{\theta}_t\}_0^\infty$ which satisfy the next two conditions. In this case, $\hat{p}_t \geq 0$ for $t = 0, 1, \ldots$.

 (i) $H_t(\hat{p}_t, \hat{x}_{t-1}, \hat{\theta}_{t-1}) = \max \{H_t(\hat{p}_t, \hat{x}_{t-1}, \theta_{t-1}) | \theta_{t-1} \in U_{t-1}\}$,
 (ii) $\hat{p}_i^t = \partial H_{t+1}(\hat{p}_{t+1}, \hat{x}_t, \hat{\theta}_t)/\partial x_i, \quad i = 1, 2, \ldots, m, t = 0, 1, \ldots$.

Here the Hamiltonian function H_t $(t = 1, 2, \ldots)$ is defined by

$$H_t(p_t, x_{t-1}, \theta_{t-1}) = \sum_{j=1}^m p_j^t f_j^{t-1}(x_{t-1}, \theta_{t-1}).$$

Again, the "only if" part in the above theorem may not always be true if we do not assume $\bar{x}_0 > 0$.

2.6.2. Consumption stream problems with infinite horizons

By using the results from the previous section, we now examine the character of optimal paths for consumption stream problems with an infinite time horizon.

As in Section 2.4, the state of the system at the beginning of period t is denoted by $x_t \in \mathbf{R}_+^m$, and the set of all feasible output vectors at the beginning of the next period is denoted by $T_t(x_t)$. Each vector in the set $T_t(x_t)$ has the form (x_{t+1}, c_{t+1}), where $x_{t+1} \in \mathbf{R}_+^m$ is the input vector for period $t+1$ and $c_{t+1} \in \mathbf{R}_+^m$ represents the part of the outputs consumed in period $t+1$. The welfare obtained in period $t+1$ is represented by $f_t(x_{t+1}, c_{t+1})$. By using this notation, the problem in this section is described as follows.

Problem V. Given non-singular m.t.cc. T_t $(t = 0, 1, \ldots)$ from $\mathbf{R}_+^m$ into $\mathbf{R}_+^{m+n}$, monotone concave gauge f_t $(t = 1, 2, \ldots)$ on $\mathbf{R}_+^{m+n}$ and an initial vector $0 < \bar{x}_0 \in \mathbf{R}_+^m$, find an "optimal path" among all the feasible paths which satisfy the following feasibility condition:

$$(x_t, c_t) \in T_{t-1}(x_{t-1}), \quad t = 1, 2, \ldots, \text{ and } x_0 = \bar{x}_0. \tag{79}$$

Here, it is assumed that there exists a period $t \geq 1$ such that $f_t(x, c)$ is not identically zero on $\mathbf{R}_+^{m+n}$.

The formulation of the above problem is not complete unless we specify the meaning of "optimal path." We may define that a feasible path is optimal when it maximizes the total welfare stock

$$\sum_{t=1}^{\infty} f_t(x_t, c_t), \tag{80}$$

among all the feasible paths. But this definition of optimality is deficient for the case of an infinite time horizon since the value of (80) may be infinite for some paths. Several authors proposed alternative definitions of optimal paths for programs with infinite horizons.[20] We may define the optimal paths as follows.

Definition of the optimal path. A feasible path $\{(\hat{x}_t, \hat{c}_t)\}_0^\infty$ [21] is:

(a) *weakly-optimal* if for every feasible path $\{(x_t, c_t)\}_0^\infty$ (starting from

---

[20]See, for example, Atsumi (1965), Gale (1967a), Halkin (1974) and Haurie (1976).

[21]For simplicity of notation, we often denote $[\{x_t\}_0^\infty, \{c_t\}_1^\infty]$ by $\{(x_t, c_t)\}_0^\infty$. In this representation, c_t with $t = 0$ has no meaning; for convention we may set $c_0 \equiv 0$.

$\bar{x}_0$), every $N \geqq 0$, and every $\epsilon > 0$ there exists a $\tau \geqq N$ such that

$$\sum_{t=1}^{\tau} f_t(x_t, c_t) \leqq \sum_{t=1}^{\tau} f_t(\hat{x}_t, \hat{c}_t) + \epsilon.$$

(b) *strongly-optimal* if for every feasible path $\{(x_t, c_t)\}_0^{\infty}$, there exists $N \geqq 0$ such that

$$\sum_{t=1}^{\tau} f_t(x_t, c_t) \leqq \sum_{t=1}^{\tau} f_t(\hat{x}_t, \hat{c}_t) \quad \text{for all } \tau \geqq N.$$

(c) *uniformly-strongly-optimal* if there exists $N \geqq 0$ such that any feasible path $\{(x_t, c_t)\}_0^{\infty}$ verifies

$$\sum_{t=1}^{\tau} f_t(x_t, c_t) \leqq \sum_{t=1}^{\tau} f_t(\hat{x}_t, \hat{c}_t) \quad \text{for all } \tau \geqq N.$$

In the above definition it is easy to observe that there is the following sequence of implications:

$$(c) \Rightarrow (b) \Rightarrow (a). \tag{81}$$

Note that, in the context of Problem V, $f_t(x_t, c_t) \geqq$ for every t. Hence, the sequence of welfare stocks, $\Sigma_{t=1}^{\tau} f_t(x_t, c_t)$, $\tau = 1, 2, \ldots$, converges to a finite number or diverges to ∞. If $\Sigma_{t=1}^{\infty} f_t(x_t, c_t) < \infty$ for all feasible paths, definition (a) of the optimality coincides with the initial definition of optimal paths (i.e., paths which maximize the total welfare stock (80) among all feasible paths).

We next examine the relationship between the optimal paths (defined by (a), (b) or (c) in the above) and the "frontier paths" for Problem V. As in Section 2.4.1, we introduce the *augmented transformation* $\mathcal{T}_t$ $(t = 0, 1, \ldots)$ defined by

$$\mathcal{T}_t(x_t, x_{m+1}^t) = \{(x_{t+1}, x_{m+1}^{t+1}) | 0 \leqq x_{m+1}^{t+1} \leqq x_{m+1}^t + f_{t+1}(x_{t+1}, c_{t+1}),$$
$$(x_{t+1}, c_{t+1}) \in T_t(x_t)\}, \tag{82}$$

for each $(x_t, x_{m+1}^t) \in \mathbf{R}_+^{m+n}$. And we define the transformation $\mathcal{A}_t$ $(t = 0, 1, \ldots)$ by

$$\mathcal{A}_t(x_0, x_{m+1}^0) = (\mathcal{T}_{t-1} \cdot \mathcal{T}_{t-2} \cdot \ldots \cdot \mathcal{T}_1 \cdot \mathcal{T}_0)(x_0, x_{m+1}^0), \tag{83}$$

for each $(x_0, x_{m+1}^0) \in \mathbf{R}_+^{m+n}$. According to Definition (83),

$$\mathcal{A}_t(\bar{x}_0, 0) = \{(x_t, x_{m+1}^t) | (x_t, x_{m+1}^t) \in (\mathcal{T}_{t-1} \cdot \ldots \cdot \mathcal{T}_1 \cdot \mathcal{T}_0)(\bar{x}_0, 0)\}.$$

In the context of Problem V, each $\mathcal{A}_t$ is a non-singular m.t.cc. from $\mathbf{R}_+^m$ into $\mathbf{R}_+^{m+n}$ $(t = 1, 2, \ldots)$. The set $\mathcal{A}_t(\bar{x}_0, 0)$ represents all the feasible output vectors in period t (starting from the initial vector $\bar{x}_0$) each of which includes the welfare stock in period t as its last component, x^t_{m+1}.

We denote by $F(\mathscr{A}_t(\bar{x}_0, 0))$ the frontier set of $\mathscr{A}_t(\bar{x}_0, 0)$; and call a feasible path $\{(x_t, c_t)\}^\infty_0$ for Problem V a *frontier path* if it passes through some point on the frontier set $F(\mathscr{A}_t(\bar{x}_0, 0))$ in every period t ($t = 1, 2, \ldots$), that is, if it satisfies the following condition.

$$\left(x_t, \sum_{\tau=1}^{t} f_\tau(x_\tau, c_\tau)\right) \in F(\mathscr{A}_t(\bar{x}_0, 0)), \quad t = 1, 2, \ldots. \tag{84}$$

We shall now prove the following lemma.

Lemma 2.12. For optimality defined by (a), (b) or (c), any optimal path for Problem V is a frontier path.

Proof. Because of relation (81), it is sufficient to prove that an optimal path in the sense of definition (a) is always a frontier path.

Suppose an optimal path $\{(\hat{x}_t, \hat{c}_t)\}^\infty_0$ is not a frontier path. Then from the definition of the frontier path by (84), there exists a period α and a number λ such that

$$\left(\lambda\hat{x}_\alpha, \lambda \sum_{t=1}^{\alpha} f_t(\hat{x}_t, \hat{c}_t)\right) \in \mathscr{A}_\alpha(\bar{x}_0, 0), \quad \lambda > 1.$$

Then, since $(\hat{x}_{\alpha+1}, \Sigma_{t=1}^{\alpha+1} f_t(\hat{x}_t, \hat{c}_t)) \in \mathscr{T}_\alpha(\hat{x}_\alpha, \Sigma_{t=1}^{\alpha} f_\tau(\hat{x}_t, \hat{c}_t))$ and $\mathscr{T}_\alpha$ is a m.t.cc.,

$$\left(\lambda\hat{x}_{\alpha+1}, \lambda \sum_{t=1}^{\alpha+1} f_t(\hat{x}_t, \hat{c}_t)\right) \in \mathscr{T}_\alpha\left(\lambda\hat{x}_\alpha, \lambda \sum_{t=1}^{\alpha} f_t(\hat{x}_t, \hat{c}_t)\right) \subset \mathscr{A}_{\alpha+1}(\bar{x}_0, 0).$$

Repeating this procedure, we have

$$\left(\lambda\hat{x}_N, \lambda \sum_{t=1}^{N} f_t(\hat{x}_t, \hat{c}_t)\right) \in \mathscr{A}_N(\bar{x}_0, 0) \quad \text{for all } N \geqq \alpha, \lambda > 1.$$

This implies there is a feasible path $\{(x'_t, c'_t)\}^\infty_0$ such that

$$\sum_{t=1}^{N} f_t(x'_t, c'_t) = \lambda \sum_{t=1}^{N} f_t(\hat{x}_t, \hat{c}_t) \quad \text{for all } N \geqq \alpha, \lambda > 1. \tag{85}$$

Next recall that each T_t is non-singular ($t = 1, 2, \ldots$), $\bar{x}_0 > 0$, and there exists a period $t' \geqq 1$ such that $f_{t'}(x, c) > 0$ when $(x, c) > 0$. Therefore, there exists a feasible path $\{(x_t, c_t)\}^\infty_0$ such that

$$f_{t'}(x_{t'}, c_{t'}) > 0.$$

Hence, if the path $\{(\hat{x}_t, \hat{c}_t)\}_0^\infty$ is an optimal path in the sense of definition (a), it can not be true that $f_t(\hat{x}_t, \hat{c}_t) = 0$ for all $t \geq 1$. This implies that there exists a period $\tau' \geq 1$ such that

$$f_{\tau'}(\hat{x}_{\tau'}, \hat{c}_{\tau'}) > 0.$$

Thus, from (85),

$$\sum_{t=1}^{N} f_t(x_t', c_t') - \sum_{t=1}^{N} f_t(\hat{x}_t, \hat{c}_t) \geq (\lambda - 1)f_{\tau'}(\hat{x}_{\tau'}, \hat{c}_{\tau'}) > 0,$$
$$\text{for all } N \geq \max\{\alpha, \tau'\},$$

which implies the path $\{(\hat{x}_t, \hat{c}_t)\}_0^\infty$ is not an optimal path in the sense of definition (a). This is a contradiction.

Therefore, we conclude that if a path is an optimal path in the sense of definition (a), it should be a frontier path.   Q.E.D.

From Theorem 2.19 and Lemma 2.12, we obtain

Theorem 2.22 (price sustainability of an optimal consumption path). Suppose $\{(\hat{x}_t, \hat{c}_t)\}_0^\infty$ is an optimal path for Problem V defined by (a), (b) or (c). Then there exists a sequence of non-negative price vectors $\{\hat{p}_t\}_0^\infty$ and a non-negative constant $\hat{p}_{m+1}$ such that

(i) $0 < \hat{p}_0 \cdot \bar{x}_0 = \hat{p}_1 \cdot \hat{x}_1 + \hat{p}_{m+1}f_1(\hat{x}_1, \hat{c}_1) = \cdots$

$$= \hat{p}_t \cdot \hat{x}_t + \hat{p}_{m+1} \sum_{\tau=1}^{t} f_\tau(\hat{x}_\tau, \hat{c}_\tau) = \cdots,$$

(ii) $\hat{p}_t \in T_t^*(\hat{p}_{t+1}), \quad t = 0, 1, \ldots.$

Here transformation T_t^* ($t = 0, 1, \ldots$) is defined by

$$T_t^*(p_{t+1}) = \{p_t \in \mathbf{R}_+^{*m} \mid p_t \cdot x_t \geq p_{t+1} \cdot x_{t+1} + \hat{p}_{m+1}f_{t+1}(x_{t+1}, c_{t+1})$$
$$\text{for all } (x_{t+1}, c_{t+1}) \in T_t(x_t), x_t \geq 0\}, \qquad (86)$$

for each $p_{t+1} \in \mathbf{R}_+^{*m}$.

Proof. From Lemma 2.12, the optimal path $\{(\hat{x}_t, \hat{c}_t)\}_0^\infty$ for Problem V should be a frontier path which satisfies condition (84). Hence, from Theorem 2.19,[22] there exists a sequence of price vectors $\{(\hat{p}_t, \hat{p}_{m+1}^t)\}_0^\infty$

---

[22]See the proof of Theorem 2.10 for the point that Theorem 2.19 can be applied even when $x_{m+1}^0 = 0$.

such that

$$0 < \hat{p}_0 \cdot \bar{x}_0 = \hat{p}_1 \cdot \hat{x}_1 + \hat{p}^1_{m+1} f_1(\hat{x}_1, \hat{c}_1) = \cdots$$

$$= \hat{p}_t \cdot \hat{x}_t + \hat{p}^t_{m+1} \sum_{\tau=1}^{t} f_\tau(\hat{x}_\tau, \hat{c}_\tau) = \cdots,$$

$$(\hat{p}_t, \hat{p}^t_{m+1}) \in \mathcal{T}^*_t(\hat{p}_{t+1}, \hat{p}^{t+1}_{m+1}), \quad t = 0, 1, \ldots,$$

$$(\hat{p}_t, \hat{p}^t_{m+1}) \geq 0, \quad t = 0, 1, \ldots.$$

Here $\mathcal{T}^*_t$ is the adjoint transformation of m.t.cc. $\mathcal{T}_t$, defined by

$$\mathcal{T}^*_t(p_{t+1}, p^{t+1}_{m+1}) = \{(p_t, p^t_{m+1}) \in \mathbf{R}^{*m+1}_+ | p_t \cdot x_t + p^t_{m+1} x^t_{m+1}$$

$$\geq p_{t+1} \cdot x_{t+1} + p^{t+1}_{m+1}(x^t_{m+1} + f_{t+1}(x_{t+1}, c_{t+1}))$$

$$\text{for all } (x_{t+1}, c_{t+1}) \in T_t(x_t), x_t \geq 0, x^t_{m+1} \geq 0), \tag{87}$$

for each $(p_{t+1}, p^{t+1}_{m+1}) \in \mathbf{R}^{*m+1}_+$. From this definition of $\mathcal{T}^*_t$, we see that

$$(p_t, p^t_{m+1}) \in \mathcal{T}^*_t(p_{t+1}, p^{t+1}_{m+1}) \quad \text{if and only if}$$

$$p^t_{m+1} \geq p^{t+1}_{m+1}, \text{ and}$$

$$p_t \cdot x_t \geq p_{t+1} \cdot x_{t+1} + p^{t+1}_{m+1} f_{t+1}(x_{t+1}, c_{t+1})$$

$$\text{for all } (x_{t+1}, c_{t+1}) \in T_t(x_t), x_t \geq 0.$$

From the above relation, we obtain

$$\hat{p}^0_{m+1} \geq \hat{p}^1_{m+1} \geq \cdots \geq \hat{p}^t_{m+1} \geq \cdots, \tag{88}$$

and

if $(p_t, p^t_{m+1}) \in \mathcal{T}^*_t(p_{t+1}, p^{t+1}_{m+1})$,

then $(p_t, p^{t+1}_{m+1}) \in \mathcal{T}^*_t(p_{t+1}, p^{t+1}_{m+1}), \quad t = 0, 1, \ldots.$

Hence, for any period $t \geq 0$,

$$(\hat{p}_\tau, \hat{p}^{t+1}_{m+1}) \in \mathcal{T}^*_\tau(\hat{p}_{\tau+1}, \hat{p}^{t+1}_{m+1}) \quad \text{for all } \tau = 0, 1, \ldots, t. \tag{89}$$

Next, recall that on optimal path $\{(\hat{x}_t, \hat{c}_t)\}^\infty_0$ there exists a period t such that $f_t(\hat{x}_t, \hat{c}_t) > 0$ (see the proof of Lemma 2.12). Denote the minimum of all such periods by N (≥ 1). Then,

$$\hat{x}^t_{m+1} = \sum_{\tau=1}^{t} f_\tau(\hat{x}_\tau, \hat{c}_\tau) \begin{cases} = 0, & t = 1, 2, \ldots, N-1, \\ > 0, & t = N, N+1, \ldots. \end{cases}$$

Further, applying Theorem 2.20 to Problem V, it should be true that

$$\hat{p}_t \cdot \hat{x}_t + \hat{p}^t_{m+1}\hat{x}^t_{m+1} = \min \{p_t \cdot \hat{x}_t + p^t_{m+1}\hat{x}^t_{m+1} | (p_t, p^t_{m+1})$$

$$\in \mathcal{T}^*_t(\hat{p}_{t+1}, \hat{p}^{t+1}_{m+1})\}, \quad t = 0, 1, \ldots.$$

Consequently, recalling (88) and (89) and setting $\hat{p}_{m+1}^{N+1} = \hat{p}_{m+1}$, we have

$\hat{p}_{m+1}^t = \hat{p}_{m+1}$ for all $t \geq N$,

$(\hat{p}_t, \hat{p}_{m+1}) \in \mathcal{T}_t^*(\hat{p}_{t+1}, \hat{p}_{m+1})$ for all $0 \leq t \leq N$,   and

$0 < \hat{p}_0 \cdot \bar{x}_0 = \hat{p}_1 \cdot \hat{x}_1 = \cdots = \hat{p}_{N-1} \cdot \hat{x}_{N-1}$

$= \hat{p}_N \cdot \hat{p}_N + \hat{p}_{m+1} \displaystyle\sum_{\tau=1}^{N} f_\tau(\hat{x}_\tau, \hat{c}_\tau) = \cdots.$

Hence the sequence of vectors $\{\hat{p}_t\}_0^\infty$ and the constant $\hat{p}_{m+1}$ satisfy conditions (i) and (ii) of Theorem 2.22.   Q.E.D.

In Theorem 2.22, it is obvious from (i) that

$$(\hat{p}_t, \hat{p}_{m+1}) \geq 0, \quad t = 0, 1, \ldots. \tag{90}$$

Corollary 2.9. In the context of Theorem 2.22, if $\hat{p}_{m+1} > 0$, then one can choose $\hat{p}_{m+1} = 1$.

Proof. If $\hat{p}_{m+1} \neq 0$, from (86) we see that

$\hat{p}_t / \hat{p}_{m+1} \in T_t^*(\hat{p}_{t+1}/\hat{p}_{m+1})$ with the new constant $\hat{p}_{m+1} = 1$.

Hence, the sequence of vectors $\{\hat{p}_t'\}_0^\infty = \{\hat{p}_t/\hat{p}_{m+1}\}$ and the new constant $\hat{p}_{m+1} = 1$ satisfy conditions (i) and (ii) of Theorem 2.22.   Q.E.D.

But, unlike the case of finite horizons (i.e., Section 2.4), we can not always assume that the constant $\hat{p}_{m+1}$ is positive. That is, we have

Corollary 2.10. In the context of Theorem 2.22, if the total welfare $\Sigma_{t=1}^\infty f_t(\hat{x}_t, \hat{c}_t)$ is not finite for the optimal path $\{(\hat{x}_t, \hat{c}_t)\}_0^\infty$, then

$$\hat{p}_{m+1} = 0, \quad \text{and} \quad \hat{p}_t \geq 0 \quad \text{for } t = 0, 1, \ldots.$$

This is an immediate consequence of (i) in Theorem 2.22. This corollary suggests that when the total welfare $\Sigma_{t=1}^\infty f_t(x_t, c_t)$ is not finite for some feasible path, then there is a possibility that there would not exist any optimal path in any definition of optimality, (a), (b) or (c). But the fact that the total welfare $\Sigma_{t=1}^\infty f_t(x_t, c_t)$ is not finite for some feasible path does not always imply that there is no optimal path.

Example 7. (An optimal path exists, and its total welfare is infinite.) Suppose each of x_t and c_t is a scalar, and

$$T_t(x_t) = \{(x_{t+1}, c_{t+1}) | 0 \le x_{t+1} \le x_t, 0 \le c_{t+1} \le x_t\}: \mathbf{R}_+ \to \mathbf{R}_+^2,$$

$$f_t(x_t, c_t) = c_t,$$

$$\bar{x}_0 = 1.$$

In this case, the optimal path is $\{(\hat{x}_t, \hat{c}_t)\}_0^\infty = \{(1, 1)\}_0^\infty$. Thus

$$\sum_{t=1}^\infty f_t(\hat{x}_t, \hat{c}_t) = \infty, \qquad \hat{p}_{m+1} = 0.$$

Example 8. (The maximum value of the total welfare is not finite, and there is no optimal path.)

Again suppose each of x_t and c_t is a scalar, and

$$T_t(x_t) = \{(x_{t+1}, c_{t+1}) | 0 \le x_{t+1} \le 2x_t,$$

$$0 \le c_{t+1} \le 2x_t - x_{t+1}\}: \mathbf{R}_+ \to \mathbf{R}_+^2,$$

$$f_t(x_t, c_t) = c_t,$$

$$\bar{x}_0 = 1.$$

In this case, one observes that any path with

$$c_t > 0 \quad \text{for a period, say, } t' \ge 1,$$

is "dominated" by some path with

$$c_t = 0 \quad \text{for } t = 1, 2, \ldots, t'.$$

Hence there is no optimal path.

Of course, the existence of a price sequence $\{\hat{p}_t\}_0^\infty$ and a constant $\hat{p}_{m+1}$ which satisfies (i) and (ii) of Theorem 2.22 does not always assure that the commodity path $\{(\hat{x}_t, \hat{c}_t)\}_0^\infty$ is the optimal path for Problem V. Though it would be difficult to obtain the general sufficient conditions for optimality, the next theorem provides one sufficient condition.

Theorem 2.23 (sufficiency theorem for Problem V). Suppose that $\{(\hat{x}_t, \hat{c}_t)\}_0^\infty$ is a feasible path for Problem V (here the assumption, $\bar{x}_0 > 0$, is not required), and that there exists a sequence of non-negative price vectors $\{\hat{p}_t\}_0^\infty$ and a non-negative constant $\hat{p}_{m+1}$ such that they together satisfy conditions (i) and (ii) of Theorem 2.22 and the next condition.

$$\lim_{t \to \infty} \hat{p}_t \cdot \hat{x}_t = 0.$$

Then the path $\{(\hat{x}_t, \hat{c}_t)\}_0^\infty$ is an optimal path for Problem V in the sense that any feasible path $\{(x_t, c_t)\}_0^\infty$ for Problem V (starting from $\bar{x}_0$)

verifies

$$\sum_{t=1}^{\infty} f_t(\hat{x}_t, \hat{c}_t) \geqq \sum_{t=1}^{\infty} f_t(x_t, c_t). \tag{91}$$

In this case, the total welfare $\Sigma_{t=1}^{\infty} f_t(\hat{x}_t, \hat{c}_t)$ is finite, and one can set $\hat{p}_{m+1} = 1$.

Proof. Recall that the sequence, $\Sigma_{\tau=1}^{t} f_\tau(x_\tau, c_\tau)$, $\tau = 1, 2, \ldots$, is non-decreasing for any feasible path $\{(x_\tau, c_\tau)\}_0^{\infty}$. Hence, the limit

$$\sum_{\tau=1}^{\infty} f_\tau(x_\tau, c_\tau) \equiv \lim_{t \to \infty} \sum_{\tau=1}^{t} f_\tau(x_\tau, c_\tau)$$

always exists. Suppose that a feasible path $\{(\hat{x}_t, \hat{c}_t)\}_0^{\infty}$, a sequence of non-negative price vectors $\{\hat{p}_t\}_0^{\infty}$ and a non-negative constant $\hat{p}_{m+1}$ together satisfy (i) and (ii) of Theorem 2.22. Then, from (i)

$$\hat{p}_t \cdot \hat{x}_t = \hat{p}_0 \cdot \bar{x}_0 - \hat{p}_{m+1} \sum_{\tau=1}^{t} f_\tau(\hat{x}_\tau, \hat{c}_\tau), \quad t = 1, 2, \ldots,$$

and hence the limit, $\lim_{t \to \infty} \hat{p}_t \cdot \hat{x}_t$, always exists; and we get

$$\hat{p}_{m+1} \sum_{\tau=1}^{\infty} f_\tau(\hat{x}_\tau, \hat{c}_\tau) = \hat{p}_0 \cdot \bar{x}_0 - \lim_{t \to \infty} \hat{p}_t \cdot \hat{x}_t.$$

Therefore, if $\lim_{t \to \infty} \hat{p}_t \cdot \hat{x}_t = 0$, then $\hat{p}_{m+1} > 0$ (since $\hat{p}_0 \cdot \bar{x}_0 > 0$), and

$$\sum_{\tau=1}^{\infty} f_\tau(\hat{x}_\tau, \hat{c}_\tau) = \hat{p}_0 \cdot \bar{x}_0 / \hat{p}_{m+1}.$$

Next, recalling the definition of transformation T_t^* by (86), from (ii) of Theorem 2.22 we have for any feasible sequence $\{(x_t, c_t)\}_0^{\infty}$ that

$$\hat{p}_0 \cdot \bar{x}_0 \geqq \hat{p}_1 \cdot x_1 + \hat{p}_{m+1} f_1(x_1, c_1),$$

$$\hat{p}_1 \cdot x_1 + \hat{p}_{m+1} f_1(x_1, c_1) \geqq \hat{p}_2 \cdot x_2 + \hat{p}_{m+1} \sum_{\tau=1}^{2} f_\tau(x_\tau, c_\tau), \ldots,$$

$$\hat{p}_t \cdot x_t + \hat{p}_{m+1} \sum_{\tau=1}^{t} f_\tau(x_\tau, c_\tau) \geqq \hat{p}_{t+1} \cdot x_{t+1} + \hat{p}_{m+1} \sum_{\tau=1}^{t+1} f_\tau(x_\tau, c_\tau), \ldots,$$

and hence,

$$\sum_{\tau=1}^{\infty} f_\tau(x_\tau, c_\tau) \leqq \hat{p}_0 \cdot \bar{x}_0 / \hat{p}_{m+1} = \sum_{\tau=1}^{\infty} f_\tau(\hat{x}_\tau, \hat{c}_\tau).$$

Therefore, we obtain (91). Since $\hat{p}_{m+1} \neq 0$, the value of $\Sigma_{\tau=1}^{\infty} f_\tau(\hat{x}_\tau, \hat{c}_\tau)$ is finite; and from Corollary 2.9, one can set $\hat{p}_{m+1} = 1$.   Q.E.D.

An optimal path in the sense of (91) is also optimal in the sense of definition (a) (i.e., weakly-optimal), but not always optimal in the sense of definitions (b) and (c). But if the optimal path in the sense of (91) is unique, then it is also the optimal path in the sense of definition (b) (i.e., strongly optimal).

The next corollary is immediately obtained by applying Corollary 2.8 to Problem V and recalling Corollary 2.6.

Corollary 2.11. In the context of Theorem 2.23, we set $\hat{p}_{m+1} = 1$. Then, the optimal path $\{(\hat{x}_t, \hat{c}_t)\}_0^\infty$ and the sequence of price vectors $\{\hat{p}_t\}_0^\infty$ together also satisfy conditions (i), (ii) and (iii) of Theorem 2.9 and conditions (iv) to (vii) of Corollary 2.6 for any $N \geq 1$. This also holds to be true in the context of Theorem 2.22 if every welfare function f_t (and f_τ, f_w) in Theorem 2.9 and Corollary 2.6 is multiplied by the constant $\hat{p}_{m+1}$.

Next, we have

Theorem 2.24 (Max-min principle for Problem V). Suppose $\{(\hat{x}_t, \hat{c}_t)\}_0^\infty$ is an optimal path for Problem V by any of the definitions (a), (b) or (c). Then there exist a non-negative constant $\hat{p}_{m+1}$ and a price path $\{\hat{p}_t\}_0^\infty$ which satisfies the feasibility condition (ii) of Theorem 2.22 such that they together satisfy the next two conditions.

(i) $\quad 0 < \hat{p}_{m+1} f_t(\hat{x}_t, \hat{c}_t) + \hat{p}_t \cdot \hat{x}_t$
$\qquad = \max \{\hat{p}_{m+1} f_t(x_t, c_t) + \hat{p}_t \cdot \hat{x}_t | (x_t, c_t) \in T_t(\hat{x}_{t-1})\}, \quad t = 1, 2, \ldots,$

(ii) $\quad 0 < \hat{p}_{m+1} f_t(\hat{x}_t, \hat{c}_t) + \hat{p}_t \cdot \hat{x}_t$
$\qquad = \min \{\hat{p}_{m+1} f_t(\hat{x}_t, \hat{c}_t) + p_t \cdot \hat{x}_t | p_t \in T_t^*(\hat{p}_{t+1})\}, \quad t = 0, 1, \ldots.$

Here the transformation T_t^* $(t = 0, 1, \ldots)$ is defined by (86). In this case, $(\hat{p}_t, \hat{p}_{m+1}) \geq 0$, $t = 0, 1, \ldots$.

Conversely, suppose that $\{(\hat{x}_t, \hat{c}_t)\}_0^\infty$ is a feasible path for Problem V (here the condition, $\bar{x}_0 > 0$, is not required), and that there exists a non-negative constant $\hat{p}_{m+1}$ and a feasible price sequence $\{\hat{p}_t\}_0^\infty$ (which satisfies condition (ii) of Theorem 2.22) such that they together satisfy the above two conditions, (i) and (ii), and the additional condition, $\lim_{t \to \infty} \hat{p}_t \cdot \hat{x}_t = 0$. Then the path $\{(\hat{x}_t, \hat{c}_t)\}_0^\infty$ is optimal for Problem V in the sense that it maximizes the welfare stock $\Sigma_{t=1}^\infty f_t(x_t, c_t)$ among all the feasible paths.

This theorem directly follows from Theorems 2.22 and 2.23 by applying the proof of Theorem 2.7. Finally, as in the case of Theorem 2.8, the next maximum principle immediately follows from Theorem 2.24.

Theorem 2.25 (maximum principle for Problem V). Suppose T_t ($t = 0, 1, \ldots$) in Problem V is represented by (63) in which, for each fixed θ_t, each of functions f_i^t and g_j^t is concave and continuously differential with respect to x_t on $\mathbf{R}_+^m$. Then, if $\{(\hat{x}_t, \hat{c}_t)\}_0^\infty$ and $\{\hat{\theta}_t\}_0^\infty$ are, respectively, an optimal path and an optimal control sequence for Problem V (by any definition of optimality (a), (b) or (c)), there exist a non-negative constant $\hat{p}_{m+1}$ and a sequence of non-negative price vectors $\{\hat{p}_t\}_0^\infty$ such that they together satisfy the next two conditions.

(i) $0 < H_t(\hat{p}_t, \hat{p}_{m+1}, \hat{x}_t, \hat{\theta}_t) = \max \{H_t(\hat{p}_t, \hat{p}_{m+1}, \hat{x}_{t-1}, \theta_{t-1}) | \theta_{t-1} \in U_{t-1}\},$

$$t = 1, 2, \ldots.$$

(ii) $\hat{p}_i^t = \partial H_{t+1}(\hat{p}_{t+1}, \hat{p}_m, \hat{x}_t, \hat{\theta}_t)/\partial x_i, \quad i = 1, 2, \ldots, m, t = 0, 1, \ldots.$

Here the Hamiltonian function H_t ($t = 1, 2, \ldots$) is defined by

$$H_t(p_t, p_{m+1}, x_t, \theta_t) = p_{m+1}f_t(f_1^{t-1}(x_{t-1}, \theta_{t-1}), \ldots,$$

$$g_n^{t-1}(x_{t-1}, \theta_{t-1})) + \sum_{i=1}^m p_i^t f_i^{t-1}(x_{t-1}, \theta_{t-1}).$$

Conversely, suppose that $\{(\hat{x}_t, \hat{c}_t)\}_0^\infty$ and $\{\hat{\theta}_t\}_0^\infty$ are, respectively, a feasible path and a feasible control sequence for Problem V in the present context (here the condition, $\bar{x}_0 > 0$, is not required). Then, if there exists a non-negative constant $\hat{p}_{m+1}$ and a sequence of non-negative price vectors $\{\hat{p}_t\}_0^\infty$ which together satisfy the above two conditions, (i) and (ii), and the next condition,

$$\lim_{t \to \infty} \hat{p}_t \cdot \hat{x}_t = 0,$$

then $\{(\hat{x}_t, \hat{c}_t)\}_0^\infty$ and $\{\hat{\theta}_t\}_0^\infty$ are an optimal path and an optimal control sequence for Problem V, respectively, in the sense that they together maximize the welfare stock $\sum_{t=1}^\infty f_t(x_t, c_t)$ among all the feasible paths.

PART II

DEVELOPMENT IN SPACE SYSTEMS
WITH CONVEX STRUCTURES

CONSTRUCTION OF MONOTONE SPACE SYSTEMS

3.1. Introduction

As explained in Sections 1.2 and 1.3, monotone space systems represent a class of space systems equipped with convex structures. In Part II, consisting of Chapters 3, 4, and 5, the optimal development in these monotone space systems will be studied. In Chapter 3, we will provide a general framework for the construction of monotone space systems. In Chapter 4, we will investigate the general character of optimal growth paths in these monotone space systems by applying the mathematical theorems from Chapter 2. Then, in Chapter 5, the turnpike behavior of the optimal growth paths will be studied by using two-region, two-good monotone space systems.

A space system is defined, as stated in Section 1.2 of Chapter 1, by three index sets: commodity index set J_i, location index set J_l and time index set J_t, and by the space transformation T_t for each t. Thus the construction of a monotone space system involves the specification of these index sets and space transformations. In this chapter we examine each of these elements, and provide a general framework for the construction of monotone space systems.

In Section 3.2, the basic elements of space systems are examined. Then, in Section 3.3, it is shown that at least three different types of monotone space systems can be constructed depending on how we combine production activities and transportation activities in a model.

In Section 3.4, monotone space systems based on composite mappings are first discussed. This representation of the space transformation will be convenient for theoretical analyses as well as for intuitive understanding of growth processes. Two concrete examples of monotone space systems are given in Section 3.5. Monotone space systems of this type are analyzed in detail in Chapters 4 and 5.

In the last section, two other types of monotone space systems are briefly discussed. It will be shown that monotone space systems based on convex-hull mappings are appropriate for the analyses of space systems in which transportation capacities are explicitly considered. Also, monotone space systems based on flow models are useful for practical planning in space systems.

3.2. Elements of monotone space systems

3.2.1. Time, commodities and locations

The three index sets: commodity index set J_i, location index set J_l and time index set J_t, which are closely interrelated with each other, are briefly discussed here.

For Part II, the discrete representation of time is adopted, and time is broken up into successive periods of equal length. The main reason for using discrete time is that we can construct practical models of space systems more easily by using discrete time than by using continuous time. In addition, we can only obtain statistics based on discrete periods. Moreover, discrete time models are more easily handled, theoretically as well as computationally, than continuous time models. The total number of plan periods, N, is assumed to be finite throughout Part II.

It is usually assumed that there exist m physically distinct *commodities* in our system, where m is a finite positive integer. Each commodity is assumed to be homogeneous qualitatively and continuously divisible quantitatively. The total number of commodities is closely related to the length of the period. For example, when the length of the period is one month and it takes one year for the construction of a factory, then eleven *intermediate products* should be introduced into the model besides the completed factory. And the number of commodities is also related to how the depreciation of durable goods is treated in the model. This point is discussed in the next Section, 3.2.2.

The definition of commodity index set is also related to that of location index set. Immobile commodities are often more conveniently defined by their locations than by their physical properties, for example, the weather in the South, a coal mine in Detroit and a highway connecting Philadelphia and New York. By definition, these commodities cannot exist in other locations. Thus, in the commodity index sets of other locations it is unnecessary to include these *immobile local commodities*. For simplicity of notation, we assume that there are the same number of local immobile commodities in each location, and hence the same commodity index set, $J_i = (1, 2, \ldots, i, \ldots m)$ is used for each location.[1] The contents of the index set for each location may differ from location to location with respect to immobile local commodities. But this does not matter since they are immobile.

Natural resources are discussed in Section 3.2.4, while man and labor are discussed in Section 3.2.5. The range of commodities considered in each model is related to the technological differences between locations. This point is discussed in Section 3.2.6.

A *location* denotes a point in space, within which every commodity can move freely: that is, we assume that every commodity can move without any transport inputs within a given location. The set of locations $J_l = (1, 2, \ldots, l, \ldots, r, \ldots, s, \ldots, n)$ constitutes the space.

A location may be a nation, a county or a town. It may be a "ring" surrounding a certain point. Or, it may be a member of a regional partition over the space. Or, it can be even a hypothetical point which denotes an industrial sector in a single-regional system in which nonshiftability of capital between different sectors is assumed. It is assumed that the total number of locations, n, is finite, and any commodity in the space system belongs to one of these locations.

3.2.2. Production activities

It is assumed that the space transformation T_t consists of two mutually exclusive groups of activities[2] – *production activities* and *transportation activities* – for each t $(= 0, 1, \ldots, N - 1)$. Thus production activities also include the *preservation activity* for the stock of each durable good, and transportation activities also include identity transportation (which transports a good from one location to itself without

---

[1]If the number of commodities is different between locations in actual models, then the obvious modifications should be considered in the following, but this does not affect the essential nature of the study.

[2]Recall that, in this book, any pair of input–output vectors (x, y) is called an *activity*. And if (x, y) is included in the graph of a certain transformation, say T, then it is called a *feasible activity* under T. T may be T_t, T_1, T_2, F_l or G_l, each of which is defined later.

any real transport inputs) within each location, and, possibly, *migra-tion activities* between locations. Transportation activities are dis-cussed in the next section.

Each production activity belongs to some location. And a production activity in a location, say l, transforms a set of inputs in location l into a set of outputs in that location after a certain length of time.[3] The properties of production activities are not discussed much here since our main concern is the spatial aspect of production. But, the *age structure* of commodities is important in relation to the mobility of commodities. Most production activities include durable capital goods as well as current inputs in their inputs. Thus, in addition to ordinary products, capital goods should be included in outputs of each activity. That is, "old" commodities appear in outputs of each activity as well as "new" commodities. We must distinguish old commodities from new commodities in two respects: old capital goods are less efficient than new capital goods, and old capital goods are less transportable than new capital goods since they may be installed in factories.

One device for the treatment of the age structure of commodities is the neoclassical one. We introduce the notion of *depreciation rate*, say δ_i, for each commodity i. And we assume that one unit of capital good i of a certain age is functionally equivalent to $(1 - \delta_i)$ units of capital good i of one unit of time earlier. But we must be careful concerning transportability since it is unrealistic to assume that capital i of one unit of age is $(1 - \delta_i)$ times less transportable than the new capital. Thus we cannot assume that there exists a common "depreciation rate" for efficiency and transportability of each commodity. But this neoclassical device is useful when we can assume that "old" com-modities are immobile and only "new" commodities are mobile.[4] In this case, only the depreciation rate for efficiency is needed.

The second approach is due to von Neumann. Capital goods at different stages of wear and tear are treated as qualitatively different goods.[5] In this treatment, there remains no ambiguity in efficiency and transportability of new and old capital. But, the number of com-modities is inevitably enlarged.

[3]The length of time that it takes for the initial inputs to be transformed into final products may be different from activity to activity. See footnote 1 in Section 3.4.1.

[4]We do not need to take the words "new" and "old" literally. They are for the convenience of transportation.

[5]For this point, see von Neumann (1945). A more detailed explanation is given in Morishima (1969, pp. 89–93).

In either approach, it is assumed that there are m commodities which are distinct with respect to quality, mobility and/or age structure.

3.2.3. *Transportation activities*

A transportation activity transforms a set of inputs, in general, over the space system into a set of outputs, in general, over the space system in a certain length of time.

Each transportation activity requires five different kinds of inputs: *transported goods, current inputs, moving facilities, route facilities*, and *terminal facilities.*[6]

As was noted in Section 3.2.1, each commodity should be assumed to be continuously divisible. For this purpose, we must introduce appropriate units of measurement for each transportation facility. For example, each highway connecting each pair of locations may be measured continuously by its number of lanes (with a certain sacrifice of reality).

And, as noted in Section 3.2.1, each commodity in the space system should belong to some location. Thus each route facility should also belong to some location. We may introduce new locations which denote the location of each facility if we are prepared for the resulting increase in the number of locations.

We shall employ the following assumption about transportation activities throughout this book:

 (i) The same facility is never used in both a transportation and a production activity in the same period.

And, depending on the problem, the following simplifying assumptions may be adopted.

 (ii) Each transportation activity transports only one transported good between only one pair of locations, and the total activity which transports a good from the origin to the final destination is one transportation activity.
(iii) The current inputs for transportation activities are supplied from the starting location.

[6]We assume that current inputs are completely used up at the end of each transportation activity.

For the explicit consideration of the capacity of transportation facilities, we need more assumptions; but this point is discussed in Section 3.5.1.

3.2.4. *Treatment of natural resources in monotone space systems*

Natural resources are commodities which are endowed by nature. For simplicity it is assumed that they cannot be produced in the system; therefore, we distinguish commodities which are given by nature from commodities which are produced in the system even though they are physically the same.

As it will become clear, a monotone space system represents a self-sustained growth system. But natural resources flow into the system from nature, that is, from outside of the system. Thus, we need some artificial device to include natural resources into a monotone space system.

For the construction of monotone space systems, the following classification of natural resources is convenient:

Group 1:   The total limitation of each resource in this group is given by the amount of the initial stock in each location.

Group 2:   Each resource in this group flows repeatedly into each location in each period.

Examples of natural resources in the first group are oil and mineral deposits in the ground. Rainfall and rivers are examples of the second group. Weather (temperature, sunshine, etc.) can also be treated as a natural resource of the second group. Unimproved land belongs to the first group (improved land should be treated as a usual durable good). Man can be treated by the same method as natural resources of the second group (see Section 3.2.5).

The natural resources of the first group can be incorporated into a monotone space system by introducing a *preservation activity* for each such natural resource as one of the production activities in each location. A preservation activity for each natural resource in the first group transforms a certain amount of that commodity into exactly the same amount of that commodity after one period of time without any other inputs. Since no costs are required for these activities, the unused portion of these resources is carried into the next period on an optimal path.

The amount of each natural resource of the second group is exogenously given in each location in each period. Hence it can be treated by the same method as durable capital goods, even if it is used as current inputs. For example, suppose the sequence of the amount of the natural resource i, for example, rainfall, in the second group in location l is given exogenously as follows:

$$\{\bar{x}_{il}^0, \bar{x}_{il}^1, \ldots, \bar{x}_{il}^t, \ldots, \bar{x}_{il}^{N-1}, \bar{x}_{il}^N\} = \{\bar{x}_{il}^t\}_0^N \tag{1}$$

where $\bar{x}_{il}^t$ is the amount of the natural resource i in location l in period t. Suppose further that we define the *rate of increase* of that resource in period t by

$$\lambda_{il}^t = (\bar{x}_{il}^{t+1}/\bar{x}_{il}^t) - 1, \quad t = 0, \quad ., \quad .., N - 1.^7 \tag{2}$$

Moreover, let us assume that if an activity in location l in period t uses x_{il}^t units of this resource as an input, then $(1 + \lambda_{il}^t) x_{il}^t$ units of this resource is included as an output of this activity. Finally, for the unemployed part of this natural resource, we introduce as one of the production activities the preservation activity which transforms a unit of this natural resource into $(1 + \lambda_{il}^t)$ units of that natural resource after one period without any other inputs. Then, when the initial amount of this natural resource is given by $\bar{x}_{il}^0$ as one of the initial conditions, the same sequence of this natural resource with (1) can be obtained on the optimal path.

3.2.5. *Man, labor, consumption and welfare functions*

As stated before, in monotone space systems man is treated by the same method as natural resources in the second group. When migration between locations is not allowed in the model, the population in each location can be given by a sequence, like (1), exogenously. Hence man can be treated in the same way as the natural resources in the

---

[7]For the rate of increase λ_{il}^t to be defined, it is necessary to assume that for each (i, l)

$\bar{x}_{il}^t > 0$ for all t,

or

$\bar{x}_{il}^t = 0$ for all t.

In the latter case, we define $\lambda_{il}^t = 0$ for all t.

second group. When migration is allowed in the model, the population in each location is a variable, and hence we cannot specify the sequence (1) beforehand. But it would be possible to specify sequence of the rate of growth, (2), exogenously. Hence, in this case also, man can be treated by the same method as natural resources of the second group. An example is shown in Section 3.5.1.

When there are several types of men in the model (depending on age, education, etc.), only the first type of men (for example, the youngest group of men) should be treated as a natural resource of the first group. The rest of the types of men should be treated in the same way as usual durable goods (like housing stocks).

Depending on a problem, it is appropriate to introduce another commodity, *labor*, into the model besides man (or, instead of man) as we see in the following.

There are typically two different types of problems in optimal growth theory in economics.[8] One is the problem of determining the optimal path for an economic system in which the objective is the maximization of the terminal stock of commodities, in its value with fixed prices or in its quantity of some bundle of goods, given some initial stock of commodities. This may be called the *final state* problem. In the other problem, the object is to maximize the utility of consumption over some time period, subject to a given initial stock of commodities. This may be called the *consumption stream problem*. The object of the analysis is quite different in each of these two problems.

In the final state problems studied in this book (for example in Section 3.5.1 and in the problems in Chapter 5), both man and labor are introduced. In these problems it is assumed that labor is a service commodity of which the inputs are man and consumption goods. And it is assumed that labor, as a service commodity, is durable for only one period, and it cannot be carried into the next period. And it is assumed that to obtain a unit of labor we must pay a fixed amount of each consumption good to each man. The fixed amount of each consumption good is not necessarily the minimum amount of each good biologically required for the subsistance of man. It is enough to assume that the amount of consumption goods paid to each unit of labor is decided beforehand by rules or mechanisms external to our

---

[8]Here we follow Chapter 17 in Hicks (1965) and Chapters 10 and 11 in Burmeister and Dobel (1970).

model. And it is also not necessary to assume that this "consumption vector" is fixed uniquely. The planner can prepare a set of consumption vectors for each period. From this set, the planner chooses a consumption vector for each period so as to maximize the value of final commodity stocks.

On the other hand, in consumption stream problems in this book, usually man is not distinguished from labor and both are called man. But, depending on the problem it is often convenient to distinguish man as a natural resource from labor as a service good (see the example in Section 3.5.2). When man and labor are distinguished from each other in consumption stream problems, one unit of man is converted into one unit of labor without any costs. But in this case, the dimension of labor is "man-period" and labor can be durable only for one period.

As formulated in Chapter 1, the objective function in the programming problem is generally represented by the function,

$$\sum_{t=1}^{N} f_t(X_t, C_t), \tag{3}$$

where $X_t = (x_l^t)_1^n \in \mathbf{R}_+^{mm}$ and $C_t = (c_l^t)_1^n \in \mathbf{R}_+^{m_c n}$, and the vector $x_l^t \in \mathbf{R}_+^m$ represents the state of the space system in location l at the beginning of the period t and the vector $c_l^t \in \mathbf{R}_+^{m_c}$ is the consumption vector in location l in period t. For the final state problem, function (3) is simply specified as follows:

$$\sum_{t=1}^{N} f_t(X_t, C_t) = f_N(X_N), \tag{4}$$

where f_N is a numerical function on $\mathbf{R}_+^{mn}$. That is, for the final state problem, $f_t(X_t, C_t) = 0$ for $t = 1, 2, \ldots, N-1$ and $m_c = 0$. In addition, it is assumed in this book that function (3) takes the following form in consumption stream problems:

$$\sum_{t=1}^{N} f_t(X_t, C_t) = \sum_{t=1}^{N} \left(\sum_{l=1}^{n} f_l^t(x_l^t, c_l^t) \right). \tag{5}$$

That is, the objective function is additive over time as well as over space. Following the economic convention, we may use the next function for the welfare function in each location.

$$f_l^t(x_l^t, c_l^t) = \frac{1}{(1+\rho)^t} x_{ml}^t u_l \left(\frac{c_l^t}{x_{ml}^t} \right), \tag{6}$$

where

u_l: a numerical function on $\mathbf{R}_+^{m_c}$ which represents the time-invariant
         utility function common to each person in location l,

x_{ml}^t: the population in location l at the beginning of period t, and

ρ: the utility discount rate where $\rho > -1$.

When migration between locations is not allowed, we can fix the population in each location in each period to some exogenously given number, $\bar{x}_{ml}^t$. But when migration between locations is allowed in the model, x_{ml}^t is a variable, and hence $x_{ml}^t = 0$ is technologically possible. But the value of function (6) cannot be defined without further assumptions when $x_{ml}^t = 0$. And when we define the value of function (6) at $(0, c_l^t)$, special care may be needed in order to guarantee the continuity at that point (in cases where global continuity is needed to ensure the existence of optimal solutions).

Special care is needed for another reason when we use (6) as the welfare function. For example, suppose u_l is a linearly homogeneous function on $\mathbf{R}_+^{m_c}$. Then $x_{ml}u_l(c_l/x_{ml}) = u_l(c_l)$ for any $(x_{ml}, c_l) \in \mathbf{R}_+^{m_c+1}$. That is, the welfare in this location is independent of the size of the population. In this case we may obtain an economically nonsensical solution. For example, suppose the space consists of two locations, $l = 1$ or 2, and there is only one consumption good, so that c_l is a scalar. Moreover, suppose the welfare function in each location is as follows:

$$x_{m1}u_1\left(\frac{c_1}{x_{m1}}\right) = x_{m1}\left(2 \cdot \frac{c_1}{x_{m1}}\right) = 2c_1$$

$$x_{m2}u_2\left(\frac{c_2}{x_{m2}}\right) = x_{m2}\left(1 \cdot \frac{c_2}{x_{m2}}\right) = c_2,$$

and that the productivity of men is higher in location 2 than in location 1. Then on the optimal path all men are allocated to location 2 and all of the consumption good is allocated to location 1. This is clearly a nonsense solution if commuting between locations is not considered.

Thus if migration between locations is allowed in the model, special care is required in the definition of the welfare function in order to avoid such problems.

In order to take into account the "scrap value" of the stock of commodities at the final period, we may use the following welfare function for the final period, N:

$$f_l^N(x_l^N, c_l^N) = \tilde{f}_l^N(x_l^N, c_l^N) + \bar{p}_l^N \cdot x_l^N,$$

where $\tilde{f}_l^N$ is an ordinary welfare function and $\bar{p}_l^N$ is a given vector of scrap value of commodities at the final period.

Further, we may add restrictions on welfares between locations. For example, in the case of two regions ($l = 1, 2$), we may add restrictions,

$$\alpha_1 \leqq f_2^t(x_2^t, c_2^t)/f_1^t(x_1^t, c_1^t) \leqq \alpha_2$$

where $0 \leqq \alpha_1 < \alpha_2$. This restriction can be handled within the framework of monotone programming by using the objective function

$$f_t(X_t, C_t) = \min\{f_1^t(x_1^t, c_1^t), f_2^t(x_2^t, c_2^t)/\alpha_1\}$$
$$+ \min\{f_2^t(x_2^t, c_2^t), \alpha_1 f_1^t(x_1^t, c_1^t)\}$$

in (5).

When we consider several types of men, say k types, in the model, we may use the following welfare function instead of (6):

$$f_l^t(x_l^t, c_l^t) = \max\left\{\sum_{j=1}^{k} \frac{w_{jl}^t}{(1+\rho)^t} x_{mjl}^t u_l^j \left(\frac{c_{mjl}^t}{x_{mjl}^t}\right) \middle| \sum_{j=1}^{k} c_{mjl}^t = c_l^t, \quad c_{mjl}^t \geqq 0, \right.$$

$$\left. j = 1, 2, \ldots, k\right\}, \qquad (7)$$

where

x_{mjl}^t: the number of men of type j in location l in period t, $j = 1, 2, \ldots, k$,

c_{mjl}^t: the total consumption vectors for the population of type j in location l in period t, $c_{mj}^t \in \mathbf{R}_+^{m_c}$,

w_{jl}^t: the weight for population type j in location l in period t,

u_l^j: the utility function for population type j in location l,

ρ: the utility discount rate where $\rho > -1$.

3.2.6. Technological differences between locations

Assume that technological knowledge is perfectly mobile over the space system. The question in this section is then whether it is possible for a technological difference in production to exist between locations. By the phrase "technological difference in production between locations," we mean the difference in the sets of activities which are technologically possible in each location.

If the distribution of natural resources is perfectly homogeneous over the space system, then there will be no technological differences between locations. On the other hand, suppose that the natural resources are not evenly distributed between locations. Then there possibly exist technological differences between locations for the industries which use these natural resources.

For example, let us examine the production of oranges in California and Pennsylvania. Suppose oranges are produced from labor which is homogeneous in two locations and from "sunshine" which (however measured) is different in the two locations. Then, when the production function of oranges in each location is represented (in the neoclassical way) by

$$y = f_l(L), \quad l = C(\text{California}) \text{ or } P(\text{Pennsylvania}),$$

where y is the amount of oranges produced from the labor input L, then f_C is different from f_P since they reflect the difference of the degree of sunshine between the two locations. On the other hand, when we explicitly take into account the degree of sunshine in the production of oranges, the production function in each location is represented as follows:

$$y = f_l(L, S), \quad l = C \text{ or } P,$$

where S is a variable which represents the degree of sunshine. In this case, there will be no difference between two functions f_C and f_P. But the degree of sunshine is actually different between the two locations. Thus the sets of activities which can be actually adopted in each location are different.

As another example, let us consider the coal extraction industry in the above two locations. We assume that coal is obtained from labor and coal ore, and the quality of coal ore is identical in both locations, but that the depth of coal ore in Pennsylvania is 10 feet and the depth of coal ore in California is 30 feet, and that the amount of coal ore in each location is unlimited. If the production function of coal is represented (in the neoclassical way) as follows in each location $l = C$, P,

$$y = f_l(L, C),$$

where y is the amount of coal obtained from labor input L and coal ore input C, then f_C is different from f_P since they reflect the difference of the depth of coal ore in each location. But, let us differentiate the coal ore by its depth as C_{10} and C_{30} where C_{30} is the amount of coal ore input in 30 feet depth. The production function for each location is now represented by

$$y = f_l(L, C_{10}, C_{30}), \quad l = C \text{ or } P.$$

In this case there is no difference between the two functions f_C and f_P. But the total amount of C_{10} in California is zero and the total amount of C_{30} is zero in Pennsylvania. Thus the set of activities which can be actually adapted differs by location.

Thus when natural resources are not evenly distributed between locations, there will be technological differences between locations in production which uses some natural resources. These differences should be reflected either in the form of the production functions or in the constraints on the total amount of each type of natural resource.

3.3. Three alternative ways for the construction of monotone space systems

In Section 3.2, the basic elements of space systems, except for space transformations, were discussed. In the remaining part of this chapter we consider the construction of space transformations, assuming we have already decided on an appropriate location index set J_l, commodity index set J_i and time index set J_t.

As noted in Section 3.2.2, it is assumed in this book that the space transformation T consists of production activities and transportation activities which are mutually exclusive of each other. Then each space system may be characterized differently depending on how these two groups of activities are combined in it. In this vein, at least three different types of space systems can be considered.

(1) As a first type, we may assume that production activities and transportation activities occur at different stages in each period. Thus, the notion of "period" is similar to that of "week" by Hicks in this case. Under this assumption, the space transformation T is represented as a *composite mapping* of two transformations – one represents production activities and the other represents transportation activities.

(2) As a second type, we may assume that these two groups of activities take place in parallel in each period and they are related to each other only at the beginning of the period. In this case, the space transformation T is represented by a *convex-hull mapping* of two transformations.

(3) As the third type, we may assume that both groups of activities are so intertwined that the performance of both activities are interrelated even within each period. Thus the space transformation T cannot be decomposed. In this case, the space transformation T takes the form of a *flow model*.

Each representation of the space transformation has its merits and faults, as will be shown later. However for the mathematical treatment as well as for the intuitive understanding of the growth process of the space system, the first method will be most convenient. Thus, the next two sections are devoted to monotone space systems based on composite mappings. In the last section, the other two types of space transformations are briefly discussed. The analysis in the remaining chapters of Part II is limited to the first type of space systems.

For simplicity of notation, it is assumed implicitly in the following discussion that the space transformation is time invariant. Thus the time subscript t is dropped from every function related to the space transformation. But if we want to consider technological changes over time, it is sufficient to put the time subscript t on each of these functions.

3.4. Monotone space systems based on composite mappings

3.4.1. Production activities, T_1

Let us assume that each of the production activities in each location has a duration of exactly one period.[9] That is, each production activity transforms a set of inputs in a location into a set of outputs in that location after one period duration. Then all of the production activities in each location can be symbolically represented as follows:

$$x_l \longrightarrow F_l(x_l), \quad l = 1, 2, \ldots, n, \tag{8}$$

where $F_l(x_l)$ is the set of all feasible output vectors for the production activities in location l after one period, given initial input vector x_l in

---

[9]As noted in footnote 3, the length of time that it takes for the initial inputs to be transformed into final products may be different from activity to activity. Then this assumption implies that we have already "standardized" production activities so that each of them is of one period duration; those of longer duration have been broken down into a set of single activities of unit time duration introducing, if necessary, intermediate products as additional commodities. Thus the commodity index set J_i should be taken to include these fictitious intermediate products. For this point, see von Neumann (1945, p. 222), and Morishima (1969, pp. 75 and 91).

that location. Thus F_l is a multivalued mapping. It is convenient for the analysis of monotone space systems based on composite mappings to assume that the domain and the range of F_l are, respectively, $\mathbf{R}_+^m$ and some subset of $\mathbf{R}_+^{m'}$, where m and m' ($\geq m$) are not necessarily the same. The reason is as follows. If we use the neoclassical treatment of the depreciation of commodities, it is not necessary to distinguish "old" commodities from "new" commodities in production activities after appropriate adjustments by depreciation rates. But they should be distinguished from each other with respect to transportation activities. For example, suppose m_1 kinds of machines are included in m kinds of inputs for production activities in each location and these machines are installed in factories and they cannot be moved between locations. Then m_1 kinds of "old" machines are included in the outputs of production activities besides m kinds of new products. And, since only "new" machines can be transported between locations, "old" machines should be differentiated from the "new" machines in the inputs for transportation activities which take place after production activities in each period. Thus, the number of output commodities of production activities should be $m + m_1$, and this is equal to m'. For details, see the model used in Chapter 5.

Hence, let us assume $x_l \in \mathbf{R}_+^m$, $F_l(x_l) \subset \mathbf{R}_+^{m'}$ and F_l is a multivalued mapping from $\mathbf{R}_+^m$ into $\mathbf{R}_+^{m'}$ in (8), where $m' \geq m$.

Using (8), the set of feasible production activities in the space system is symbolically represented as follows:

$$X \longrightarrow T_1(X),$$

where

$$T_1(X) = \{Y \mid Y = (y_l)_1^n \text{ for some } y_l \in F_l(x_l),$$
$$l = 1, 2, \ldots, n \text{ where } (x_l)_1^n = X\}. \tag{9}$$

In (9), X is a vector in $\mathbf{R}_+^{mn}$ and represents the state of the space system at the beginning of a period. $T_1(X)$ is a subset of $\mathbf{R}_+^{m'n}$ and represents the set of all feasible states of the space system after a one-period duration of production activities, given initial input vector X. Hence, T_1 is a multivalued mapping from $\mathbf{R}_+^{mn}$ into $\mathbf{R}_+^{m'n}$.

3.4.2. Transportation activities, T_2

Let us assume that each of the transportation activities is such that it uses a set of commodities over the space system at the end of each period and it obtains a set of outputs over the space at the beginning of the next period. Thus we are assuming that the time required for each transportation activity can be neglected.[10] Then, the set of all possible transportation activities in the space system can be represented symbolically as follows:

$$Y \longrightarrow T_2(Y).\tag{10}$$

Here Y is a vector in $\mathbf{R}_+^{m'n}$ and it represents the state of the space system at the end of the production activities in a period. $T_2(Y)$ is a subset in $\mathbf{R}_+^{mn}$ and it shows the set of all feasible states of the space system after the performance of transportation activities given transportation input vector Y. After being transported, "old" machines and "new" machines are added together in each location since they are functionally equivalent as an input commodity for production activities (after appropriate adjustment by depreciation rates). Thus, each vector in $T_2(Y)$ is an mn-dimensional vector. Hence, T_2 is a multivalued mapping from $\mathbf{R}_+^{m'n}$ into $\mathbf{R}_+^{mn}$.

If we want to specify transportation activities in more detail than in representation (10), we need some assumptions. For example, if we do not consider transportation facilities explicitly and if we adopt assumption (iii) in Section 3.2.3, transformation T_2 takes a simple form.[11] In this case, each transportation activity transforms a set of inputs in a location, say l, to a set of outputs in another location, say s, as follows:

$$y_{ls} \longrightarrow z_{ls}, \quad l, s = 1, 2, \ldots, n,\tag{11}$$

where $y_{ls} \in \mathbf{R}_+^{m'}$, $z_{ls} \in \mathbf{R}_+^{m}$. And the set of feasible transportation activi-

---

[10]Under this assumption we can treat in a correct manner the amount of current inputs required for each transportation activity since its dimension does not include time directly, for example, one gallon of gasoline per each movement of one ton of steel from Philadelphia to New York.

Moreover, we are not double-counting the function of some capital in production activities and transportation activities because of assumption (i) in Section 3.2.3.

But, if we take this assumption literally, that is, if we assume that the transportation activity does not require time, then the capacity of each transportation facility cannot be correctly treated since it is defined per unit of time. Thus, if we want to consider the capacity of transportation facilities under this method of the construction of space systems, that is, under the composite mapping, some conventions are needed.

In the other two methods which are discussed in Section 3.6, transportation capacities can be handled correctly.

[11]This simplification may be appropriate in some cases, for example in international trade problems. Even in interregional problems, there are very few papers in which transportation facilities are seriously considered.

ties from location l to location s is represented as follows:

$$y_{ls} \longrightarrow G_{ls}(y_{ls}), \quad l, s = 1, 2, \ldots, n, \tag{12}$$

where G_{ls} is a multivalued mapping from $\mathbf{R}_+^{m'}$ into $\mathbf{R}_+^{m}$. Then the set of feasible transportation activities which originate from location l is represented as follows:

$$y_l \longrightarrow G_l(y_l) = \Big\{ Z_{l\cdot} | Z_{l\cdot} = (z_{l1}, \ldots, z_{ls}, \ldots, z_{ln})$$

$$\text{for some } z_{ls} \in G_{ls}(y_{ls}), \sum_{s=1}^{n} y_{ls} = y_l, y_{ls} \geqq 0, s = 1, \ldots, n \Big\},$$

$$l = 1, 2, \ldots, n, \tag{13}$$

where $y_l \in \mathbf{R}_+^{m'}$, $Z_{l\cdot} \in \mathbf{R}_+^{mn}$, $G_l(y_l) \in \mathbf{R}_+^{mn}$. Hence G_l is a multivalued mapping from $\mathbf{R}_+^{m'}$ into $\mathbf{R}_+^{mn}$. Using (13), $T_2(Y)$ in (10) is represented as follows:

$$T_2(Y) = \Big\{ Z | Z \in \sum_{l=1}^{n} G_l(y_l), \quad \text{where } (y_l)_1^n = Y \Big\}. \tag{14}$$

3.4.3. Space transformation, T

Now, given the state X_t of the space system at the beginning of period t, the set of all feasible states at the end of this period is given by,

$$T_2 \cdot T_1(X_t) = \{ Z_t | Z_t \in T_2(Y_t) \quad \text{for some } Y_t \in T_1(X_t) \}. \tag{15}$$

Thus $T_2 \cdot T_1$ represents the space transformation T. That is, in this case, the space transformation T is represented as the composite mapping of two transformations T_1 and T_2 where the former represents production activities over the space system and the latter transportation activities. Hence, T, which is defined by

$$T = T_2 \cdot T_1, \tag{16}$$

is a multivalued mapping from $\mathbf{R}_+^{mn}$ into $\mathbf{R}_+^{mn}$.

Summarizing the above discussion, our space system is symbolically represented as follows:

$$x_t \longrightarrow T(X_t), \quad t = 0, 1, \ldots, N - 1, \tag{17}$$

where, from (15) and (16)

$$T(X_t) = T_2 \cdot T_1(X_t)$$

$$= \{ X_{t+1} | X_{t+1} \in T_2(Y_t) \quad \text{for some } Y_t \in T_1(X_t) \}, \tag{18}$$

and, from (9)

$$T_1(X_t) = \{Y_t | Y_t = (y_l)_1^n \quad \text{for some } y_l^t \in F_l(x_l^t), \quad l = 1, \ldots, n,$$
$$\text{where } (x_l^t)_1^n = X_t\}, \tag{19}$$

and, if we use (11) to (14) with their simplifying assumptions,

$$T_2(Y_t) = \left\{ Z_t | Z_t \in \sum_{l=1}^n G_l(y_l^t) \quad \text{where } (y_l^t)_1^n = Y_t \right\}, \tag{20}$$

in which

$$G_l(y_l^t) = \left\{ Z_{l\cdot}^t | Z_{l\cdot}^t = (z_{l1}^t, \ldots, z_{ls}^t, \ldots, z_{ln}^t) \quad \text{for some } z_{ls}^t \in G_{ls}(y_{ls}^t), \right.$$

$$\left. \sum_{s=1}^n y_{ls}^t = y_l^t, \, y_{ls}^t \geqq 0, \quad s = 1, 2 \ldots, n \right\}, \quad l = 1, 2, \ldots, n. \tag{21}$$

In (17) to (21),

$$X_t, \, Z_{l\cdot}^t \in \mathbf{R}_+^{mn}, \, Y_t \in \mathbf{R}_+^{m'n}, \, x^t, \, z_{ls}^t \in \mathbf{R}_+^m, \, y_l^t, \, y_{ls}^t \in \mathbf{R}_+^{m'}, \quad l, s = 1, \ldots, n. \tag{22}$$

Hence,

$$T: \mathbf{R}_+^{mn} \longrightarrow \mathbf{R}_+^{mn}, \qquad T_1: \mathbf{R}_+^{mn} \longrightarrow \mathbf{R}_+^{m'n}, \qquad T_2: \mathbf{R}_+^{m'n} \longrightarrow \mathbf{R}_+^{mn},$$

$$F_l: \mathbf{R}_+^m \longrightarrow \mathbf{R}_+^{m'}, \qquad G_l: \mathbf{R}_+^{m'} \longrightarrow \mathbf{R}_+^{mn}, \qquad G_{ls}: \mathbf{R}_+^{m'} \longrightarrow \mathbf{R}_+^m,$$

$$l, s = 1, 2, \ldots, n. \tag{23}$$

The space transformation T which is obtained above may be depicted as in Figure 3.1.

It may be noted that, in this representation of the space transformation, T_1 should include such activities as the preservation of each commodity, and T_2 should include identity transportation activities which transport commodities from a location to itself without any additional inputs. These are illustrated in the next section.

3.4.4. Monotone space systems

We have not yet specified the form of T_1 and T_2. But if we give the form of mapping $F_l(l = 1, \ldots, n)$ in (19), T_1 is determined. Thus if we further give the form of T_2, the space transformation is determined by (18). Moreover, if we use (20) and (21) with their simplifying assumptions, it is sufficient to specify the form of $G_{ls}(l, s = 1, \ldots, n)$ in order to determine T_2.

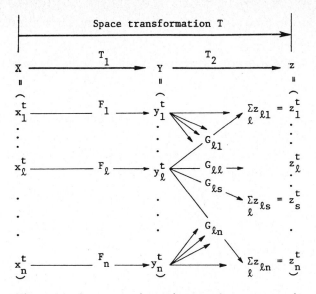

Figure 3.1. Space transformation based on a composite mapping.

Our objective in this chapter is the construction of monotone space systems. And the following theorem tells us how to construct monotone space systems in the framework of (17) to (21).

Theorem 3.1

(i) If T_1 and T_2 in (18) are m.t.cc. respectively, then so is T.

(ii) If each $F_l(l = 1, \ldots, n)$ in (19) is a m.t.cc., then so is T_1.

(iii) If each of $G_l(l = 1, \ldots, n)$ in (20) is a m.t.cc., then so is T_2.

(iv) If each of $G_{ls}(l = 1, \ldots, n)$ in (21) is a m.t.cc., then so is G_l.

The proof is given in Appendix D.1. To obtain an appropriate function F_l which is a m.t.cc., we may use neoclassical type production functions, Leontief type input–output models or von Neumann type models. Similarly, an appropriate G_{ls} of a m.t.cc. can be obtained by using any one of these models.

In the next section, two concrete examples of monotone space systems based on composite mappings are given. The first example is for a final state problem. And in Example 2, a model for a consumption stream problem is obtained by slightly modifying the first example.

3.5. Examples of monotone space systems based on composite mappings

3.5.1. Example 1, two-location von Neumann space system for a final state problem

We here construct a simple multisectoral space system based on the von Neumann model of production.[12]

Our space consists of two locations (regions or countries), $l = r$ and s. There are six distinct commodities i in each location:

$i = 1$:   good 1,
   2:   good 2,
   3:   labor,
   4:   factory 1 (for good 1),
   5:   factory 2 (for good 2),
   6:   man.

This model is an example of a final state problem. Thus, as explained in Section 3.2.5, man is considered as a natural resource of the second group and labor is a service commodity. It is assumed that factories and labor cannot be moved between locations. And man can migrate between locations without transport inputs. But man cannot commute between locations during a single time period. The population growth rate in each location is constant over time. Moreover, for simplicity, it is assumed that the production of any commodity in this space system requires one unit of time which defines a single period. As to the depreciation of each commodity, we use the neoclassical treatment and assume for simplicity that good 1 and good 2 do not depreciate but that factory 1 and factory 2 depreciate at constant rates in each period, whether they are used or not. Also, we assume that current inputs for transportation activities are supplied from the starting location.

Given the above assumptions, we consider the following basic activities in the space system:

Basic production activities, $j(l)$, $j = 1, 2, \ldots, 10$, $l = r, s$
   1(l):   production of one unit of good 1 in location l.
   2(l):   production of one unit of good 2 in location l.

---

[12]For the von Neumann model of production, see von Neumann (1945) and, for example, Morishima (1969).

3(l): production of one unit of labor in location l.[13]
4(l): construction of one unit of factory 1 in location l.
5(l): construction of one unit of factory 2 in location l.
6(l): preservation of one unit of good 1 in location l.
7(l): preservation of one unit of good 2 in location l.
8(l): preservation of one unit of factory 1 in location l.
9(l): preservation of one unit of factory 2 in location l.
10(l): preservation of one unit of man in location l.

Basic transportation activities, j(rs), j(rr), j(sr), j(ss)

1(rs), 1(sr): transportation of one unit of good 1 from r to s, from s to r.

2(rs), 2(sr): transportation of one unit of good 2 from r to s, from s to r.

3(rs), 3(sr): migration of one unit of man from r to s, from s to r.

1(rr), 1(ss): transportation of one unit of good 1 from r to r, from s to s.

2(rr), 2(ss): transportation of one unit of good 2 from r to r, from s to s.

3(rr), 3(ss): transportation of one unit of labor from r to r, from s to s.

4(rr), 4(ss): transportation of one unit of factory 1 from r to r, from s to s.

5(rr), 5(ss): transportation of one unit of factory 2 from r to r, from s to s.

6(rr), 6(ss): transportation of one unit of man from r to r, from s to s.

The preservation activity for each commodity should be included as one of the production activities since some part of each commodity may be unused. However for labor, this preservation activity is not needed since labor is a service and cannot be preserved for more than one period.

Among the basic transportation activities, $j(rr)$, and $j(ss)$, $j = 1, \ldots, 6$, represent *identity transportation*, that is, they transport each good from one location to itself without any real transport inputs.

---

[13]We do not need to assume that this activity is literally the production of labor. It may just be assumed that the planner decides that each unit labor service is paid a certain fixed amount of each good. For this point, see Section 3.2.5.

The matrices in Table 3.1 describe basic production activities in each location. In these matrices:

a^l_{ij} = the quantity of commodity i technologically required[14] at the beginning of the period per unit level of operation of the basic activity j in location l.

b^l_{ij} = the quantity of commodity i obtained at the end of the corresponding period per unit level of operation of the basic activity j in location l.

λ^l_i = the rate of depreciation of commodity i per one period in location l, $i = 4, 5$.

λ^l_6 = the rate of population growth in location l.

Matrix A_l represents the input coefficients of the basic production activities in location l, each column being the input vector of a particular basic production activity in location l, $l = r, s$. And matrix B_l represents the output coefficient of the basic production activities in location l, each column being the output vector of a particular basic production activity in location l, $l = r, s$.

Let us denote the jth column of A_l and B_l by a^l_j and b^l_j, respectively. Then the basic production activity $j(l)$ is symbolically represented as follows:

$$a^l_j \longrightarrow b^l_j.$$

That is, activity $j(l)$ transforms the set of inputs a^l_j which is given at the beginning of a period into a set of outputs b^l_j at the end of the corresponding period.

We put the following restrictions on these coefficients:

$$\left.\begin{array}{l} A_l \geqq 0,\ B_l \geqq 0, \\ 1 > \lambda^l_i \geqq 0, \quad i = 1, \ldots, 5, \quad \lambda^l_6 > -1, \\ \text{no column of } A_l \text{ is a zero vector.} \end{array}\right\} \quad l = r, s. \qquad (24)$$

Using the notation of Section 3.4, the set of production activities in each location in period t is represented as follows under the assumption of free disposability of inputs and outputs.

$$x^t_l \rightarrow F_l(x^t_l) = \{y^t_l | x^t_l \geqq A_l q^t_l,\ 0 \leqq y^t_l \leqq B_l q^t_l$$
$$\text{for some } q^t_l \geqq 0\}, \quad l = r, s. \qquad (25)$$

---

[14]For activity 3(l), coefficient a^l_{ij} does not necessarily reflect technological requirements. For this point, see Section 3.2.5.

Table 3.1. Basic production activities.

$A_l = [a_{ij}^l]$: production input matrix, $l = r,s$

Input					Activity					
	1(l)	2(l)	3(l)	4(l)	5(l)	6(l)	7(l)	8(l)	9(l)	10(l)
1. Good 1	a_{11}^l	a_{12}^l	a_{13}^l	a_{14}^l	a_{15}^l	1	0	0	0	0
2. Good 2	a_{21}^l	a_{22}^l	a_{23}^l	a_{24}^l	a_{25}^l	0	1	0	0	0
3. Labor	a_{31}^l	a_{32}^l	0	a_{34}^l	a_{35}^l	0	0	0	0	0
4. Factory 1	a_{41}^l	0	0	0	0	0	0	1	0	0
5. Factory 2	0	a_{51}^l	0	0	0	0	0	0	1	0
6. Man	0	0	1	0	0	0	0	0	0	1

$B_l = [b_{ij}^l]$: production output matrix

Input					Activity					
	1(l)	2(l)	3(l)	4(l)	5(l)	6(l)	7(l)	8(l)	9(l)	10(l)
1. Good 1	1	0	0	0	0	1	0	0	0	0
2. Good 2	0	1	0	0	0	0	1	0	0	0
3. Labor	0	0	1	1	0	0	0	0	0	0
4. Factory 1	$(1-\lambda_4^l)a_{41}^l$	0	0	0	0	0	0	$1-\lambda_4^l$	0	0
5. Factory 2	0	$(1-\lambda_5^l)a_{51}^l$	0	0	1	0	0	0	$1-\lambda_5^l$	0
6. Man	0	0	$1+\lambda_6^l$	0	0	0	0	0	0	$1+\lambda_6^l$

Hence, the set of feasible production activities in this space system in period t, given the state X_t of the space system at the beginning of period t, is represented as follows:

$$X_t \to T_1(X_t) = \left\{ Y_t \,|\, Y_t = (y_r^t, y_s^t) \quad \text{for some } y_l^t \in F_l(x_l^t), \right.$$

$$\left. l = r, s, (x_r^t, x_s^t) = X_t \right\}$$

$$= \left\{ Y_t \,|\, X_t \geq \begin{pmatrix} A_r & 0 \\ 0 & A_s \end{pmatrix} \begin{pmatrix} q_r^t \\ q_s^t \end{pmatrix}, 0 \leq Y_t \leq \begin{pmatrix} B_r & 0 \\ 0 & B_s \end{pmatrix} \begin{pmatrix} q_r^t \\ q_s^t \end{pmatrix} \right.$$

$$\left. \text{for some } q_l^t \geq 0, \quad l = r, s \right\}, \qquad (26)$$

where

$$X_t, Y_t \in \mathbf{R}_+^{6 \times 2}, x_l^t, y_l^t \in \mathbf{R}_+^6, A_l, B_l : 6 \times 10 \text{ matrices}, q_l^t \in \mathbf{R}_+^{10},$$

$$l = r, s. \qquad (27)$$

Under assumption (24), the transformation F_l defined by (25) is clearly a m.t.cc. from $\mathbf{R}_+^6$ into $\mathbf{R}_+^6$. Thus, from (ii) in Theorem 3.1, the transformation T_1 defined by (26) is a m.t.cc. from $\mathbf{R}_+^{6 \times 2}$ into $\mathbf{R}_+^{6 \times 2}$.

Next, the matrices given in Table 3.2 describe the basic transportation activities. Matrix V_{rs} represents the input coefficients of the basic transportation activities from location r to location s, and W_{rs} is the corresponding output matrix. To obtain the basic transportation activities from location s to location r, simply change the superscript on these coefficients to sr. We then get V_{sr} and W_{sr}.

Matrices V_{ll} and W_{ll} represent, respectively, the input coefficients and output coefficients of the identity transportation activities.

We put the following restrictions on these coefficients:

$$V_{rs} \geq 0, \; W_{rs} \geq 0,$$
$$V_{sr} \geq 0, \; W_{sr} \geq 0,$$
$$v_{11}^{rs} = 1 + v_{11}'^{rs}, \; v_{22}^{rs} = 1 + v_{22}'^{rs}, \; v_{11}'^{rs} \geq 0, \; v_{22}'^{rs} \geq 0, \qquad (28)$$
$$v_{11}^{sr} = 1 + v_{11}'^{sr}, \; v_{22}^{sr} = 1 + v_{22}'^{sr}, \; v_{11}'^{sr} \geq 0, \; v_{22}'^{sr} \geq 0,$$

Using the notation in Section 3.4, the set of transportation activities, for example, from location r to s at the end of period t is represented as follows (under the assumption of free disposability of inputs and outputs):

Table 3.2. Basic transportation activities.

Transportation activities from region r to region s

$V_{rs} = (v_{ij}^{rs})$

Input	Activity		
	1(rs)	2(rs)	3(rs)
1. Good 1	v_{11}^{rs}	v_{12}^{rs}	0
2. Good 2	v_{21}^{rs}	v_{22}^{rs}	0
3. Labor	v_{31}^{rs}	v_{32}^{rs}	0
4. Factory 1	0	0	0
5. Factory 2	0	0	0
6. Man	0	0	1

$W_{rs} = (w_{ij}^{rs})$

Input	Activity		
	1(rs)	2(rs)	3(rs)
1. Good 1	1	0	0
2. Good 2	0	1	0
3. Labor	0	0	0
4. Factory 1	0	0	0
5. Factory 2	0	0	0
6. Man	0	0	1

Identity transportation activities

$V_{ll} = (v_{ij}^{ll})$, $l = r, s$

i	j					
	1(ll)	2(ll)	3(ll)	4(ll)	5(ll)	6(ll)
1.	1	0	0	0	0	0
2.	0	1	0	0	0	0
3.	0	0	1	0	0	0
4.	0	0	0	1	0	0
5.	0	0	0	0	1	0
6.	0	0	0	0	0	1

$W_{ll} = (w_{ij}^{ll})$, $l = r, s$

i	j					
	1(ll)	2(ll)	3(ll)	4(ll)	5(ll)	6(ll)
1.	1	0	0	0	0	0
2.	0	1	0	0	0	0
3.	0	0	1	0	0	0
4.	0	0	0	1	0	0
5.	0	0	0	0	1	0
6.	0	0	0	0	0	1

$$y_{rs}^t \to G_{rs}(y_{rs}^t) = \{x_{rs}^{t+1}|y_{rs}^t \geqq V_{rs}\xi_{rs}^t, 0 \leqq x_{rs}^{t+1} \leqq W_{rs}\xi_{rs}^t$$

$$\text{for some } \xi_{rs}^t \geqq 0\}.^{15} \qquad (29)$$

And the set of transportation activities "originating", for example, from location r at the end of period t is represented as follows:

$$y_r^t \to G_r(y_r^t) = \left\{ X_{r.}^{t+1}|X_{r.}^{t+1} = (x_{rr}^{t+1}, x_{rs}^{t+1}) \quad \text{for some } x_{rl}^{t+1} \in G_{rl}(y_{rl}^t), \right.$$

$$l = r, s, \sum_{l=r,s} y_{rl}^t = y_r^t, y_{rl}^t \geqq 0, \quad l = r, s \bigg\}$$

$$= \left\{ x_{r.}^{t+1}|y_r^t \geqq (V_{rr}V_{rs}) \begin{pmatrix} \xi_{rr}^t \\ \xi_{rs}^t \end{pmatrix}, 0 \leqq x_{r.}^{t+1} \right.$$

$$\leqq \begin{pmatrix} W_{rr} & 0 \\ 0 & W_{rs} \end{pmatrix} \begin{pmatrix} \xi_{rr}^t \\ \xi_{rs}^t \end{pmatrix} \quad \text{for some } \xi_{rr}^t \geqq 0, \xi_{rs}^t \geqq 0 \right\}. \qquad (30)$$

Hence, the set of feasible transportation activities in this space system in period t, given the state Y_t of the space system at the end of production activities is represented as follows:

$$Y_t \to T_2(Y_t) = \left\{ X_{t+1}|X_{t+1} \in \sum_{l=r, s} G_l(Y_l^t), (Y_r^t, Y_s^t) = Y_t \right\}$$

$$= \left\{ X_{t+1}|0 \leqq X_{t+1} \leqq \begin{pmatrix} \sum\limits_{l=r,s} W_{lr}\xi_{lr}^t & \sum\limits_{l=r,s} V_{rl}\xi_{rl}^t \\ \sum\limits_{l=r,s} W_{ls}\xi_{ls}^t & \sum\limits_{l=r,s} V_{sl}\xi_{sl}^t \end{pmatrix} \right.$$

$$\leqq Y_t \quad \text{for some } \xi_{lr}^t, \xi_{ls}^t, \xi_{rl}^t, \xi_{sl}^t \geqq 0, \quad l = r, s \bigg\}, \qquad (31)$$

where

$$Y_t, X_{t+1} \in \mathbf{R}_+^{6\times 2}, y_r^t, y_s^t \in \mathbf{R}_+^6,$$

$$\xi_{rs}^t, \xi_{sr}^t \in \mathbf{R}_+^3, \xi_{ll}^t \in \mathbf{R}_+^6, \quad l = r, s, V_{rs}, V_{sr}, \qquad (32)$$

$$W_{rs}, W_{sr}: 6 \times 3, V_{ll}, W_{ll}: 6 \times 6, \quad l = r, s.$$

Under assumption (28), the transformation T_2 defined by (31) is a m.t.cc. from $\mathbf{R}_+^{6\times 2}$ into $\mathbf{R}_+^{6\times 2}$.

---

[15]Since the present model is a final state problem, it is not necessary to consider the flow of consumption goods outside the space system. Hence the outputs of transportation activities directly become input commodities in the next period. Thus we use x^t and X^t instead of z^t and Z^t, respectively.

From (26) and (31), the space transformation T is defined as follows:

$$X_t \to T(X_t),$$

where

$$T(X_t) = T_2 \cdot T_1(X_t)$$
$$= \{X_{t+1} | X_{t+1} \in T_2(Y_t) \quad \text{for some } Y_t \in T_1(X_t)\}$$

$$= \left\{ X_{t+1} \Big| 0 \leqq X_{t+1} \leqq \begin{pmatrix} \displaystyle\sum_{l=r,s} W_{lr}\xi_{lr}^t \\ \displaystyle\sum_{l=r,s} W_{ls}\xi_{ls}^t \end{pmatrix}, \begin{pmatrix} \displaystyle\sum_{l=r,s} V_{rl}\xi_{rl}^t \\ \displaystyle\sum_{l=r,s} V_{sl}\xi_{sl}^t \end{pmatrix} \right.$$

$$\leqq \begin{pmatrix} B_r & 0 \\ 0 & B_s \end{pmatrix} \begin{pmatrix} q_r^t \\ q_s^t \end{pmatrix}, \begin{pmatrix} A_r & 0 \\ 0 & A_s \end{pmatrix} \begin{pmatrix} q_r^t \\ q_s^t \end{pmatrix} \leqq X_t$$

$$\left. \text{for some } \xi_{rs}^t, \xi_{sr}^t \geqq 0, \xi_{ll}^t, q_l^t \geqq 0, l = r, s \right\}. \tag{33}$$

As noted before, T_1 and T_2 are m.t.cc. from $\mathbf{R}_+^{6\times2}$ into $\mathbf{R}_+^{6\times2}$. Hence, from (i) in Theorem 3.1, the space transformation T defined by (33) is a m.t.cc. from $\mathbf{R}_+^{6\times2}$ into $\mathbf{R}_+^{6\times2}$.

This is the simplest kind of monotone space system in one sense. We may give alternate techniques for each basic production and transportation activity. For example, we may specify that each basic production activity j can be any pair of input vector $a_j \in \mathbf{R}_+^6$ and output vector $b_j \in \mathbf{R}_+^6$ such that

$$(a_j, b_j) \in D_j$$

where D_j is a compact convex nonempty set in $\mathbf{R}_+^{12}$ for which $(0, b_j) \in D_j$ only if $b_j = 0$. Then our space system is still a monotone space system. But this is not discussed further.

Finally, if we want to prohibit migration, then we need only to remove activities $3(rs)$ and $3(sr)$ from the basic transportation activities.

3.5.2. Example 2, two-location space system for a consumption stream problem

Using the framework of the previous example, we now construct a space system exemplifying a consumption stream problem. The previous example was a final state problem. Thus, labor and man were distinguished from each other and a fixed amount of each consumption good was paid for each unit of labor.

It was stated in Section 3.2.5 that generally man is not distinguished from labor in consumption stream problems. However, the model constructed here is a special exception. Note that if we do not distinguish labor from man, then man is a durable capital good. And for the construction of monotone space systems based on composite mappings of two transformations, T_1 and T_2, recall that we assume transportation activities do not require time. Hence, as noted in footnote 10, the requirement of transportation activities for an input commodity, man, is not correctly specified. In other words, some part of the function of total man is double counted in production and transportation activities in the same period. But if we distinguish labor from man and assume that labor is a service commodity which is used only for current inputs, then this double counting is avoided. The total amount of labor produced in a period is consumed in the subsequent transportation activities and production activities. Thus, we distinguish labor from man in monotone space systems based on composite mappings, even if they are used for consumption stream problems.

The framework of this model is exactly the same as in example 1, except that we now put the following additional condition on the coefficients of matrix A_l:

$$a^l_{13} = a^l_{23} = 0, \quad l = r, s \tag{34}$$

This is necessary because the amount of consumption is determined after the end of all activities – both production activities and transportation activities – in each period.

Equations (24) through (33) remain valid for our present problem, keeping in mind that (34) is assumed in these equations. And the transformations T_1, T_2 and T are m.t.cc., respectively, assuming (34). But, in equations (29) to (33), each variable which has the subscript (or the superscript) $t+1$ should be replaced by z or Z and the subscript (or the superscript) t should be attached, since we determine the amount of consumption after transportation activities.

3.6. Two other types of monotone space systems

3.6.1. Monotone space systems based on convex-hull mappings

As noted in footnote 10, it is difficult to treat the capacity of transportation facilities in monotone space systems based on composite mappings since the time required for each transportation activity is not considered explicitly in these models. We shall show that the space system discussed in this section is more appropriate for the analysis of those space systems in which transportation facilities are explicitly considered.

For the construction of monotone space systems based on convex-hull mappings, we assume that production activities and transportation activities in the space system take place in parallel in each period in the following sense.

At the beginning of each period, the stock of commodities, X, over the space system is divided into two parts as follows:

$$X = X^1 + X^2$$
$$\equiv (x_l^1)_1^n + (x_l^2)_1^n \tag{35}$$

where X^1 is a part of X allocated to the production activities and X^2 is that part allocated to the transportation activities.

The production activities in location $l (l = 1, \ldots, n)$ use x_l^1 as the initial input in that period, and obtain a set of outputs at the end of that period as a function of x_l^1. Hence the set of production activities in each location in a period may be symbolically represented as follows:

$$x_l^1 \to F_l(x_l^1), \quad l = 1, 2, \ldots, n. \tag{36}$$

Relation (35) implies that within each period the production activities in each location take place independently from both production activities in other locations and transportation activities over the space system. In other words, production activities in location l in a period do not use commodities which are transported from other locations into that location by transportation activities in the same period. Conversely, transportation activities over the space system in a period do not use the outputs of production activities over the space system in that same period.[16] This assumption is not so unrealistic when the length of each period is taken to be relatively short.

Using (35), the set of feasible production activities in the space system in each period, given X^1, is represented as follows:

$$X^1 \to T_1(X^1) = \{Y^1 | Y^1 = (y_l^1)_1^n \quad \text{for some } y_l^1 \in F_l(x_l^1),$$
$$l = 1, 2, \ldots, n \text{ where } (x_l^1)_1^n = X^1\}. \tag{37}$$

[16]To satisfy the assumption of the independency between production activities and transportation activities within each period, it is sufficient to assume that: (i) every production activity in each location is of the point-input, point-output type; and (ii) each of them has the same duration of one period. These are, of course, not necessary conditions.

And the set of feasible transportation activities over the space system is symbolically represented as follows:

$$X^2 \to T_2(X^2), \tag{38}$$

where $T_2(X^2)$ is the set of all feasible output vectors of transportation activities, given initial input vector X^2, after one period.

Using (37) and (38), the space transformation T is described as follows:

$$X \to T(X)$$

where

$$T(X) = (T_1 \vee T_2)X$$
$$\equiv \{Z | Z = Y^1 + Y^2 \quad \text{for some } Y^1 \in T_1(X^1), \, Y^2 \in T_2(X^2),$$
$$X^1 + X^2 = X, \, X^1, \, X^2 \geqq 0\}$$

for each $X \in \mathbf{R}_+^{mn}$.[17]

Transformation T which is defined above may be called the *convex-hull mapping* of T_1 and T_2. T_1, T_2 and T are each multivalued mappings from $\mathbf{R}_+^{mn}$ into $\mathbf{R}_+^{mn}$. And, from B-property 21, when T_1 and T_2 are both m.t.cc., then T is also a m.t.cc.

To specify the transportation activities in more detail than equation (38), we need some simplifying assumptions. An explicit treatment of the capacity of transportation facilities in a more realistic dynamic model is a quite complicated problem and it is beyond the scope of this book.[18] Thus, for the introduction of transportation facilities into the model, we need some strong assumptions besides assumption (i) in Section 3.2.3. The following is one possible set of assumptions:

(a) The required time for transportation of commodities between any pair of locations in the space system is relatively small compared to the length of the period.

(b) Each transportation activity takes place "uniformly" over each period.

---

[17] It may be noted that when we divide X into X^1 and X^2, it is enough to consider the case in which all the transportation facilities are included in X^2 because of assumption (i) in Section 3.2.3.

[18] Network problems by ordinary definitions are not dynamic analysis since they consider only steady state flows. Dynamic transportation problems, in which the level of transportation activities changes over time, transportations require time and transportation capacities are explicitly considered, are quite complicated problems.

(c) "Transported goods" during a period are not transported again[19] in the same period and are not used for current inputs for transportation activities during that period. But, facilities for transportation can be used repeatedly in the same period.

Assumption (b) means, for example, that when x units of commodity i are transported from location l to location s during period t and one period is equal to k hours, we assume x/k units of commodity i are sent from location l to s in each hour during period t.

From assumptions (a) and (b), transportation flows over the space system converge rapidly to the "steady state flow" in each period. Thus the methods of static transportation analysis, originated by Koopmans and Reiter (1951), can be safely used within each period. But the analysis of transportation in space systems is beyond the scope of this book.

The space transformation T which is defined by (39) may be depicted as in Figure 3.2.

3.6.2. *Monotone space systems based on flow models*

The preceding two types of space systems are based on point-input, point-output type activity models originated by von Neumann (1945), and are appropriate for the theoretical analysis of dynamic systems. But the von Neumann model of production requires that the length of each period be relatively short since it is assumed that every activity lasts exactly one period and thus no transactions between behavioral units can exist (in the case of our space system, between locations) within each period.

On the other hand, we may base our theoretical framework on the static activity model of Koopmans (1954), (or on the static Leontief input–output model which is a special case of Koopmans model). Examples of applications of Koopmans type static activity model to single-period programming in space systems are to be found in Isard (1958), Moses (1960) and Stevens (1958). Vietorisz (1967) offers an extension of these models to multiperiod programming. And Bhatia

---

[19]If a commodity is transported from location l to r and this same commodity is transported again from r to s in the same period, the total activity which transports that commodity from l to s is defined as one transportation activity in that period. Thus this assumption is reasonable.

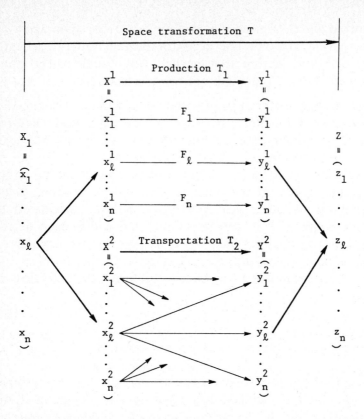

Figure 3.2. Space transformation based on a convex-hull mapping.

(1961) develops an application of dynamic Leontief input–output model to a multiperiod programming problem in a space system. They may be called *flow models*. In these models, it is assumed that the whole system is in static equilibrium within each period. Hence the level of transportation activities over the space system as well as the level of the production activity in each location is decided simultaneously in each period. Compared to von Neumann type models, these flow models are more appropriate when the length of each period is relatively long (for example, a year). Thus, they are appropriate for practical analyses. But, since all the relations of the space system are decided simultaneously in each period in these flow models, the intuitive understanding as well as the theoretical treatment of the dynamic process of the space system may be difficult. Thus, we do not investigate space systems based on flow models in this book.

Chapter 4

EFFICIENT GROWTH IN MONOTONE SPACE SYSTEMS

4.1. Introduction

Monotone space systems were defined in Section 1.2, and three types of monotone space systems were discussed in Chapter 3. The purpose of this chapter is to investigate the general character of optimum growth paths in these monotone space systems by applying the mathematical theorems obtained in Chapter 2.

As noted in Section 3.3, monotone space systems based on composite mappings will be the most appropriate framework for theoretical analysis as well as for intuitive understanding of the growth process among the three types of monotone space systems. Hence we limit the discussion in this chapter to monotone space systems based on composite mappings as represented by Equations (17) to (21) in Section 3.4.2.

Given a monotone space system, a *final state problem* is defined when the objective function is specified for the final period. A *consumption stream problem* is defined when the welfare function is specified for each period. It is assumed in this chapter that both the objective function and the welfare function in each period are additive over all locations. Thus Equation (5) in Chapter 3 is the general form of the objective function in consumption stream problems in this chapter. And, to avoid repetition, a final state problem is treated as a special case of a consumption stream problem in its mathematical representation, even though their economic meanings are quite different. That is, we may construct an objective function for final state problems by simply setting $f_l^t(x_l^t, c_l^t) \equiv 0$ for $l = 1, 2, \ldots, n$, $t = 1, 2, \ldots, N - 1$ in Equation (5) of Chapter 3.

Further, the welfare function $f_l^t(x_l^t, c_l^t)$ in each location and in each period will, at first, be limited to a monotone concave gauge. Though it may be too restrictive in consumption stream problems to assume that each welfare function is a monotone concave gauge, we shall show that a typical consumption stream problem can be handled within this limitation. In the last section of this chapter, we briefly examine the modifications of our results when the welfare function is replaced by a generalized monotone concave gauge.

In Section 4.2.1, the primal problem in this chapter is formulated as Problem I(a). An intuitive diagram of the problem is presented. Then, the dual problem for I(a) is formulated in Section 4.2.2 as Problem I(b). It is shown there that the dual problem can be formulated quite automatically by using Figure 4.1.

Then, in Section 4.3, the duality theorem, existence theorem and max-min principle for the pair of problems, I(a) and I(b), are obtained, respectively, by applying the results in Chapter 2. The interpretations of these theorems are given in the following two sections.

The duality between the commodity system and the price system is important for the theoretical understanding of the efficient growth process in a monotone space system. Thus, we devote Section 4.4 to an interpretation of dual variables.

In Section 4.5, stepwise procedures for optimal paths are explained by using the max-min principle. First, in Section 4.5.1, the selection of optimal activities and price vectors in each period is divided into three stages – production, transportation and consumption. We then employ the max-min principle to explain how optimal activities and price vectors are determined in each stage. In Section 4.5.2 we show that while the whole program can be divided into numbers of smaller single step programs, this method will encounter the difficulties inherent in "singular control problems." Singular control problems are interesting theoretically as well as computationally in programming in space systems, and they are discussed in detail in this section.

Finally, in Section 4.6, we briefly examine what modifications of the previous results are needed when we replace each welfare function with a generalized monotone concave gauge.

4.2. Problem formulation

4.2.1. Primal problem

According to the assumptions given in the previous section, the primal problem in this chapter is summarized as follows:

Problem I(a)

$$\text{Maximize } \sum_{t=0}^{N-1} \sum_{l=1}^{n} f_l^{t+1}(x_l^{t+1}, c_l^{t+1}) \tag{1}$$

subject to $(X_{t+1}, [f_l^{t+1}(x_l^{t+1}, c_l^{t+1})]_1^n) \in \mathcal{T}_t(X_t), \quad t = 0, 1, \ldots, N-1,$
and $X_0 = \bar{X}_0 \geq 0,$ (2)

where

$$\mathcal{T}_t(X_t) = T_{3t} \cdot T_2 \cdot T_1(X_t) = \{(X_{t+1}, [f_l^{t+1}(x_l^{t+1}, c_l^{t+1})]_1^n) |$$
$$(X_{t+1}, [f_l^{t+1}(x_l^{t+1}, c_l^{t+1})]_1^n) \in T_{3t}(Z_t)$$
$$\text{for some } Z_t \in T_2(Y_t) \text{ for some } Y_t \in T_1(X_t)\}, \tag{3}$$

$$T_1(X_t) = \{Y_t | Y_t = (y_l^t)_1^n \text{ for some } y_l^t \in F_l(x_l^t),$$
$$l = 1, \ldots, n, \text{ where } (x_l^t)_1^n = X_t\}, \tag{4}$$

$$T_2(Y_t) = \{Z_t | Z_t = (z_l^t)_1^n \in \sum_{l=1}^{n} G_l(y_l^t) \text{ where } (y_l^t)_1^n = Y_t\}, \tag{5}$$

$$G_l(Y_l^t) = \{Z_l^t | Z_l^t = (z_{l1}^t, \ldots, z_{ls}^t, \ldots, z_{ln}^t)$$
$$\text{for some } z_{ls}^t \in G_{ls}(y_{ls}^t), \sum_{s=1}^{n} y_{ls}^t = y_l^t, y_{ls}^t \geq 0,$$
$$s = 1, \ldots, n\}, \quad l = 1, 2, \ldots, n, \tag{6}$$

$$T_{3t}(Z_t) = \{(X_{t+1}, [f_l^{t+1}(x_l^{t+1}, c_l^{t+1})]_1^n) |$$
$$\text{for some } (x_l^{t+1}, f_l^{t+1}(x_l^{t+1}, c_l^{t+1})) \in g_l^t(z_l^t),$$
$$l = 1, 2, \ldots, n, \text{ where } (z_l^t)_1^n = Z_t\}, \tag{7}$$

$$g_l^t(z_l^t) = \{(x_l^{t+1}, f_l^{t+1}(x_l^{t+1}, c_l^{t+1})) |$$
$$\text{for some } x_l^{t+1} + (c_l^{t+1}, 0) \leq z_l^t, x_l^{t+1} \geq 0, c_l^{t+1} \geq 0\},$$
$$l = 1, 2, \ldots, n. \tag{8}$$

The dimension of each vector is:

$$X_t = (x_l^t)_1^n \in \mathbf{R}_+^{mn}, x_l^t \in \mathbf{R}_+^m,$$
$$C_t = (c_l^t)_1^n \in \mathbf{R}_+^{m_c n}, c_l^t \in \mathbf{R}_+^{m_c},$$
$$Y_t = (y_l^t)_1^n \in \mathbf{R}_+^{m'n}, y_l^t \in \mathbf{R}_+^{m'}, y_{ls}^t \in \mathbf{R}_+^{m'},$$
$$Z_t = (z_l^t)_1^n \in \mathbf{R}_+^{mn}, z_{ls}^t \in \mathbf{R}_+^m, Z_l^t \in \mathbf{R}_+^{mn},$$
$$l = 1, 2, \ldots, n, \quad t = 0, 1, \ldots, N.$$

For an intuitive understanding of the above problem, Figure 4.1(a) will be helpful. This figure illustrates the operation of the problem in

(a) Primal relation

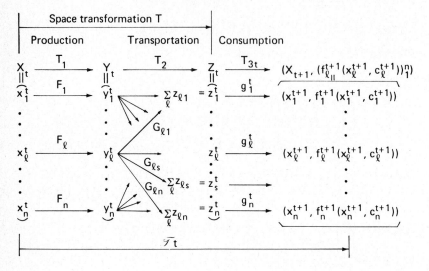

(b) Dual relation

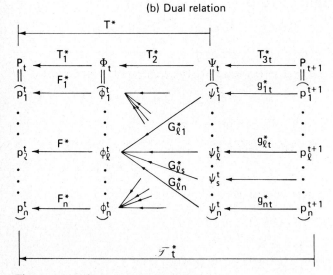

Figure 4.1. Primal and dual relations.

each period. The state of the space system at the beginning of period t is denoted by $X_t = (x_l^t)_1^n$ where each vector x_l^t represents the stocks of commodities in location l at that time.

First, production activities take place in each location l by using x_l^t as the production input vector in that location. Production activities in location l transform a set of inputs denoted by x_l^t into a set of outputs denoted by y_l^t after one unit of time under the production function F_l in this location. The state of the space system at the end of the production activities is denoted by $Y^t = (y_l^t)_1^n$. This operation is represented by the transformation T_1 which is defined by (4), and it is depicted in the left half of Figure 4.1(a).

Second, outputs of production activities in location l, which are represented by y_l^t, are distributed between locations by transportation activities starting from this location as defined by (6). Thus, all the transportation activities over the space system are defined by (5). The state of the space system at the end of the transportation activities over the space system is denoted by $Z_t = (z_l^t)_1^n$ where z_l^t represents the stocks of commodities in location l at the end of transportation activities over the space system. This operation is depicted in the middle of Figure 4.1(a).

Finally, each vector z_l^t is divided into two parts, x_l^{t+1} and $c_{l.}^{t+1}$ where c_l^{t+1} represents the consumption vector in location l in the next period and x_l^{t+1} becomes the production input vector in location l in the next period. The two parts have the relation, $x_l^{t+1} + (c_l^{t+1}, 0) = z_l^t$, where $0 \in \mathbf{R}^{m-m_c}$ and $m - m_c$ is the number of non-consumable commodities.[1] A pair of vectors (x_l^{t+1}, c_l^{t+1}) represents the result in location l after the termination of all activities over the space system in period t, including consumption decisions. This result is evaluated by the welfare function $f_l^{t+1}(x_l^{t+1}, c_l^{t+1})$ in this location. The consumption decision and the evaluation procedure are represented by a fictitious transformation g_l^t defined by (8). Function g_l^t transforms vector z_l^t into vector $(x_l^{t+1}, f_l^{t+1}(x_l^{t+1}, c_l^{t+1}))$. Thus, the consumption decision and the evaluation procedure over the space system are represented by transformation T_{3t} defined by (7), and depicted in the right half of Figure 4.1(a).

Hence, after the termination of all activities over the space system in period t, we obtain the total welfare $\Sigma_{l=1}^n f_l^{t+1}(x_l^{t+1}, c_l^{t+1})$ and the input vector $X_{t+1} = (x_l^{t+1})_1^n$ for the next period.

---

[1]For this point, see Section 1.2.

As previously stated, the final state problem is treated in this chapter as a special case of the consumption stream problem in its mathematical representation. That is, when

$$f_l^t(x_l^t, c_l^t) \equiv 0 \quad \text{for } l = 1, 2, \ldots, n, \quad t = 1, 2, \ldots, N-1, \tag{9}$$

in Problem I(a), we have a final state problem. In this case we don't need to consider the consumption decision in each period. Hence, the third transformation T_{3t} is omitted and the output vector of transportation activities becomes directly the input vector of production activities in the next period, that is,

$$Z_t = X_{t+1}. \tag{10}$$

Keeping in mind these differences, we represent both problems by Problem I(a).

We put the following restrictions on Problem I(a).

A1. Each F_l, $l = 1, 2, \ldots, n$, is a m.t.cc. from $\mathbf{R}_+^m$ into $\mathbf{R}_+^{m'}$ and each G_{ls}, $l, s = 1, 2, \ldots, n$, is a m.t.cc. from $\mathbf{R}_+^{m'}$ into $\mathbf{R}_+^m$.

A2. Each F_l, $l = 1, 2, \ldots, n$, and G_{ll}, $l = 1, 2, \ldots, n$, is non-singular.

A3. Each $f_l^t(x_l^t, c_l^t)$, $l = 1, 2, \ldots, n$, $t = 1, 2, \ldots, N$, is a monotone concave gauge on $\mathbf{R}_+^{m+m_c}$. Not all of $f_l^N(x_l^N, c_l^N)$, $l = 1, 2, \ldots, n$, are identically zero.

From Theorem 3.1, under A1, each G_l which is defined by (6) is a m.t.cc. from $\mathbf{R}_+^m$ into $\mathbf{R}_+^{m'}$ for $l = 1, 2, \ldots, n$; and T_1, T_2, T_3 and $\mathcal{T}_t$ which are defined by, respectively, (4), (5), (7) and (3) are m.t.cc. from $\mathbf{R}_+^{mn}$ into $\mathbf{R}_+^{m'n}$, from $\mathbf{R}_+^{m'n}$ into $\mathbf{R}_+^{mn}$, from $\mathbf{R}_+^{mn}$ into $\mathbf{R}_+^{mn+n}$ and from $\mathbf{R}_+^{mn}$ into $\mathbf{R}_+^{mn+n}$, respectively.

In order to state the optimality conditions in terms of auxiliary variables, we need to require that the space transformation $T = T_2 \cdot T_1$ be non-singular. It follows from (3), (7) and (8) that when T is non-singular, a strictly positive X_{t+1} is obtained from a strictly positive X_t under $\mathcal{T}_t$.

From B-definition 6, the non-singularity of T means that given a positive amount of every commodity in every location at the beginning of a period, we can obtain a positive amount of every commodity in every location at the end of that period. This condition does not require that every commodity be able to be produced in every location, nor that some commodities be increased in some locations. The condition simply requires that each commodity in each location can be produced or preserved in that location or can be transported from other locations. If some commodity in some location cannot be obtained by any of the above methods, then it is not necessary to include it in the model. Thus, if T is singular, it can be made non-singular without losing any economic significance merely by restricting the range of T.

Under A2, T_1 and T_2 (and thus T) are non-singular. But A2 is not the necessary condition for non-singularity. It is only one possible sufficient condition, which appears reasonable.[2]

The latter half of A3 is needed in order to make the problem meaningful. The first half of A3 is needed in order to make the problem a typical monotone program.

Since the function, $f_l^t(x_l^t, c_l^t) \equiv 0$, is a monotone concave gauge, and since $f_l^N(x_l^N, c_l^N) = p_l^N \cdot x_l^N$ is a monotone concave gauge when p_l^N is a vector in $\mathbf{R}_+^{*m}$, Problem I(a) includes a typical final state problem as a special case under A3.

Moreover, when u_l is a generalized monotone concave gauge[3] on $\mathbf{R}_+^{mc}$ in the definition of the function $f_l^t(x_l^t, c_l^t)$ in (6) of Chapter 3 and when the value of that function at $x_{ml}^t = 0$ is defined as in Lemma 2.7 in Chapter 2, then the function $f_l^t(x_l^t, c_l^t)$ is a monotone concave gauge on $\mathbf{R}_+^{m+mc}$. Hence Problem I(a) also includes a typical consumption stream problem as a special case under A3.

Since it is difficult to assume that the welfare function in each period is time invariant, the welfare function f_l^t includes a time superscript. For example, all f_l^t, $t = 1, 2, \ldots, N-1$ are identically zero in the final state problem. For another example, the welfare function in a typical consumption stream problem has a time discount in each period. Hence transformations g_l^t, T_{3t} and $\mathcal{T}_t$ each possess time index t.

All other transformations are assumed to be time invariant for simplicity of notation.

4.2.2. Adjoint transformations and the dual problem

The duality between the commodity system and the price system is the key to understanding the efficient growth process of a monotone space

---

[2] A2 means that each commodity can be produced or can be preserved in each location, and that transportation activities include an identity transportation activity for each commodity in each location.

[3] For the definition of a generalized monotone concave gauge, see Section 2.5.

system. To state the dual of Problem I(a), we must first introduce the necessary dual variables. This can be done quite automatically by replacing each variable in Figure 4.1(a) by an arbitrary new variable. Figure 4.1(b) shows the result. One point may need explanation. Observe that we did not introduce the dual variable corresponding to each $f_l^{t+1}(x_l^{t+1}, c_l^{t+1})$. This is because Lemma 2.5 allows us to set the price for a unit of welfare equal to unity on the optimal path.

The meaning of each dual variable is discussed in Section 4.4 in detail. But for the present we shall simply refer to each dual variable as the "price" of the corresponding commodity. The validity of this interpretation is shown in Section 4.4.

The existence of functional relations between dual variables can be identified by reversing the direction of each arrow in Figure 4.1(a). The result is shown in Figure 4.1(b). The functional relations between these variables are determined by the simple rule that any feasible activities in the space system cannot obtain positive profits.

Since the price relation is determined by moving backwards through time, let us start from P_{t+1}. First, given the price vector p_l^{t+1} in location l at the beginning of period $t + 1$, the adjoint transformation g_l^* gives the set of "transportation output price vectors"[4] in location l for which no feasible consumption activities in location l obtain positive profits. That is,

$$g_l^{*t}(p_l^{t+1}) = \{\psi_l^t \in \mathbf{R}_+^{*m} | \psi_l^t \cdot z_l^t \geq p_l^{t+1} \cdot x_l^{t+1} + f_l^{t+1}(x_l^{t+1}, c_l^{t+1})$$
$$\text{for all } x_l^{t+1} + (c_l^{t+1}, 0) \leq z_l^t, x_l^{t+1} \geq 0, c_l^{t+1} \geq 0, z_l^t \geq 0\}. \quad (11)$$

As noted before, the price of one unit of welfare is considered to be unity.

By using (11), the set of transportation output price vectors $\Psi_t = (\psi_l^t)_1^n$ over the space system for which no feasible consumption activities in any location obtain positive profits under given price vector $P_{t+1} = (p_l^{t+1})_1^n$ is given as follows:

$$T_{3t}^*(P_{t+1}) = \{\Psi_t | \Psi_t = (\psi_l^t)_1^n \quad \text{for some } \psi_l^t \in g_l^{*t}(p_l^{t+1}),$$
$$l = 1, 2, \ldots, n, \text{ where } (p_l^{t+1})_1^n = P_{t+1}\}.$$

This procedure is depicted on the right half of Figure 4.1(b).

---

[4]"Transportation output price vectors" are the price vectors for the output commodities of transportation activities, which are also the price vectors for the input commodities of consumption activities.

Next, given a transportation output price vector ψ_s^t in location s, the adjoint transformation G_{ls}^* gives the set of "transportation input price vectors" in location l for which no feasible transportation activities from location l to location s obtain positive profits. Thus, G_{ls}^* is defined as follows:

$$G_{ls}^*(\psi_t^s) = \{\phi_l^t \in \mathbf{R}_+^{*m'} | \phi_l^t \cdot y_{ls}^t \geqq \psi_s^t \cdot z_{ls}^t \quad \text{for all } (y_{ls}^t, z_{ls}^t) \in \text{graph } G_{ls}\}. \tag{12}$$

But, since the transportation input price vector ϕ_l^t in location l is the input price vector not only for the transportation activities from location l to location s but also for all the transportation activities starting from l, the real transportation input price vectors in location l are given by the intersection of sets, $G_{ls}^*(\psi_s^t)$, $s = 1, 2, \ldots, n$. That is, given the transportation output price vector Ψ_t over the space system, the set of transportation input price vectors in location l at which no feasible transportation activities starting from this location obtain positive profits is given as follows:

$$G_l^*(\Psi_t) = \{\phi_l^t | \phi_l^t \in G_{l1}^*(\psi_1^t) \cap \cdots \cap G_{ls}^*(\psi_s^t) \cap \cdots \cap G_{ln}^*(\psi_n^t)$$
$$\text{where } (\psi_s^t)_1^n = \Psi_t\}.$$

Therefore, the set of transportation input price vectors Φ_t over the space system for which no feasible transportation activities in the space system obtain positive profits under given transportation output price vector Ψ_t over the space system is given as follows:

$$T_2^*(\Psi_t) = \{\Phi_t | \Phi_t = (\phi_l^t)_1^n \quad \text{for some } \phi_l^t \in G_e^*(\Psi_t), l = 1, \ldots, n\}.$$

This procedure is depicted in the middle of Figure 4.1(b).

Finally, given a transportation input price vector ϕ_l^t (that is, "a production output price vector") in location l, the adjoint transformation F_l^* gives the set of "production input price vectors" in location l for which no feasible production activities in location l obtain positive profits. Thus F_l^* is defined as follows:

$$F_l^*(\phi_l^t) = \{p_l^t \in \mathbf{R}_+^m | p_l^t \cdot x_l^t \geqq \phi_l^t \cdot y_l^t \quad \text{for all } (x_l^t, y_l^t) \in \text{graph } F_l\}. \tag{13}$$

Thus, the set

$$T_1^*(\Phi_t) = \{P_t | P_t = (p_l^t)_1^n \quad \text{for some } p_l^t \in F_l^*(\phi_l^t),$$
$$l = 1, 2, \ldots, n \text{ where } (\phi_l^t)_1^n = \Phi_t\}$$

gives all of the production input price vectors over the space system at which no feasible production activities in any location obtain positive profits under the given production output price vector Φ_t over the space system.

By using these adjoint transformations, the dual problem for Problem I(a) is stated as follows:

Problem I(b)

Minimize $P_0 \cdot \bar{X}_0,$ (14)

subject to $P_t \in \mathcal{T}_t^*(P_{t+1}),$ $t = 0, 1, \ldots, N-1$ and $P_N = 0,$ (15)

where

$$\mathcal{T}_t^*(P_{t+1}) = T_1^* \cdot T_2^* \cdot T_{3t}^*(P_{t+1}) = \{P_t | P_t \in T_1^*(\Phi_t) \quad \text{for some}$$
$$\Phi_t \in T_2^*(\Psi_t) \text{ for some } \Psi_t \in T_{3t}^*(P_{t+1})\}, \quad (16)$$

$$T_1^*(\Phi_t) = \{P_t | P_t = (p_l^t)_1^n \text{ for some } p_l^t \in F_l^*(\phi_l^t),$$
$$l = 1, 2, \ldots, n, \text{ where } (\phi_l^t)_1^n = \Phi_t\}, \quad (17)$$

$$T_2^*(\Psi_t) = \{\Phi_t | \Phi_t = (\phi_l^t)_1^n \quad \text{for some } \phi_l^t \in G_l^*(\Psi_t), \quad l = 1, \ldots, n\}, \quad (18)$$

$$G^*(\Psi_t) = \{\phi_l^t | \phi_l^t \in G_{l1}^*(\psi_1^t) \cap \cdots \cap G_{ls}^*(\psi_s^t) \cap \cdots \cap G_{ln}(\psi_n^t),$$
$$\text{where } (\psi_s^t)_1^n = \Psi_t\}, \quad (19)$$

$$T_{3t}^*(P_{t+1}) = \{\Psi_t | \Psi_t = (\psi_l^t)_1^n \quad \text{for some } \psi_l^t \in g_l^{*t}(p_l^{t+1}),$$
$$l = 1, 2, \ldots, n, \text{ where } (p_l^{t+1})_1^n = P_{t+1}\}, \quad (20)$$

in which $F_l^*(l = 1, 2, \ldots, n),$ $G_{ls}^*(l, s = 1, 2, \ldots, n)$ and $g_l^{*t}(l = 1, 2, \ldots, n)$ are, respectively, defined by (13), (12) and (11).

The dimension of each vector is:

$$P_t = (p_l^t)_1^n \in \mathbf{R}_+^{*mn}, p_l^t \in \mathbf{R}_+^{*m},$$
$$\Psi_t = (\psi_l^t)_1^n \in \mathbf{R}_+^{*mn}, \psi_l^t \in \mathbf{R}_+^{*m},$$
$$\Phi_t = (\phi_l^t)_1^n \in \mathbf{R}_+^{*m'n}, \phi_l^t \in \mathbf{R}_+^{*m'},$$
$$l = 1, 2, \ldots, n, t = 0, 1, \ldots, N.$$

Under A1 to A3, $g_l^{*t},$ $G_{ls}^*,$ $G_l^*,$ $F_l^*,$ $T_1^*,$ $T_2^*,$ T_{3t}^* and $\mathcal{T}_t^*$ $(l, s = 1, 2, \ldots, n, t = 0, 1, \ldots, N-1)$ are m.t.cc., respectively, and they are non-singular[5] except possibly for $G_{ls}^*(l \neq s)$.

Roughly speaking, the objective of Problem I(b) is to determine a sequence of price vectors $\{\hat{P}_t\}_0^N$ which minimizes the value $P_0 \cdot \bar{X}_0$ of

---

[5]For the definition of non-singularity, see B-definition 6.

the initial stocks of commodities $\bar{X}_0$ among the set of sequences of price vectors $\{P_t\}_0^N$ which do not allow positive profits for any activities in the space system throughout the planning period.

Note also from (11) and (20) that when $f_l^t(x_l^t, c_l^t) \equiv 0$ for $l = 1, 2, \ldots, n; t = 1, 2, \ldots, N - 1$, in the final state problem, we can set

$$\Psi_t = P_{t+1}, \quad t = 0, 1, \ldots, N - 1, \tag{21}$$

and adjoint transformations T_{3t}^* $(t = 0, 1, \ldots, N - 1)$ can be omitted in Problem I(b).

4.3. Duality and related theorems

In this section, the duality theorem, existence theorem and max-min principle are obtained, respectively, for the pair of problems I(a) and I(b) by applying the theorems in Chapter 2. The detailed interpretations of these theorems and the usefulness of these theorems in the analysis of the optimal growth path for the problem are discussed in Sections 4.4 and 4.5.

4.3.1. Duality theorem

Suppose a pair of sequences of vectors $\{\hat{X}_t\}_0^N = \{(\hat{x}_i^t)_1^n\}_0^N$ and $\{\hat{C}_t\}_1^N = \{(\hat{c}_i^t)_1^n\}_1^N$ maximizes the objective function (1) in I(a) among all feasible paths. Then we know from (3) to (7) that at least one pair of sequences of vectors $\{\hat{Y}_t\}_0^{N-1} = \{(\hat{y}_i^t)_1^n\}_0^{N-1}$ and $\{\hat{Z}_t\}_0^{N-1} = \{(\hat{z}_i^t)_1^n\}_0^{N-1}$ corresponds to them. Let us call a quadruplet of such sequences of vectors $[\{\hat{X}_t\}_0^N, \{\hat{Y}_t\}_0^{N-1}, \{\hat{Z}_t\}_0^{N-1}, \{\hat{C}_t\}_1^N]$ *a solution to I(a)*. Similarly, from (15) to (20), we know at least one pair of sequences of vectors $\{\hat{\Phi}_t\}_0^{N-1} = \{(\hat{\phi}_i^t)_1^n\}_0^{N-1}$ and $\{\hat{\Psi}_t\}_1^{N-1} = \{(\hat{\psi}_i^t)_1^n\}_1^{N-1}$ corresponds to each optimal path $\{\hat{P}_t\}_0^N = \{(\hat{p}_i^t)\}_0^N$ for I(b). Thus, a triplet of such sequences of vectors $[\{\hat{P}_t\}_0^N, \{\hat{\Phi}_t\}_0^{N-1}, \{\hat{\Psi}_t\}_0^{N-1}]$ may be called *a solution to I(b)*.

Then, by applying Theorem 2.5 to the pair of problems I(a) and I(b), we get the following theorem.

Theorem 4.1 (duality theorem). Under A1 to A3, suppose $[\{\hat{X}_t\}_0^N, \{\hat{Y}_t\}_0^{N-1}, \{\hat{Z}_t\}_0^{N-1}, \{\hat{C}_t\}_1^N] \equiv [\{(\hat{x}_i^t)_1^n\}_0^N, \{(\hat{y}_i^t)_1^n\}_0^{N-1}, \{(\hat{z}_i^t)_1^n\}_0^{N-1}, \{(\hat{c}_i^t)_1^n\}_1^N]$ is a solution to I(a) and $[\{\hat{P}_t\}_0^N, \{\hat{\Phi}_t\}_0^{N-1}, \{\hat{\Psi}_t\}_0^{N-1}] \equiv [\{(\hat{p}_i^t)_1^n\}_0^N, \{(\hat{\phi}_i^t)_1^n\}_0^{N-1},$

$\{(\hat{\psi}_t^t)_1^n\}_0^{N-1}]$ is a solution to I(b). Then, we have:

(i) $\mu = \sum_{\tau=1}^{N} \sum_{l=1}^{n} f_l^\tau(\hat{x}_l^\tau, \hat{c}_l^\tau) = \sum_{\tau=1}^{t} \sum_{l=1}^{n} f_l^\tau(\hat{x}_l^\tau, \hat{c}_l^\tau) + \hat{\Psi}_t \cdot \hat{Z}_t$

$= \sum_{\tau=1}^{t} \sum_{l=1}^{n} f_l^\tau(\hat{x}_l^\tau, \hat{c}_l^\tau) + \hat{\Phi}_t \cdot \hat{Y}_t = \sum_{\tau=1}^{t} \sum_{l=1}^{n} f_l^\tau(\hat{x}_l^\tau, \hat{c}_l^\tau) + \hat{P}_t \cdot \hat{X}_t$

$= \hat{\Psi}_0 \cdot \hat{Z}_0 = \hat{\Phi}_0 \cdot \hat{Y}_0 = \hat{P}_0 \cdot \bar{X}_0, \quad t = 1, 2, \ldots, N-1$, where

μ is a non-negative real number.

(ii-1) $\max\{\hat{\Phi}_t \cdot Y_t | Y_t \in T_1(\hat{X}_t)\} = \hat{\Phi}_t \cdot \hat{Y}_t = \min\{\Phi_t \cdot \hat{Y}_t | \Phi_t \in T_2^*(\hat{\Psi}_t)\}$,
$t = 0, 1, \ldots, N-1$.

(ii-2) $\max\{\hat{\Psi}_t \cdot Z_t | Z_t \in T_2(\hat{Y}_t)\} = \hat{\Psi}_t \cdot \hat{Z}_t = \min\{\Psi_t \cdot \hat{Z}_t | \Psi_t \in T_{3t}^*(\hat{P}_{t+1})\}$,
$t = 0, 1, \ldots, N-1$.

(ii-3) $\sum_{l=1}^{n} f_l^{t+1}(\hat{x}_l^{t+1}, \hat{c}_l^{t+1}) + \hat{P}_{t+1} \cdot \hat{X}_{t+1} = \max\{\sum_{l=1}^{n} f_l^{t+1}(x_l^{t+1}, c_l^{t+1}) +$

$\hat{P}_{t+1} \cdot X_{t+1} | (X_{t+1}, [f_l^{t+1}(x_l^{t+1}, c_l^{t+1})]_1^n) \in T_{3t}(\hat{Z}_t)\}$,
$t = 0, 1, \ldots, N-1$.

$\sum_{l=1}^{n} f_l^{t+1}(\hat{x}_l^{t+1}, \hat{c}_l^{t+1}) + \hat{P}_{t+1} \cdot \hat{X}_{t+1} = \sum_{l=1}^{n} f_l^{t+1}(\hat{x}_l^{t+1}, \hat{c}_l^{t+1}) +$

$\min[P_{t+1} \cdot \hat{X}_{t+1} | P_{t+1} \in T_1^*(\hat{\Phi}_{t+1})], \quad t = 0, 1, \ldots, N-2.$

(iii) $\hat{P}_t \geq 0$, $\hat{\Phi}_t \geq 0$ and $\hat{\Psi}_t \geq 0$ for $t = 0, 1, \ldots, N-1$, $\hat{P}_N = 0$. And if
$f_l^N(x_l^N, c_l^N) \neq 0$, then $\hat{p}_l^t \geq 0$, $\hat{\phi}_l^t \geq 0$ and $\hat{\psi}_l^t \geq 0$ for $t = 0, 1, \ldots, N-1$.

Conversely, under A1 to A3, if a pair of feasible paths, $[\{\hat{X}_t\}_0^N, \{\hat{Y}_t\}_0^{N-1}$, $\{\hat{Z}_t\}_0^{N-1}, \{\hat{C}_t\}_1^N]$ for I(a) and $[\{\hat{P}_t\}_0^N, \{\hat{\Phi}_t\}_0^{N-1}, \{\hat{\Psi}_t\}_0^{N-1}]$ for I(b), satisfy the condition $\sum_{\tau=1}^{N} \sum_{l=1}^{n} f_l^\tau(\hat{x}_\tau^t, \hat{c}_\tau^t) = \hat{P}_0 \cdot \bar{X}_0$, then they are solutions to I(a) and I(b), respectively.

Parts (i), (ii), the first half of (iii) and the last half of the above theorem are directly obtained from Theorem 2.9 by observing that each transformation T_1, T_2 and T_3 in I(a) is a space transformation in each stage within each period. The latter half of (iii) can be proved in a manner quite similar to (iii) in Theorem 2.9. Hence their proofs are omitted.

Part (i) says that on the optimal path the sum of the total value of the stocks of commodities in the space system and the total accumulation of welfare in the space system is preserved not only in each period but also in any stage of activity within each period. The economic interpretation of dual variables is given in Section 4.4. The interpretation of (ii) is made simpler by the max-min principle, and its discussion is postponed until Section 4.5. Part (iii) says that price vectors in each location are semipositive in each stage of activities. And, (i) and the last half of the theorem together say that when there are optimal paths for I(a) and I(b), a necessary and sufficient condition for optimality is the equality between the total accumulated welfare in the final period and the value of the initial stock of commodities.

4.3.2. Existence theorem

To obtain the existence theorem for Problem I, first we need the next lemma, which is proved in Appendix D.2.

Lemma 4.1

(i) When each function $f_l^t(x_l^t, c_l^t)$, $l = 1, 2, \ldots, n$, satisfies the Lipschitz condition, then the function $f_t(X_t, C_t) = \Sigma_{l=1}^n f_l^t(x_l^t, c_l^t)$ also satisfies this condition.

(ii) When each $F_l(x_l^t)$, $l = 1, 2, \ldots, n$, satisfies the Lipschitz condition, then T_1, which is defined by (4), also satisfies this condition.

(iii) When each G_l, $l = 1, 2, \ldots, n$, satisfies the Lipschitz condition, then T_2, which is defined by (5), also satisfies this condition.

(iv) When each G_{ls}, $s = 1, 2, \ldots, n$, satisfies the Lipschitz conditions, then G_l, which is defined by (6), also satisfies this condition.

(v) The transformation T_3' defined by

$$T_3'(Z_t) = \{(X_{t+1}, C_{t+1}) | (X_{t+1}, C_{t+1}) = ((x_l^{t+1})_1^n, (c_l^{t+1})_1^n)$$
$$\text{for some } x_l^{t+1} + (c_l^{t+1}, 0) \le z_l^t, \ x_l^{t+1} \ge 0, \ c_l^{t+1} \ge 0,$$
$$l = 1, 2, \ldots, n \text{ where } (z_l^t)_1^n = Z_t\}$$

satisfies the Lipschitz condition.

Hence, by applying Lemma 2.3 and Theorem 2.10, we obtain the following theorem.

Theorem 4.2 (existence theorem). If A1 to A3 hold, then we have:

(i) There always exists an optimal path for I(a).

(ii) An optimal path exists for I(b) under either one of the next two conditions:

(a) $\bar{X}_0 > 0$.

(b) Each F_l, G_{ls} and f_l^t, $l, s = 1, 2, \ldots, n$, $t = 1, 2, \ldots, N$, satisfies the Lipschitz condition.

Proof. Problem I(a) is equivalent to the following problem.

$$\text{Maximize} \sum_{l=1}^{n} f_t(X_t, C_t) \equiv \sum_{t=1}^{N} \sum_{l=1}^{n} f_l^t(x_l^t, c_l^t)$$

subject to $(X_t, C_t) \in T_3' \cdot T_2 \cdot T_1(X_t)$, $t = 1, 2, \ldots, N$,

and $X_0 = \bar{X}_0$.

Here T_1 and T_2 are defined by (4) and (5), respectively, and T_3' is defined in Lemma 4.1. Under A3, f_t is a monotone concave gauge, $t = 1, 2, \ldots, N$, and considering 3.1 one easily sees that under A1 and A2 the transformation $T_3' \cdot T_2 \cdot T_1$ is a non-singular m.t.cc. Hence, under A1 to A3, Problem I(a) is a special case of Problem II(a) in Chapter 2. Therefore, from Theorem 2.10 and Lemma 4.1, one immediately obtains Theorem 4.2.   Q.E.D.

Of course, the function $f_l^t(x_l^t, c_l^t) \equiv 0$, satisfies the Lipschitz condition. Hence, for the final state problem, only the condition on f_l^N is needed. Moreover, when we use function (6) in Chapter 3 with a generalized monotone concave gauge u_l, $f_l^t(x_l^t, c_l^t)$ satisfies the Lipschitz condition if, for example, u_l has an upper bound on $\mathbf{R}_+^m$ or u_l is a linearly homogeneous function. And when each F_l and G_{ls} is of the Leontief type or von Neumann type, then they naturally satisfy the Lipschitz condition. When F_l is based on a neoclassical type production function, then F_l satisfies the Lipschitz condition if the marginal productivity of the original production function has an upper bound on $\mathbf{R}_+^m$.

4.3.3. Max-min principle

The next theorem states the max-min principle for Problem I(a) in its most detailed form.

Theorem 4.3 (max-min principle).   If A1 to A3 hold, then under either one of the two conditions (a) or (b) in Theorem 4.2, a feasible path $[\{\hat{X}_t\}_0^N, \{\hat{Y}_t\}_0^{N-1}, \{\hat{Z}_t\}_0^{N-1}, \{\hat{C}_t\}_1^N] \equiv [\{(\hat{x}_l^t)_1^n\}_0^N, \{(\hat{y}_l^t)_1^n\}_0^{N-1}, \{(\hat{z}_l^t)_1^n\}_0^{N-1}, \{(\hat{c}_l^t)_1^n\}_1^N]$ for I(a) is a solution to that problem if and only if there exists a feasible path $[\{\hat{P}_t\}_0^N, \{\hat{\Phi}_t\}_0^{N-1}, \{\hat{\Psi}_t\}_0^{N-1}] \equiv [\{(\hat{p}_l^t)_1^n\}_0^N, \{(\hat{\phi}_l^t)_1^n\}_0^{N-1},$ $\{(\hat{\psi}_l^t)_1^n\}_0^{N-1}]$ for I(b) such that they together satisfy the next three conditions:

(i) $\max\{\hat{\phi}_l^t \cdot y_l^t | y_l^t \in F_l(\hat{x}_l^t)\} = \hat{\phi}_l^t \cdot \hat{y}_l^t = \min\{\phi_l^t \cdot \hat{y}_l^t | \phi_l^t \in G_l^*(\hat{\Psi}_t)\},$

$l = 1, 2, \ldots, n, \ t = 0, 1, \ldots, N-1.$

(ii) $\sum_{l=1}^n \max\{\hat{\Psi}_t \cdot Z_l^t | Z_l^t \in G_l(\hat{y}_l^t)\} = \hat{\Psi}_t \cdot \hat{Z}_t = \sum_{l=1}^n \min\{\psi_l^t \cdot \hat{z}_l^t | \psi_l^t \in$

$g_l^{*t}(\hat{p}_l^{t+1})\}, \quad t = 0, 1, \ldots, N-1.$

(iii) $\max\{f_l^{t+1}(x_l^{t+1}, c_l^{t+1}) + \hat{p}_l^{t+1} \cdot x_l^{t+1} | (x_l^{t+1}, f_l^{t+1}(x_l^{t+1}, c_l^{t+1})) \in g_l^t(z_l^t)\}$

$= f_l^{t+1}(\hat{x}_l^{t+1}, \hat{c}_l^{t+1}) + \hat{p}_l^{t+1} \cdot \hat{x}_l^{t+1}$

$= f_l^{t+1}(\hat{x}_l^{t+1}, \hat{c}_l^{t+1}) + \min\{p_l^{t+1} \cdot \hat{x}_l^{t+1} | p_l^{t+1} \in F_l^*(\hat{\phi}_l^{t+1})\},$

$l = 1, 2, \ldots, n, \ t = 0, 1, \ldots, N-2;$ and

$\max\{f_l^N(x_l^N, c_l^N) | (x_l^N, f_l^N(x_l^N, c_l^N)) \in g_l^{N-1}(\hat{z}_l^{N-1})\} = f_l^N(\hat{x}_l^N, \hat{c}_l^N),$

$l = 1, 2, \ldots, n.$

In this case, this price path is a solution for I(b).

The necessity of the above three conditions follows from Theorem 4.1 since they are just a restatement of (ii) in that theorem. For example, the left hand side of (i) is obtained from (ii-1) in Theorem 4.1 as follows. Since $\hat{y}_l^t$ is in the set $F_l(\hat{x}_l^t)$ from the definition of a feasible path for I(a), it follows that max $\{\hat{\phi}_l^t \cdot y_l^t | y_l^t \in F_l(\hat{x}_l^t)\} \geq \hat{\phi}_l^t \cdot \hat{y}_l^t$. Thus $\sum_{l=1}^n \max\{\hat{\phi}_l^t \cdot y_l^t | y_l^t \in F_l(\hat{x}_l^t)\} = \max\{\hat{\Phi}_t \cdot Y_t | Y_t \in T_1(\hat{X}_t)\} \geq \sum_{l=1}^n \hat{\phi}_l^t \cdot \hat{y}_l^t = \hat{\Phi}_t \cdot \hat{Y}_t$. But the left hand side of (ii-1) in Theorem 4.1 requires that max $\{\hat{\Phi}_t \cdot Y_t | Y_t \in T_1(\hat{X}_t)\} = \hat{\Phi}_t \cdot \hat{Y}_t$. Hence, we must have $\max\{\hat{\phi}_l^t \cdot y_l^t | y_l^t \in F_l(\hat{x}_l^t)\} = \hat{\phi}_l^t \cdot \hat{y}_l^t$ for each $l = 1, 2, \ldots, n$. All other conditions are also obtained directly from (ii) in Theorem 4.1. Finally, the sufficiency follows from Theorem 2.12.

Since the above theorem tells us how optimal activities and price vectors are determined in each stage of each period, we can determine the solution step by step. A detailed interpretation of this theorem is given in Section 4.5.

4.4. Interpretations of dual variables

Recall that the dual variable corresponding to each commodity variable was called the "price" of that commodity. In this section we justify this interpretation.

Let us define a family of functions $\mu_t(X_t)$, $t = 0, 1, \ldots, N - 1$, by
$\mu_t(X_t) = \max \Sigma_{\tau=t+1}^{N} \Sigma_{l=1}^{n} f_l^{\tau}(x_l^{\tau}, c_l^{\tau})$, subject to

$$(X_{\tau+1}, [f_l^{\tau+1}(x_l^{\tau+1}, c_l^{\tau+1})]_1^n) \in \mathcal{T}_\tau(X_\tau), \quad \tau = t, t + 1, \ldots, N - 1,$$

where $X_\tau = (x_l^\tau)_1^n$, and where $\mathcal{T}_\tau$ is the transformation defined in Problem I(a). By this definition, $\mu_t(X_t)$ represents the maximum total welfare obtainable from input vector X_t in $N - t$ periods.

Suppose that $\{\hat{X}_t\}_0^N = \{(\hat{x}_t)_1^n\}_0^N$ is an optimal path for I(a) and $\{\hat{P}_t\}_0^N = \{(\hat{p}_l^t)_1^n\}_0^N$ is an optimal path for I(b). Then, from (40a) in Chapter 2, we see that $\mu_t(\hat{X}_t)$ and $\hat{P}_t$ have the following relation (when $\hat{x}_{il}^t = 0$, we must use (40b) in Chapter 2):

$$\frac{\partial \mu_t^-(\hat{X}_t)}{\partial x_{il}^t} \geq \hat{p}_{il}^t \geq \frac{\partial \mu_t^+(\hat{X}_t)}{\partial x_{il}^t}, \quad i = 1, 2, \ldots, m, l = 1, 2, \ldots, n. \tag{22}$$

Especially when the function $\mu_t(X_t)$ is smooth at $\hat{X}_t$, that is, continuously differentiable at $\hat{X}_t$, the above relation is rewritten as follows:

$$\frac{\partial \mu_t(\hat{X}_t)}{\partial x_{il}^t} = \hat{p}_{il}^t, \quad i = 1, 2, \ldots, m, l = 1, 2, \ldots, n. \tag{23}$$

We can replace $(\hat{X}_t, x_{il}^t, \hat{p}_{il}^t)$ by $(\hat{Y}_t, y_{il}^t, \hat{\phi}_{il}^t)$ or $(\hat{Z}_t, z_{il}^t, \hat{\psi}_{il}^t)$ in (22) and (23). Hence the economic interpretation of the dual variables is clear. For simplicity, let us suppose that function $\mu_t(X_t)$ is smooth at $\hat{X}_t$, that is, Equation (23) holds. Then $\hat{p}_{il}^t$ measures the marginal contribution of the stock of commodity i in location l at the beginning of period t to the maximum possible value of the objective function at the final period. That is, when the unit for each commodity is chosen to be sufficiently small, $\hat{p}_{il}^t$ measures how much the total accumulated welfare in the final period can be increased if one extra unit of commodity i is added in location l at the beginning of period t on the optimal path. Similarly, $\hat{\phi}_{il}^t$ and $\hat{\psi}_{il}^t$ measure how much the total accumulated welfare in the final period can be increased if one extra unit of commodity i is added in location l, respectively, at the end of production activities and at the end of transportation activities in period t. Thus in economics, each dual variable is generally called the "*(shadow) price*" of the corresponding commodity – in our present problem, the "price in terms of welfare."[6]

---

[6]Besides "price," several other terms are used for dual variables in mathematical programming. For example, dual variables for natural resources are often called "*(scarcity) rent*" and the dual variable for capital is called "*quasi-rent*" in single-

Now let us examine the spatial aspect of dual variables. First, the spatial price relation of transportable commodities is considered. For simplicity let us assume that each of basic transportation activities in the space system is of the following type:

$$_i y_{ls}^t = (y_{1i}^{ls} \ldots, y_{i-1,i}^{ls}, 1 + y_{ii}^{ls}, y_{i+1,i}^{ls}, \ldots, y_{m',i}^{ls})$$

$$\downarrow^{i}$$

$$\to {_i z_{ls}^t} = (0, \ldots, 0, 1, 0, \ldots, 0).$$

That is, each basic transportation activity originating in location l and ending in location s transports one unit of each commodity i from l to s by using a set of current transport inputs. The corresponding price vectors are $\phi_l^t = (\phi_{1l}^t, \ldots, \phi_{il}^t, \ldots, \phi_{m'l}^t)$ and $\psi_s^t = (\psi_{1s}^t, \ldots, \psi_{is}^t, \ldots, \psi_{ms}^t)$, where ϕ_l^t is the price vector for the inputs of transportation activities starting from location l, and ψ_s^t is the price vector for the outputs of transportation activities ending in location s. Then, if $(_i y_{ls}^t, {_i z_{ls}^t})$ is a feasible transportation activity for I(a) and if (ϕ_l^t, ψ_s^t) is a feasible price vector for I(b), Equation (12) requires that

$$\phi_{il}^t + \sum_{j=1}^{m'} \phi_{jl}^t y_{ji}^{ls} \geqq \psi_{is}^t, \tag{24}$$

that is,

$$(\text{price of } i \text{ in } l) + \binom{\text{transport cost for one}}{\text{unit of } i \text{ from } l \text{ to } s} \geqq \binom{\text{price of}}{i \text{ in } s}.$$

Furthermore, suppose $(\hat{\phi}_l^t, \hat{\psi}_s^t)$ is the price vector on the optimal path for I(b) and $(_i \hat{y}_{ls}^t, {_i \hat{z}_{ls}^t})$ is a basic transportation activity which is used with positive activity level on the optimal path for I(a), then we must have

$$\hat{\phi}_{il}^t + \sum_{j=1}^{m'} \hat{\phi}_{jl}^t \hat{y}_{ji}^{ls} = \hat{\psi}_{is}^t \tag{25}$$

since no activities with negative profits can be used on the optimal path.[7] That is, if any transportation takes place, then the difference of

---

period programming. As Stevens points out (1958, p. 85), in order to explain the relation between price, rent and quasi-rent in dynamic models, it is convenient to introduce another term, "royalty." For a detailed discussion on the relationship between these terms, see Fujita (1972, pp. 195–199).

[7]This is easily seen from (i) in Theorem 4.1. By definition of adjoint transformation G_{ls}^* by (12), any feasible transportation activity cannot obtain positive profits. Hence, if some transportation activity gets negative profits, condition $\hat{\Psi}_t \cdot \hat{Z}_t = \hat{\Phi}_t \cdot \hat{Y}_t$ which is required in (i) of Theorem 4.1 cannot be satisfied.

prices between two locations should be exactly the same as the transport cost. And, as long as $\sum_{j=1}^{m} \hat{\phi}_{jl}^{t} \hat{y}_{ji}^{ls} > 0$, (25) implies that $\hat{\psi}_{is}^{t} > \hat{\phi}_{il}^{t}$. Thus, any feasible transportation activities for commodity i from s to l obtain negative profits, and hence there is no transportation of commodity i from s to l (i.e., there is no "cross-hauling") on the optimal path. These are well known conditions of spatial price equilibrium.[8]

Next, let us examine the spatial price relation of immobile commodities, for example, natural resources. We may assume that each immobile commodity is a transportable good with infinite transport inputs, and we may apply spatial equilibrium condition (24). But, since the transport cost for immobile commodities is infinite by definition, condition (24) is satisfied for any pair of prices $(\phi_{il}^{t}, \psi_{is}^{t})$. And, since by definition, immobile commodities are not transported between locations on the optimal path, condition (25) need not be satisfied. Hence, spatial equilibrium conditions (24) and (25) yield no useful information to us in the case of immobile commodities. The situation is similar for many capital goods such as factories and installed machinery. For these capital goods, transportation costs are very high, and thus condition (24) will always be satisfied and condition (25) need not be satisfied.

In reality, when the programming is short-run, any values can be assumed by dual variables corresponding to immobile commodities and "virtually-immobile commodities" depending on the initial conditions and the objective function. On the other hand, suppose the planning period is long-run. Then we may naturally expect that there will be some kind of spatial price relation between these commodities, since in this case the initial condition and the form of the objective function may not be crucial factors in shaping the optimal growth path except at the initial and the terminal periods. This point is discussed in detail in Chapter 5.

Though it is difficult to determine any definite spatial relation between the prices of immobile commodities, it may not be impossible to find some spatial relations between the rent on immobile commodities. The reason is that the rent on a resource (or capital) in a given period is decided by the value of the outputs newly produced by the marginal unit of that resource in that period, and the value of the outputs are spatially related when they are mobile goods. In particular, when the

---

[8]See, for example, Samuleson (1952).

problem is simple, spatial relations of rent on resources (or capital) can be investigated effectively by using the notion of "*location rent.*" A typical example in static problems is the von Thünen model of agricultural locations. The relation between the dual variables of a linear programming problem and location rent was investigated successfully by Stevens (1961) in the formulation of a transportation problem.

Finally, suppose that the amounts of a commodity i in location r and location s are, respectively, $\hat{x}^t_{ir}$ and $\hat{x}^t_{is}$ on the optimal path. Then $\hat{p}^t_{ir} > \hat{p}^t_{is}$ implies that a part of commodity i in location s would be moved to location r if this commodity could be moved freely at that time. And $\hat{p}^t_{ir} = \hat{p}^t_{is}$ implies there would be no changes in these amounts even if commodity i could be moved freely between these two locations at that time. Hence the "spatial variation of prices" at any time measures the inefficiency of the space system at that time due to the immobilities of some resources and transport inputs required for the movement of commodities (though it will be difficult to construct an exact measure of spatial variation of prices, since $\hat{p}^t_{il}$ stays constant only for a small variation of $\hat{x}^t_{il}$).

4.5. Stepwise procedures for optimal paths

4.5.1. Stepwise procedures

In this section we use Theorem 4.3 to examine how optimal activities and price vectors are determined in each period. The main concern of this section is not to examine the computational procedures but rather to gain an understanding of the general nature of optimal paths by means of Theorem 4.3.

First, the determination of optimal activities and price vectors in each period is divided into three stages – production, transportation and consumption. We then consider how optimal activities and price vectors are decided in each stage.

Suppose that the state of the space system $\hat{X}_t = (\hat{x}^t_l)^n_1$ at the beginning of period t on the optimal path is already known. Then recall from the left side equality of (i) in Theorem 4.3 that each location chooses its production output vector $\hat{y}^t_l$ among the set $F_l(\hat{x}^t_l)$ so as to maximize the value $y^t_l \cdot \hat{\phi}^t_l$ of its production outputs under some price vector $\hat{\phi}^t_l$. This is depicted in Figure 4.2(a).

Also recall from the left side of equality (ii) in Theorem 4.3 that after

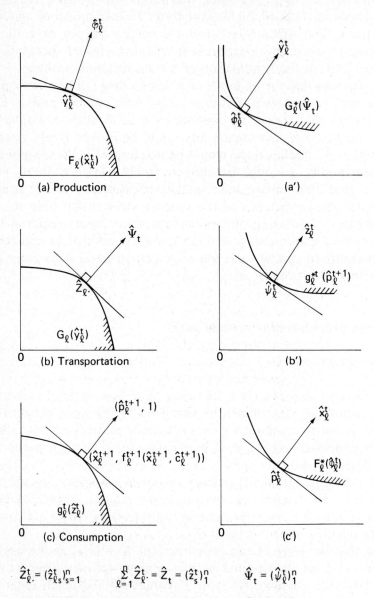

$$\hat{Z}^t_\ell = (\hat{z}^t_{\ell s})^n_{s=1} \qquad \sum_{\ell=1}^n \hat{Z}^t_\ell = \hat{Z}_t = (\hat{z}^t_s)^n_1 \qquad \hat{\Psi}_t = (\hat{\psi}^t_\ell)^n_1$$

Figure 4.2. Stepwise procedures for the optimal path.

$\hat{y}_l^t$ is determined by the above procedure, now location l should choose its transportation output vector $\hat{Z}_l^t. = (\hat{z}_{ls}^t)_{s=1}^n$ over the space system among the set $G_l(\hat{y}_l^t)$ so as to maximize the value $Z_l^t. \cdot \hat{\Psi}_t$ of its transport outputs under some price vector $\hat{\Psi}_t = (\hat{\psi}_l^t)_1^n$. This is depicted in Figure 4.2(b). From (6), we see that the determination of the optimum $\hat{Z}_l^t.$ may be understood in two steps. First, the transport input vector $\hat{y}_l^t$ is divided into $(\hat{y}_{l1}^t, \ldots, \hat{y}_{ls}^t, \ldots, \hat{y}_{ln}^t)$. Then the optimum transport output vector $\hat{z}_{ls}^t$ from l to s is chosen among set $G_{ls}(\hat{y}_{ls}^t)$ so as to maximize the value $z_{ls}^t \cdot \hat{\psi}_s^t$ $(s = 1, 2, \ldots, n)$. But in reality, the vectors $(\hat{y}_{l1}^t, \ldots, \hat{y}_{ls}^t, \ldots, \hat{y}_{ln}^t)$ and $\hat{Z}_l^t. = (\hat{z}_{l1}^t, \ldots, \hat{z}_{ls}^t, \ldots, \hat{z}_{ln}^t)$ are determined simultaneously since we do not know how to obtain vector $(\hat{y}_{l1}^t, \ldots, \hat{y}_{ls}^t, \ldots, \hat{y}_{ln}^t)$ by itself.

The sum $\Sigma_{l=1}^n \hat{Z}_l^t. = (\Sigma_{l=1}^n \hat{z}_{ls}^t)_{s=1}^n = (\hat{z}_s^t)_{s=1}^n = \hat{Z}_t$ is the final result of all such transportation activities over the space system. Then, as seen from (5), vector $\hat{Z}_t$ naturally maximizes the value $\hat{\Psi}_t \cdot Z_t$ among the set $T_2(\hat{Y}_t)$, as is required in (ii) of Theorem 4.3.

Finally, recall from the first equality in (iii) of Theorem 4.3 that the commodity vector z_l^t in each location l after the transportation activities over the space system is divided into two vector components, $\hat{x}_l^{t+1}$ and $\hat{c}_l^{t+1}$, so as to maximize the value $f_l^{t+1}(x_l^{t+1}, c_l^{t+1}) + \hat{p}_l^{t+1} \cdot x_l^{t+1}$ for some price vector $\hat{p}_l^{t+1}$. This is depicted in Figure 4.2(c). Then the total welfare obtained in the space system is $\Sigma_{l=1}^n f_l^{t+1}(\hat{x}_l^{t+1}, \hat{c}_l^{t+1})$, and vector $\hat{X}_t = (\hat{x}_l^{t+1})_1^n$ constitutes the production input vector in period $t + 1$.

Conversely, suppose that we somehow know the production input price vector $\hat{P}_{t+1} = (\hat{p}_l^{t+1})_1^n$ in period $t + 1$ corresponding to the optimal path. Then, recall from the first equality in (ii) of Theorem 4.3 that the input price vector $\hat{\psi}_l^t$ for consumption input commodities in each location l in period t is chosen from the set $g_l^{*t}(\hat{p}_l^{t+1})$ so as to minimize the value $\psi_l^t \cdot \hat{z}_l^t$ of the consumption inputs under some vector $\hat{z}_l^t$. This is shown in Figure 4.2(b'). Then, as seen from (20), the vector $\hat{\Psi}_t = (\hat{\psi}_l^t)^n$ naturally minimizes the value $\Psi_t \cdot \hat{Z}_t$ among the set $T_{3t}^*(\hat{P}_{t+1})$, as is required in (ii) of Theorem 4.3.

Next, observe from the second equality in (i) of Theorem 4.3 that after $\hat{\psi}_l^t$ is determined by the above procedure, the transportation input price vector (that is, the production output price vector) $\hat{\phi}_l^t$ is chosen in location l so as to minimize the value $\phi_l^t \cdot \hat{y}_l^t$ of transport inputs in that location under some vector $\hat{y}_l^t$. This is shown in Figure 4.2(a').

Finally, observe from the second equality in (iii) of Theorem 4.3 that the production input price vector $\hat{p}_l^t$ in each location l is decided so as to minimize the value $p_l^t \cdot \hat{x}_l^t$ of production inputs for some vector $\hat{x}_l^t$. This is depicted in Figure 4.2(c').

In reality, a pair of vectors $\hat{Y}_t$ and $\hat{\Phi}_t$, $\hat{Z}_t$ and $\hat{\Psi}_t$, and $\hat{X}_{t+1}$ and $\hat{P}_{t+1}$ are, respectively, decided simultaneously by iterations so as to satisfy, respectively, conditions (i), (ii) and (iii) in Theorem 4.3. If the whole sequence of vectors satisfy these conditions in each period, then we have a solution.

4.5.2. Singular control in space systems

In the previous section, we treated the optimal-path problem as if the optimum activity and the corresponding price vector could be selected uniquely in each stage by the maximum principle. Moreover, all the diagrams in Figure 4.2 are depicted as if each supporting hyperplane has a unique tangential point. But this will not be true for many problems in space systems.

In production decisions and consumption decisions this problem is not so important. For example, if the frontier curve of the set $F_l(\hat{x}_l^l)$ is strictly concave, we obtain a unique optimal output vector $\hat{y}_l^l$ by the max-min principle. And the strict concavity of the frontier curve of the feasible production set $F_l(\hat{x}_l^l)$ is theoretically possible. For example, if the number of the basic production activities is infinite, this concavity condition may hold. Even if the frontier curve of the set $F_l(\hat{x}_l^l)$ is not strictly concave, this problem may not be so important. The reason is that in the case of production decisions there is no theoretical reason why the production output price vector $\hat{\phi}_l^l$ assumes such a value so as to give multiple maximizing points more than accidentally. Even if this happens, it is likely to occur only in a few periods.

But the situation is entirely different in the case of transportation decisions. For example, as long as we assume constant returns to scale in transportation activities, the frontier curve of the feasible transportation set $G_l(\hat{y}_l^l)$ is never strictly concave even if we assume an infinite number of basic transportation activities.

Let us start from the simplest example. Suppose our space system consists of two locations – Philadelphia ($= r$) and New York ($= s$). At first, let us suppose that the transportation of steel ($= i$) does not require any extra inputs, and that 10 units of steel are produced in Philadelphia in a period on the optimal path. Then the feasible set for the steel allocation between two locations can be depicted as in Figure 4.3(a). Now suppose the optimal path requires that 6 units of steel are distributed to New York and 4 units are retained in Philadelphia. Then,

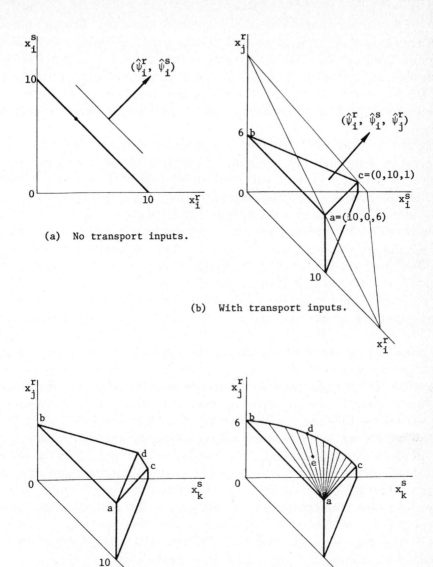

(a) No transport inputs.

(b) With transport inputs.

(c) With alternative transportation   (d) With infinite transportation
     techniques.   techniques.

Figure 4.3. Feasible sets for transport outputs.
Note: Subscripts *i*, *j* and *k* indicate steel, gasoline and oil, respectively, while
superscripts *r* and *s* represent Philadelphia and New York, respectively.

to sustain this optimal allocation, the corresponding transport output prices, $\hat{\psi}_i^r$ and $\hat{\psi}_i^s$, should be exactly the same, as shown in condition (ii) of Theorem 4.3. But the trouble is that any efficient point on the feasible allocation set obtains the same revenue. Thus the max-min principle is unable to select the optimal point $(4, 6)$. This kind of situation is called a *singular control problem* in optimal control theory.[9]

Unfortunately, this problem is not an exceptional case for transportation decisions. For example, in the next period it may be necessary to produce 11 units of steel in Philadelphia, and the optimal allocation may be $(5, 6)$. Here again we have the same trouble. That is, the allocation of the product between more than two locations (including possibly, the production site) need not be an exceptional case on the optimal path in a space system with multiple commodities.

The predicament does not change even when we consider the transport inputs. For example, let us introduce the current transport input commodity, gasoline $(= j)$ into the previous example. Suppose that 0.5 units of gasoline are required for the transportation of each unit of steel from Philadelphia to New York, and that Philadelphia has 6 units of gasoline at the end of the production in a period. Then the transport input point is given by point a in Figure 4.3(b) and the feasible set for transportation outputs is as depicted in this figure. The efficient allocation set is then the line segment $\overline{ac}$. Hence whichever point on the efficient set $\overline{ac}$ (except for extreme points a and c) is required on the optimal path, we have a singular control problem.

Let us introduce alternative transportation techniques into the above example. Since it is not reasonable to assume that steel and gasoline are substitutable for each other as current transport inputs, let us consider the transportation of oil $(= k)$, instead of steel, from Philadelphia to New York. Thus the original transportation technique is to transport one unit of oil from Philadelphia to New York with 0.5 units of gasoline, and the alternative transportation technique is to transport one unit of oil from Philadelphia to New York with 0.3 units of gasoline and 0.1 units of oil. When Philadelphia has 10 units of oil and 6 units of gasoline initially, the feasible transport output set is depicted as in Figure 4.3(c). The efficient set is given by the triangular surface acd.

---

[9]For the singular control problem in optimal control theory, see, for example, Athans and Falb (1966).

Figure 4.3(d) shows the case with an infinite number of basic transportation activities. The initial transport input point is denoted by point *a*. As we see in the figure, as long as we assume constant returns to scale in the transportation activities, the efficient set is not strictly concave even if there exist an infinite number of basic transportation activities. For example, the line segment $\overline{ad}$ is part of the efficient set. Thus the hyperplane (that is, the price system) which sustains point *e* gives the same revenue to any point on the line segment $\overline{ad}$. Thus we have a singular control problem.

It is interesting to note that the singular control problem would almost be impossible to observe if the optimal path were actually calculated. For example, in the case of Figure 4.3(a), we see that the total amount of steel in Philadelphia is allocated to both locations only when transport output prices $\hat{\psi}_i^r$ and $\hat{\psi}_i^s$ are exactly the same. Thus, even small computational errors in the values of prices bring about extreme allocations. Therefore, the calculated optimal growth path may exhibit "zigzag" movements, which tends to lead one to the wrong conclusion that the spatial allocation of commodities must have a strong "all-or-none" character.

How do we avoid this singular control problem in actual calculations? This task may not be impossible if we are prepared to perform a large-scale computation in each step. That is, instead of performing the computation for each stage of activities (i.e., production, transportation and consumption) we solve the following problem simultaneously for each period.

$$\max \sum_{l=1}^{n} f_l^{t+1}(x_l^{t+1}, c_l^{t+1}) + \hat{\Phi}_{t+1} \cdot Y_{t+1},$$

s.t. $Y_{t+1} \in T_1(X_{t+1}),$

$\qquad (X_{t+1}, [f_l^{t+1}(x_l^{t+1}, c_l^{t+1})]_1^n) \in T_{3t}(Z_t),$

$\qquad Z_t \in T_2(\hat{Y}_t).$

Since, as noted before, the singular control problem is less likely to occur in production and consumption than in transportation, the above problem is less likely to have the singular control problem. And the optimal transportation schedule is automatically determined from the solution to the above problem.

4.6. In the case of generalized monotone programming

Up to this point we have assumed that each welfare function and each transformation in the problem are linearly homogeneous, and have shown that a fairly broad class of problems can be treated within this framework.

As long as we do not permit increasing returns to scale in production and transportation, the assumption of first degree homogeneity in production and transportation technologies may not be a bad assumption, since decreasing returns to scale arise mostly from the fixedness of some production factors (especially from the limitations on natural resources). But the fixedness of some production factors can be treated quite naturally within the framework of monotone programming, as was explained in Section 3.2.4. Thus, in the case of final state problems, the previous problem formulation was fairly general.

But, in the case of consumption stream problems, considering only linearly homogeneous welfare functions may be far too restrictive. Thus, let us briefly examine what modifications are needed to the previous results when assumption A3 is replaced by the following:

A3'. Each $f_l^t(x_l^t, c_l^t)$, $l = 1, 2, \ldots, n$, $t = 1, 2, \ldots, N$, is a generalized monotone concave gauge on $\mathbf{R}_+^{m+m_c}$, and not all of $f_l^N(x_l^N, c_l^N)$, $l = 1, 2, \ldots, n$, are identically zero.

As shown below the previous results remain basically unchanged under this assumption.[10] There is no change in the formulation of the primal problem, but the assumptions are different. Hence, let us designate the problem with assumptions A1, A2 and A3' in Problem I(a) as Problem II(a). A slight modification is needed in the dual problem. First, the adjoint transformation (11) is now changed as follows:

$$g_l^{*t}(p_l^{t+1}, p_{\eta,l}^{t+1}) = \{(\psi_l^t, p_{\eta,l}^t) | \psi_l^t \cdot z_l^t + p_{\eta,l}^t \eta$$
$$\geq p_l^{t+1} \cdot x_l^{t+1} + p_{\eta,l}^{t+1} \eta + \eta f_l^{t+1}\left(\frac{x_l^{t+1}}{\eta}, \frac{c_l^{t+1}}{\eta}\right) \quad \text{for all}$$
$$x_l^{t+1} + (c_l^{t+1}, 0) \leq z_l^t, x_l^{t+1} \geq 0, c_l^{t+1} \geq 0, \eta > 0\}, \quad (11')$$

In this definition, η is a fictitious scalar variable introduced to change function $f_l^{t+1}(x_l^{t+1}, c_l^{t+1})$ which is not linearly homogeneous into a

---

[10]We may also replace A1 with A1': Each of F_l and G_{ls}, $l, s = 1, 2, \ldots, n$, is a generalized m.t.cc. By using theorems in Section 2.5 in Chapter 2, we can show that the previous results remain basically unchanged even under A1', A2 and A3'.

linearly homogeneous function $\eta f_{t+1}(x_{t+1}/\eta, c_{t+1}/\eta)$. $p_{\eta,l}^{t+1}$ denotes the fictitious price variable corresponding to η in location l (for these points, see Section 2.5 in Chapter 2).

The dual problem can now be rewritten as follows:

Problem II(b).

$$\text{Minimize } P_0 \cdot \bar{X}_0 + \sum_{l=1}^{n} p_{\eta,l}^0,$$

subject to

$$P_t \in T_1^*(\Phi_t) = \{P_t | P_t = (p_l^t)_1^n \quad \text{for some } p_l^t \in F_l^*(\phi_l^t),$$
$$l = 1, 2, \ldots, n, \text{ where } (\phi_l^t)_1^n = \Phi_t\},$$

$$\Phi_t \in T_2^*(\Psi_t) = \{\Phi_t | \Phi_t = (\phi_l^t)_1^n \quad \text{for some } \phi_l^t \in G_l^*(\Psi_t),$$
$$l = 1, 2, \ldots, n\},$$

$$G_l^*(\Psi_t) = \{\phi_l^t | \phi_l^t \in G_{l1}^*(\psi_1^t) \cap \cdots \cap G_{ls}^*(\psi_s^t) \cap \cdots \cap G_{ln}^*(\psi_n^t),$$
$$\text{where } (\psi_s^t)_1^n = \Psi_t\},$$

$$(\Psi_t, (p_{\eta,l}^t)_1^n) \in T_{3t}^*(P_{t+1}, (p_{\eta,l}^{t+1})_1^n)$$
$$= \{(\Psi_t, (p_{\eta,l}^t)_1^n) | (\Psi_t, (p_{\eta,l}^t)_1^n) = ((\psi_l^t)_1^n, (p_{\eta,l}^t)_1^n)$$
$$\text{for some } (\psi_l^t, p_{\eta,l}^t) \in g_l^{*t}(p_l^{t+1}, p_{\eta,l}^{t+1})\}, \quad t = 0, 1, \ldots, N,$$

given

$$P_N = 0, \qquad p_{\eta,l}^N = 0, \quad l = 1, 2, \ldots, n,$$

where g_l^{*t}, $l = 1, 2, \ldots, n$, is defined by (11') and F_l^*, $l = 1, 2, \ldots, n$, and G_{ls}^*, $l, s = 1, 2, \ldots, n$, are defined, respectively, by (13) and (12).

For this pair of problems, II(a) and II(b), Theorem 4.1 (duality theorem) needs some slight modifications. First of all, condition (i) in Theorem 4.1 is rewritten using Theorem 2.14 as follows:

(i')
$$\mu = \sum_{\tau=1}^{N} \sum_{l=1}^{n} f_l^\tau(\hat{x}_l^\tau, \hat{c}_l^\tau) = \sum_{\tau=1}^{t} \sum_{l=1}^{n} f_l^\tau(\hat{x}_l^\tau, \hat{c}_l^\tau) + \hat{\Psi}_t \cdot \hat{Z}_t + \sum_{l=1}^{n} \hat{p}_{\eta,l}^t$$

$$= \sum_{\tau=1}^{t} \sum_{l=1}^{n} f_l^\tau(\hat{x}_l^\tau, \hat{c}_l^\tau) + \hat{\Phi}_t \cdot \hat{Y}_t + \sum_{l=1}^{n} \hat{p}_{\eta,l}^t$$

$$= \sum_{\tau=1}^{t} \sum_{l=1}^{n} f_l^\tau(\hat{x}_l^\tau, \hat{c}_l^\tau) + \hat{P}_t \cdot \hat{X}_t + \sum_{l=1}^{n} \hat{p}_{\eta,l}^t$$

$$= \hat{\Psi}_0 \cdot \hat{Z}_0 + \sum_{l=1}^{n} \hat{p}_{\eta,l}^0 = \hat{\Phi}_0 \cdot \hat{Y}_0 + \sum_{l=1}^{n} \hat{p}_{\eta,l}^0$$

$$= \hat{P}_0 \cdot \bar{X}_0 + \sum_{l=1}^{n} \hat{p}_{\eta,l}^0.$$

Condition (ii-2) is replaced by:

(ii-2′) $\max\{\hat{\Psi}_t \cdot Z_t | Z_t \in T_2(\hat{Y}_t)\} + \sum_{l=1}^{n} \hat{p}_{\eta,l}^t$

$$= \hat{\Psi}_t \cdot \hat{Z}_t + \sum_{l=1}^{n} \hat{p}_{\eta,l}^t$$

$$= \min\{\Psi_t \cdot \hat{Z}_t + \sum_{l=1}^{n} p_{\eta,l}^t | (\Psi_t, (p_{\eta,l}^t)_1^n) \in T_{3t}^*(\hat{P}_{t+1}, (\hat{p}_{\eta,l}^{t+1})_1^n)\}.$$

Finally, the equality between the total welfare in the final period and the value of the initial stock of commodities in the latter half of Theorem 4.1 is replaced by:

$$\sum_{\tau=1}^{N} \sum_{l=1}^{n} f_l^{\tau}(\hat{x}_l^{\tau}, \hat{c}_l^{\tau}) = \hat{P}_0 \cdot \bar{X}_0 + \sum_{l=1}^{n} \hat{p}_{\eta,l}^0.$$

According to the above modifications, condition (ii) in Theorem 4.3 (max-min principle) is changed as follows:

(ii′) $\sum_{l=1}^{n} \max\{\hat{\Psi}_t \cdot Z_l^t | Z_l^t \in G_l(\hat{y}_l^t)\} + \sum_{l=1}^{n} \hat{p}_{\eta,l}^t$

$$= \hat{\Psi}_t \cdot \hat{Z}_t + \sum_{l=1}^{n} \hat{p}_{\eta,l}^t$$

$$= \sum_{l=1}^{n} \min\{\psi_l^t \cdot \hat{z}_l^t + p_{\eta,l}^t | (\psi_l^t, p_{\eta,l}^t) \in g_l^{*t}(\hat{p}_l^{t+1}, \hat{p}_{\eta,l}^{t+1})\}.$$

It then follows from the definition of g_l^{*t} in (11′) that on the optimal path we must have:

$$\hat{p}_{\eta,l}^t \geqq \hat{p}_{\eta,l}^{t+1}, \quad l = 1, 2, \ldots, n, \ t = 0, 1, \ldots, N - 1.$$

Thus, from (i′), we get

$$\sum_{l=1}^{n} f_l^{t+1}(\hat{x}_l^{t+1}, \hat{c}_l^{t+1}) + \hat{P}_{t+1} \cdot \hat{X}_{t+1} \geqq \hat{P}_t \cdot \hat{X}_t.$$

That is, if we neglect the payment $\hat{p}_{\eta,l}^t$ to the fictitious variable η_l, $l = 1, 2, \ldots, n$, then activities on the optimal path may obtain positive profits.

Chapter 5

OPTIMUM GROWTH IN TWO-REGION, TWO-GOOD MONOTONE SPACE SYSTEMS: FINAL STATE PROBLEMS

5.1. Introduction

One of the distinguishing properties of a space system, compared to a point system, is the possible existence of physically identical commodities in different locations. Hence, a key theoretical topic in the study of space systems is the price relationship between identical commodities at different locations.

In particular, when we are programming the optimum growth of a spatial system, the main concern is how to allocate investment goods between locations at each time period. The optimal allocation of investment goods between locations is one which maximizes the "allocation revenue."

To illustrate, let us assume that there is only one kind of investment good, I, which can be moved without transport inputs between the two locations, region r and region s. And once a unit of investment good is invested in a region, it becomes one unit of (immobile) capital in that region. Then, on the optimal path, the total amount of the newly produced investment good, $I(t)$ in each period t, is allocated between the two regions according to the following rule (assuming there are no further institutional and/or political constraints):

$$\text{Maximize}_{0 \le \theta \le 1} [\theta p^r(t+1) + (1-\theta)p^s(t+1)]I(t),$$

where θ is the proportion of the investment good allocated to region r and p^l ($l = r, s$) is the price of the capital in region l.[1]

---

[1] For the derivation of this allocation rule, see Section 5.2.3. This allocation rule holds even when there are many kinds of investment goods. The essential character of the discussion in this section does not change even when transport inputs are required for the allocation of investment goods. See Section 5.3.3.

According to this rule, no matter how small the difference between the price of capital, all of the investment good is allocated to the region with the higher price in each period. The investment good can only be allocated to both regions when the price of capital is exactly the same in each region. Thus, one may conclude that the optimal spatial allocation of investment goods has a strong "all or none" character. That is, throughout most of the plan period, the investment good will be entirely allocated to a single region in each period (since it is unlikely that the price of capital will be exactly the same in both regions over a long period).

Almost all existing analytical results are consistent with the above conclusion. For example, see Rahman (1963), Sakashita (1967) and Ohtsuki (1971). Indeed, these authors reach the much stronger conclusion that all of the investment good is allocated to the same single region throughout most of the plan period when there are no further institutional and/or political constraints. But we must be careful to note that their conclusions are obtained only for the case of single-good systems.[2] If we consider a multicommodity system, the opposite result can be possible.

The purpose of this chapter is to show that in certain spatial systems the investment good should be allocated between regions in a "balanced manner" throughout most of the plan period. To establish this, we must show that the price of capital is exactly the same in both regions in each period for most of the plan period. However, this is impossible to attain in single-good systems (except for some uninteresting cases). Also it is difficult to attain capital price equalization in a short-run analysis since in this case the optimal path is crucially affected by the initial condition and the objective function. Hence, to establish the necessity of "balance" between regions for the efficient growth of the spatial system, the analysis should be conducted in a multicommodity system with a long-range planning horizon.

In this chapter an emphasis is put on graphical representation of the problem rather than mathematical strictness. For this purpose, one of the simplest multicommodity systems without transport inputs for commodity movement is analyzed in Section 5.2. This simplification

---

[2]The all or none character of regional investment allocation in these papers stems basically from two factors: (1) there is only one tradable commodity in their models, (2) they assume production functions with constant returns to scale. The latter point is discussed in Chapter 6.

enables us to obtain an intuitive understanding of the problem by geometrical means. The extension of the results to the space system with transport inputs for commodity movement is given in Section 5.3.

Before going to the actual analysis, it may be helpful to give a brief outline of our approach. At first, the primal problem is formulated in Section 5.2.1, and the dual problem is given in Section 5.2.2. Then, by applying the max-min principle from Chapter 2 to this pair of problems, the stepwise procedures for the optimal path are obtained in Section 5.2.3. As one might expect from the "turnpike analyses" of economics, the maximum balanced growth path shall play a very important role in our later discussion. Thus we examine in Section 5.2.4 how to obtain geometrically the maximum balanced growth path for the space system. The so-called "von Neumann facet" is derived in Section 5.2.5. Then it is shown in Section 5.2.6 that the optimum path must remain most of the time in a small neighborhood of this facet, and we obtain turnpike behavior of the optimal path by examining the movement of the optimal path on the von Neumann facet. Finally, in Section 5.2.7, the desired price equalization result is obtained by analyzing the properties of the dual optimal path which enables the commodity path to stay most of the time in a small neighborhood of the von Neumann facet and the maximum balanced growth ray. Throughout this section, we concentrate on the geometrical representation of the analysis. For this purpose, a fictitious third region, the "ideal region," is introduced. By observing the movement of the corresponding optimal path in the ideal region we can acquire a graphical understanding of the movement of the optimal path in the space system.

The analysis in Section 5.3 is parallel to that in Section 5.2, though the analysis and the graphical representation of the problem both become more difficult.

5.2. Optimum growth in a space system without transport inputs

5.2.1. Problem formulation

Suppose that the space system consists of two regions $l = r, s$, and that there are three kinds of commodities besides labor service and men. These goods are capital, an investment good and a material good. We assume that the investment good and the material good can be moved without any transport inputs between regions at the end of each period, and that capital is perfectly immobile. Once a unit of investment good is invested in a region, it becomes a unit of capital in that location. And, for simplicity, we neglect the depreciation of capital.

Each production activity in each location transforms in each period a set of capital, material good and labor service into a new set of investment good and material good, plus the original amount of capital in that location.

One unit of labor service is obtained by paying a certain fixed amount of material good to one unit of man, which may be determined subjectively by the planner. But it is assumed that men do not become a scarce resource in either region throughout the planning time. Thus, we can essentially drop men from consideration. Furthermore we assume that payment to labor service takes place simultaneously with the utilization of that service. Hence the amount of labor service necessary for a production activity is economically equivalent to the amount of material good paid to labor service. Therefore we do not need to consider labor service explicitly, and each production activity in each location can be represented as follows:

$$x_l = (x_1^l, x_2^l) \rightarrow y_l = (x_1^l, y_1^l, y_2^l), \quad l = r, s,$$

where

x_1^l: the amount of the capital used in region l,
x_2^l: the amount of the material good used in region l,
y_1^l: the amount of the investment good produced in region l,
y_2^l: the amount of the material good produced in region l.

In this representation of production activities, x_2^l includes also the payment to labor service. As an input commodity for production activities, it is not necessary to distinguish the newly produced investment good from capital. But as outputs of production activities, they should be distinguished from each other since only the newly produced investment good can be moved between regions. Each production activity represented above is restricted by the production technology in each location. That is,

$$(y_1^l, y_2^l) \in F_l(x_1^l, x_2^l), \quad l = r, s,$$

where F_l is a set-valued production function in location l. We now assume that:

A1. F_l is a non-singular m.t.cc. from $\mathbf{R}_+^2$ into $\mathbf{R}_+^2$, $l = r, s$.

The non-singularity of F_l means that positive amounts of both commodities can be produced if (x_1^l, x_2^l) is strictly positive. Production functions F_l, $l = r, s$, are assumed to be time invariant.

The following is an example of the production function F_l which satisfies A1:

$$F_l(x_1^l, x_2^l) = \{(y_1^l, y_2^l) | 0 \le y_1^l \le f_I^l(x_{1I}^l, x_{2I}^l),$$
$$0 \le y_2^l \le f_M^l(x_{1M}^l, x_{2M}^l) \quad \text{for some}$$
$$0 \le x_{iI}^l + x_{iM}^l \le x_i^l, x_{iI}^l, x_{iM}^l \ge 0, \quad i = 1, 2\}$$

where f_I^l and f_M^l are neoclassical production functions for the investment good sector and for the material good sector in location l, respectively. We can also obtain a production function F_l by using von Neumann models of production or Leontief input–output models (see Section 2.2 in Chapter 2).

The problem is to determine the production plan and the allocation plan of commodities in each period so as to maximize an objective function which is a linear function of the stock of commodities in the final period, N. That is:

Problem I(a)

Maximize $\displaystyle\sum_{l=r,s} \sum_{i=1,2} \bar{p}_i^l(N) x_i^l(N),$

$$\left.\begin{aligned}
&\text{subject to } 0 \le x_1^l(t+1) \le x_1^l(t) + \theta_1^l(t)(y_1^r + y_1^s)_t, \\
&0 \le x_2^l(t+1) \le \theta_2^l(t)(y_2^r + y_2^s)_t, \quad l = r, s, \\
&\text{for some } \theta_i^r(t) + \theta_i^s(t) = 1, \theta_i^r(t), \theta_i^s(t) \ge 0, \quad i = 1, 2, \\
&\text{and } (y_1^l, y_2^l)_t \in F_l(x_1^l, x_2^l)_t, \quad l = r, s,
\end{aligned}\right\} \quad (1)\dagger$$

for $t = 0, 1, \dots, N - 1$, given

$$(x_1^l, x_2^l)_0 = (\bar{x}_1^l(0), \bar{x}_2^l(0)), \quad l = r, s.$$

Or, in a compact form:

Maximize $\bar{P}_N \cdot X_N,$

subject to $X_{t+1} \in T(X_t), \quad t = 0, 1, \dots, N - 1, \quad \text{and } X_0 = \bar{X}_0,$ (2)

where

$$X_t = ((x_1^r, x_2^r), (x_1^s, x_2^s))_t,$$
$$\bar{P}_N = ((\bar{p}_1^r, \bar{p}_2^r), (\bar{p}_1^s, \bar{p}_2^s))_N \quad \text{and}$$
$$T(X_t) = \{X_{t+1} = ((x_1^r, x_2^r), (x_1^s, x_2^s))_{t+1} | \text{subject to (1)}\}. \quad (3)$$

†In this chapter, $(\cdot)_t$ means that all the variables in the parentheses are functions of time t. For example, $(y_1^r + y_1^s)_t$ means $(y_1^r(t) + y_1^s(t))_t$.

We next assume:

A2. $\bar{P}(N) \geq 0$; that is, $\bar{P}(N)$ is semipositive.

Because of A1, the transformation defined by (3) is a non-singular m.t.cc. from $\mathbf{R}_+^4$ into $\mathbf{R}_+^4$. For convenience in the analysis below, let us decompose the space transformation T into two transformations, T_1 and T_2. T_1 represents the production activities, and T_2 shows the transportation (i.e., allocation) activities. They are defined respectively as follows:

$$T_1(X_t) = T_1((x_1^r, x_2^r), (x_1^s, x_2^s))_t$$
$$= \{Y_t = ((x_1'^r, y_1^r, y_2^r), (x_1'^s, y_1^s, y_2^s))_t| \quad \text{for some}$$
$$(y_1^l, y_2^l)_t \in F_l(x_1^l, x_2^l)_t, 0 \leq x_1'^l(t) \leq x_1^l(t), \quad l = r, s\} \tag{4}$$

$$T_2(Y_t) = T_2((x_1'^r, y_1^r, y_2^r), (x_1'^s, y_1^s, y_2^s))_t$$
$$= \{X_{t+1} = ((x_1^r, x_2^r), (x_1^s, x_2^s))_{t+1}| \quad \text{for some}$$
$$0 \leq x_1^l(t+1) \leq x_1'^l(t) + \theta_1^l(t)(y_1^r + y_1^s)_t,$$
$$0 \leq x_2^l(t+1) \leq \theta_2^l(t)(y_2^r + y_2^s)_t, \quad l = r, s,$$
$$\theta_i^r + \theta_1^s = 1, \theta_i^r, \theta_i^s \geq 0, \quad i = 1, 2\}. \tag{5}$$

The new variable $x_1'^l$ is introduced to represent the free disposability of capital. $T_1(X_t)$ denotes the set of feasible production output vectors in the space system at the end of period t, given the stock X_t of commodities at the beginning of the period. Assumption A1 implies that T_1 is a non-singular m.t.cc. from $\mathbf{R}_+^4$ into $\mathbf{R}_+^6$. Given the production output vector Y_t, the possible allocation of commodities between regions is designated by the set $T_2(Y_t)$. Clearly, T_2 is a non-singular m.t.cc. from $\mathbf{R}_+^6$ into $\mathbf{R}_+^4$.

Using these functions, the space transformation defined by (3) can be represented as the following composite mapping:

$$T(X_t) = T_2 \cdot T_1(X_t)$$
$$= \{X_{t+1}|X_{t+1} \in T_2(Y_t) \quad \text{for some } Y_t \in T_1(X_t)\}. \tag{6}$$

The space transformation T may be represented diagrammatically as in Figure 5.1(a). In this figure, the direction of each arrow shows the direction of the functional relations between variables. For example, (y_1^r, y_2^r) is produced from (x_1^r, x_2^r) through F_r, and y_1^r can be invested in location s. The expression "(or)" in this figure means that one specific unit of the corresponding good can be used either way but not both.

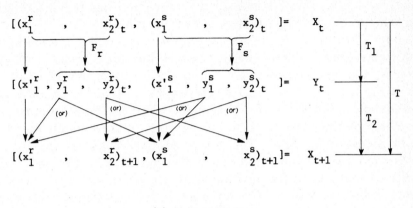

(a) Primal relation

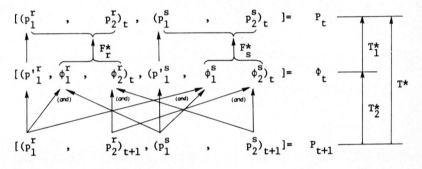

(b) Dual relation

Figure 5.1. Primal and dual relations.

The production possibilities in each location may be depicted by applying the diagrammatic techniques which are developed in Koopmans (1964).

First, the input space of each location is normalized so that

$$x_1^l + x_2^l = 1, \quad x_1^l, x_2^l \geqq 0, \quad l = 1, 2.$$

For this normalized input space, we get the (normalized) *technology set*,

$$S_l = \{z_l = ((x_1^l, x_2^l), (x_1^{l\prime} + y_1^l, y_2^l)) | x_1^l + x_2^l = 1,$$
$$x_1^l, x_2^l \geqq 0, (y_1^l, y_2^l) \in F_l(x_1^l, x_2^l), 0 \leqq x_1^{l\prime} \leqq x_1^l\}, \tag{7}$$

for each region $l = r, s$.

Since the production function F_l has the property of constant returns to scale, this technology set represents all the information about the production possibilities in each location.

Figure 5.2(a) shows the technology set in region r. Each point in this figure represents an input–output combination $[(x_1^r, x_2^r), (x_1'^r + y_1^r, y_2^r)]$; 0_1^r and 0_2^r represent, respectively, points $[(0, 1), (0, 0)]$ and $[(1, 0), (0, 0)]$. If a point is in the set S_r, it represents a feasible input–output combination in region r. For example, given an input combination designated by the point "a", the amount "ab" represents the output of the original capital good. The newly produced investment good is represented by the amount beyond the point "b". The set "bcd" represents that part of the outputs which can be moved freely between regions. The amount "bd" represents the maximum amount of material good which can be obtained from the inputs denoted by the point "a", and this is equal to the amount "ae". The set "$abde$" is a part of the technology set only because of the assumption of the free disposability of outputs. The convexity of the set S_r follows from assumption A1. Also from A1, the amounts "bc" and "bd" are nonzero for each input combination except possibly for 0_1^r and 0_2^r.

Likewise, Figure 5.2(b) shows the technology set for region s.

5.2.2. Dual problem

The importance of the duality between the commodity system and the price (rent) system in locational programming has been pointed out by many authors, especially by Stevens (1958, 1961, 1968). Moreover, in multiperiod programming, duality is important practically as well as theoretically. The whole program can be divided into a number of single step programs by introducing dual variables and dual trans-formations.

To obtain the dual system, each variable in Figure 5.1(a) is replaced by an arbitrary new variable. Figure 5.1(b) shows the result. The implication of each dual variable is clear, for example, $p_1^r(t)$ is the (efficiency) price of one unit of capital in region r at the beginning of period t. The value of $p_1^r(t)$ shows how much the value of the objective function can be increased when one unit of capital is increased in region r on the optimal path at the beginning of period t. As another example, $\phi_1^r(t)$ is the price of the newly produced investment good in region r at the end of the production activities in period t.

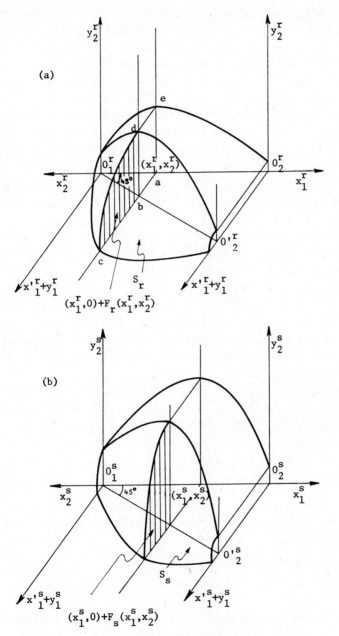

Figure 5.2. Normalized technology set in each region.

The appropriate functional relations between dual variables is obtained by reversing the direction of each arrow in Figure 5.1(a), and the result is shown in Figure 5.1(b). The functional relations between these dual variables are determined by the simple rule that no feasible activities[3] in the space system can obtain positive profits.

The direction of the arrow shows the direction of the appropriate inequality. For example, $\phi_1^r(t) \geqq p_1^r(t+1)$, that is, the price of the newly produced investment good in region r in period t should be no less than the price of the capital in region r in period $t+1$. Corresponding to each "(or)" in Figure 5.1(a), there is an "(and)" in Figure (b). This means that *both* functional relations must be satisfied. For example, $\phi_1^r(t) \geqq p_1^r(t+1)$ and $\phi_1^r(t) \geqq p_1^s(t+1)$. As another example, $\phi_1^s(t) \geqq p_1^s(t+1)$ and $\phi_1^s(t) \geqq p_1^r(t+1)$. Hence, we must have $\phi_1^r(t)$, $\phi_1^s(t) \geqq \max(p_1^r(t+1), p_1^s(t+1))$.

But double arrows do not always imply "and" or "or". For example, in Figure 5.1(a) the two arrows leading from $x_1^r(t)$ indicate that each unit of $x_1^r(t)$ contributes simultaneously to the production of the two sets of commodities, (y_1^r, y_2^r) and x_1^{r}. Hence since $x_1^r(t)$ contributes to the production of the two different sets of commodities, it should receive payment from each. Thus, each price vector (p_1^r, p_2^r) for inputs and each price vector $(p_1^{r}, \phi_1^r, \phi_2^r)$ for outputs should satisfy the next relation in each period so as not to allow positive profits for any production activities in region r.

$$(p_1^r, p_2^r)_t \in (p_1^{r}, 0)_t + F_r^*(\phi_1^r, \phi_2^r)_t,^4$$

where F_r^* is the adjoint transformation of F_r and is defined by

$$F_l^*(\phi_1^l, \phi_2^l) = \{(\Delta p_1^l, p_2^l) | \Delta p_1^l x_1^l + p_2^l x_2^l \geqq \phi_1^l y_1^l + \phi_2^l y_2^l$$
$$\text{for all } (y_1^l, y_2^l) \in F_l(x_1^l, x_2^l), (x_1^l, x_2^l) \geqq 0\}, \tag{8}$$

for each $(\phi_1^l, \phi_2^l) \in \mathbf{R}_+^2$, $l = r, s$. In (8), the new dual variable Δp_1^l represents the rent for each unit of capital in region l.

Using Figure 5.1(b), the adjoint transformations of T_1 and T_2 are respectively defined as follows:

$$T_2^*(P_{t+1}) = T_2^*((p_1^r, p_2^r), (p_1^s, p_2^s))_{t+1}$$
$$= \{\Phi_t = ((p_1^{r}, \phi_1^r, \phi_2^r), (p_1^{s}, \phi_1^s, \phi_2^s))_t | \quad \text{for some}$$

---

[3]Here, feasible activities mean those activities which are technologically feasible and are independent of the amount of the stocks of commodities in any time (refer to Section 2.2.1).

[4]Note that $F_r^*(\phi_1^r, \phi_2^r)_t$ is a subset of $\mathbf{R}_+^{*2}$. Hence, the sign, "+", in the right hand side of this equation denotes the vectorial sum of the point $(p_1^{r}, 0)_t$ and the set $F_r^*(\phi_1^r, \phi_2^r)_t$.

$$p_1^{l'}(t) \geqq p_1^l(t+1), \quad l = r, s,$$
$$\phi_1^r(t), \phi_1^s(t) \geqq \max (p_1^r, p_1^s)_{t+1},$$
$$\phi_2^r(t), \phi_2^s(t) \geqq \max (p_2^r, p_2^s)_{t+1}\}, \tag{9}$$

$$T_1^*(\Phi_t) = T_1^*(\phi_r, \phi_s)_t = T_1^*((p_1^{r'}, \phi_1^r, \phi_2^r), (p_1^{s'}, \phi_1^s, \phi_2^s))_t$$
$$= \{P_t = ((p_1^r, p_2^r), (p_1^s, p_2^s))_t | \quad \text{for some}$$
$$(p_1^l, p_2^l)_t \in (p_1^{l'}, 0)_t + F^*(\phi_1^l, \phi_2^l)_t, \quad l = r, s\}. \tag{10}$$

Given the prices P_{t+1} of goods at the beginning of period $t + 1$, $T_2^*(P_{t+1})$ gives the set of prices for the "allocation input goods" (i.e., the production output goods) at the end of period t which do not allow positive profits for any feasible allocation activities. And $T_1^*(\Phi_t)$ is the set of prices for production input goods at the beginning of period t which do not allow positive profits for any feasible production activities under given production output prices Φ_t.

From assumption A1, T_2^* and T_1^* are, respectively, non-singular m.t.cv. from $\mathbf{R}_+^{*4}$ into $\mathbf{R}_+^{*6}$ and from $\mathbf{R}_+^{*6}$ into $\mathbf{R}_+^{*4}$. Hence the adjoint transformation T^* of the transformation T is defined by

$$T^*(P_{t+1}) = \{P_t | P_t \in T_1^*(\Phi_t) \quad \text{for some } \Phi_t \in T_2^*(P_{t+1})\}, \tag{11}$$

and this is a non-singular m.t.cv. from $\mathbf{R}_+^{*4}$ into $\mathbf{R}_+^{*4}$. Taking into account the fact that the adjoint transformation F_1^* has the property of "free additivity" (i.e., property (vi*) in the definition of adjoint transformation in Chapter 2), T^* can also be rewritten as follows by using (9) and (10).

$$T^*(P_{t+1}) = \{P_t = ((p_1^r, p_2^r), (p_1^s, p_2^s))_t | \quad \text{for some}$$
$$(p_1^l, p_2^l)_t \in (p_1^l, 0)_{t+1} + F_1^*(\max (p_1^r, p_1^s), \max (p_2^r, p_2^s))_{t+1},$$
$$l = r, s\}. \tag{12}$$

Hence, the dual problem for Problem I(a) can be formulated as follows:

Problem I(b)

Minimize $\sum\limits_{l=r,s} \sum\limits_{i=1,2} p_i^l(0)\bar{x}_i^l(0)$

subject to $(p_1^l, p_2^l)_t \in (p_1^l, 0)_{t+1} + F_1^*(\max (p_1^r, p_1^s), \max (p_2^r, p_2^s))_{t+1}$,
$t = 0, 1, \ldots, N - 1$,

given

$(p_1^l, p_2^l)_N = (\bar{p}_1^l(N), \bar{p}_2^l(N)), \quad l = r, s.$

Or, in a compact form:

$$
\left.
\begin{aligned}
&\text{Minimize } P_0 \cdot \bar{X}_0 \\
&\text{subject to } P_t \in T^*(P_{t+1}), \quad t = 0, 1, \ldots, N-1, \\
&\text{and } P_N = \bar{P}_N.
\end{aligned}
\right\}
\tag{13}
$$

Hence the problem here is to minimize the value of the initial inputs subject to the condition that the sequence of prices $\{P_t\}_0^N$ should not allow positive profits for any feasible activities in the space system throughout the plan period.

5.2.3. Stepwise procedures for the optimal path

In addition to A1 and A2, we now assume that:

A3. At least one of the next two conditions is satisfied in Problem I(a):

(a) $\bar{X}(0) > 0$,
(b) Both F_r and F_s satisfy the Lipschitz condition.[5]

Given A1 to A3, it follows from Theorem 4.1 and Theorem 4.2 (equivalently, from Theorem 2.7) that a feasible path $[\{\hat{X}_t\}_0^N, \{\hat{Y}_t\}_0^{N-1}] = [\{(\hat{x}_1^r, \hat{x}_2^r)_t, (\hat{x}_1^s, \hat{x}_2^s)_t\}_0^N, \{(\hat{x}_1^{\prime r}, \hat{y}_1^r, \hat{y}_2^r)_t, (\hat{x}_1^{\prime s}, \hat{y}_1^s, \hat{y}_2^s)_t\}_0^{N-1}]$ for Problem I(a) is an optimal path for that problem if and only if there exists a feasible path $[\{\hat{P}_t\}_0^N, \{\hat{\Phi}_t\}_0^{N-1}] = [\{(\hat{p}_1^r, \hat{p}_2^r)_t, (\hat{p}_1^s, \hat{p}_2^s)_t\}_0^N, \{(p_1^{\prime r}, \hat{\phi}_1^r, \hat{\phi}_2^r)_t, (\hat{p}_1^{\prime s}, \hat{\phi}_1^s, \hat{\phi}_2^s)_t\}_0^{N-1}]$ for Problem I(b) such that they together satisfy the following conditions (and in this case, the price path is an optimal path for Problem I(b)):

$$
\max\{\Phi_t \cdot Y_t \mid Y_t \in T_1(\hat{X}_t)\} = \hat{\Phi}_t \cdot \hat{Y}_t = \min\{\Phi_t \cdot \hat{Y}_t \mid \Phi_t \in T_2^*(\hat{P}_{t+1})\},
$$
$$
t = 0, 1, \ldots, N-1. \tag{14}
$$

$$
\max\{\hat{P}_{t+1} \cdot X_{t+1} \mid X_{t+1} \in T_2(\hat{Y}_t)\}
$$
$$
= \hat{P}_{t+1} \cdot \hat{X}_{t+1} = \min\{P_{t+1} \cdot \hat{X}_{t+1} \mid P_{t+1} \in T_1^*(\hat{\Phi}_{t+1})\},
$$
$$
t = 0, 1, \ldots, N-1 \text{ for the first equality, and}
$$
$$
t = -1, 0, \ldots, N-1 \text{ for the second equality.} \tag{15}
$$

where T_1, T_2, T_1^* and T_2^* are defined, respectively, by (4), (5), (10) and (9).

[5]For a description of the Lipschitz condition, see Section 2.3.3.

From (9) and the second equality in (14), it can be seen that

$$\left.\begin{array}{l} \hat{p}_1^{\,l}(t) = \hat{p}_1^{\,l}(t+1), \quad l = r, s \\[4pt] \hat{\phi}_1^{\,r}(t) = \hat{\phi}_1^{\,s}(t) = \max{(\hat{p}_1^{\,r}, \hat{p}_1^{\,s})}_{t+1} \\[4pt] \hat{\phi}_2^{\,r}(t) = \hat{\phi}_2^{\,s}(t) = \max{(\hat{p}_2^{\,r}, \hat{p}_2^{\,s})}_{t+1} \end{array}\right\}, \quad t = 0, 1, \ldots, N-1. \tag{16}$$

This is an expected result since the investment good and the material good can be moved freely. Also, from (16) and the second equality in (15), the optimal values of $(p_1^{\,l}, p_2^{\,l})_t$, $l = r, s$, are determined for a given $(\hat{p}_1^{\,l}, \hat{p}_2^{\,l})_{t+1}$ and $(\hat{x}_1^{\,l}, \hat{x}_2^{\,l})_t$, $l = r, s$, so as to:

Minimize $p_1^{\,l}(t)\hat{x}_1^{\,l}(t) + p_2^{\,l}(t)\hat{x}_2^{\,l}(t)$,

subject to $(p_1^{\,l}, p_2^{\,l})_t \in (\hat{p}_1^{\,l}, 0)_{t+1} + F_l^*(\max{(p_1^{\,r}, p_1^{\,s})}, \max{(p_2^{\,r}, p_2^{\,s})})_{t+1}.$
$$\tag{17}$$

Figure 5.3 depicts this operation in period $t + 1$.

Dually, from (4), (16) and the first equality in (14), the optimal production plan in period t is determined in each location as follows for a given $(\hat{x}_1^{\,l}, \hat{x}_2^{\,l})_t$ and $(\hat{p}_1^{\,l}, \hat{p}_2^{\,l})_{t+1}$.

Maximize $[\max{(\hat{p}_1^{\,r}, \hat{p}_1^{\,s})}_{t+1}]y_1^{\,l}(t) + [\max{(\hat{p}_2^{\,r}, \hat{p}_2^{\,s})}_{t+1}]y_2^{\,l}(t)$,

subject to $(y_1^{\,l}, y_2^{\,l})_t \in F_l(\hat{x}_1^{\,l}, \hat{x}_2^{\,l})_t.$
$$\tag{18}$$

And, since the transformation of capital stock in each step requires no extra inputs, it follows that without loss of generality we can specify that on the optimal path the following condition holds in each location and in each period:

$$\hat{x}_1^{\,l}(t) = \hat{x}_1^{\,l}(t). \tag{19}$$

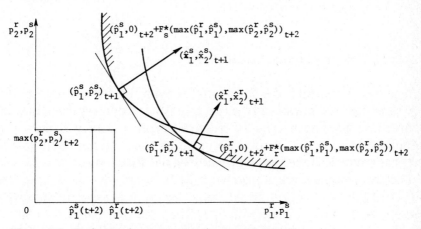

Figure 5.3. Optimal price determination in period $t + 1$.

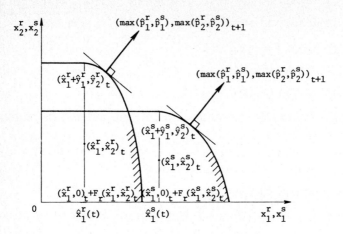

Figure 5.4. Optimal production plan in period t.

Figure 5.4 depicts this production decision in period t.

Finally, from (5), (16) and the first equality of (15), the allocation plan in each period t is determined so as to

Maximize $(\hat{p}_i^s + \theta_i^t(\hat{p}_i^r - \hat{p}_i^s))_{t+1}(\hat{y}_i^r + \hat{y}_i^s)_t$

subject to $1 \geqq \theta_i^t(\equiv \theta_i^r(t)) \geqq 0,$ (20)

for $i = 1, 2$.

Note that according to this allocation rule, all of the newly produced goods i $(i = 1, 2)$ are distributed to the region with the higher price. Thus, the all-or-none character of the optimal allocation is avoided only when $\hat{p}_i^r(t + 1) = \hat{p}_i^s(t + 1)$, $i = 1, 2$.

5.2.4. Maximum balanced growth path

When we investigate the long-run behavior of the optimal path for Problem I(a), it is naturally expected from the turnpike analyses in economics that the maximum balanced growth path will play a very important role. In this section we examine geometrically how to obtain the maximum balanced growth path for the space system.

For this purpose, we introduce a fictitious third region, which may be called the *ideal region*. In this ideal region, it is assumed that we can use any technology of region r or s. Then the (normalized) technology set S of the ideal region, which we designate as the *ideal technology set*, is obtained as the *convex hull* of S_r and S_s. That is,

$$S = \{z|z = \alpha z_r + (1-\alpha)z_s \quad \text{for some}$$
$$0 \leq \alpha \leq 1, z_r \in S_r \text{ and } z_s \in S_s\}, \tag{21}$$

where $z = ((x_1, x_2), (x'_1 + y_1, y_2))$. For definitions of z_r and z_s see (7).

From (7), it follows easily that set S can be equivalently defined as:

$$S = \{((x_1, x_2), (x'_1 + y_1, y_2))|(y_1, y_2) \in F_r(x'_1, x'_2)$$
$$+ F_s(x^s_1, x^s_2),^6 \, 0 \leq x'_1 \leq x'_1 + x^s_1, x_i = x'_i + x^s_i \quad \text{for some}$$
$$x'_1 + x'_2 + x^s_1 + x^s_2 = 1, x^l_i \geq 0, i = 1, 2, l = r, s\}. \tag{22}$$

When we add the restrictions $x'_1 + x'_2 = 1$ and $x^s_1 + x^s_2 = 0$ in (22), S equals S_r. Similarly, when $x'_1 + x'_2 = 0$ and $x^s_1 + x^s_2 = 1$ in (22), S equals S_s. Thus the ideal technology set S includes not only activities in region r and region s, but also the convex combination of any activity in r and any activity in s.

When the technology set of one region is completely included in the technology set of the other region, the ideal technology set is equal to that of the latter region. But this is not an interesting case for the analysis in space systems. Hence, we employ the next assumption:

A4. $\Delta S \equiv \{z|z \in S, z \notin S_r, z \notin S_s\} \neq \emptyset$

Under this assumption, the ideal technology set S has points which are not contained in either S_r or S_s. And this assumption implies, of course, that

$$S_r \not\supset S_s \quad \text{and} \quad S_s \not\supset S_r. \tag{23}^7$$

Figure 5.5 shows how to obtain S. First, the technology sets S_r in Figure 5.2(a) and S_s in Figure 5.2(b) are put together in Figure 5.5. Then S is obtained as the convex hull of these two sets. The shaded portion of the set S corresponds to ΔS. Note that each point in the newly produced set ΔS is a convex combination of two independent activities. Thus the surface of ΔS is not strictly convex.

It is important to know precisely what the ideal region represents.

---

[6]Note that $F_r(x'_1, x'_2) + F_s(x^s_1, x^s_2)$ represents the vectorial sum of two sets $F_r(x'_1, x'_2)$ and $F_s(x^s_1, x^s_2)$.

[7]Note that A4 implies (23), but (23) does not always imply A4.

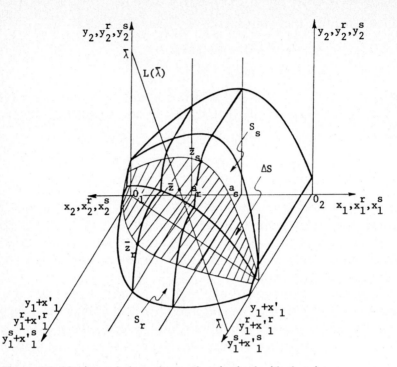

Figure 5.5. Maximum balanced growth point in the ideal region.

The technology in this region shows what can be done when the capital can be moved freely between regions. But, under the assumption of immobility of capital, this is a fictitious region.[8]

The *maximum growth factor* $\bar{\lambda}$ and the *maximum balanced growth vector* $\bar{X}$ (which is a four dimensional vector) in the space system are

[8]When the stocks of goods in each region at the beginning of a certain period are given by (x_1^r, x_2^r) and (x_1^s, x_2^s), respectively, the sets of newly produced goods in each region are given by $F_r(x_1^r, x_2^r)$ and $F_s(x_1^s, x_2^s)$, respectively. Then any point

$$(y_1, y_2) \in F_r(x_1^r, x_2^r) + F_s(x_1^s, x_2^s)$$

can be obtained in the ideal region when the stocks of goods are given by $(x_1^r + x_1^s, x_2^r + x_2^s)$ in this region. But the converse is not true. That is, when (y_1, y_2) is produced from (x_1, x_2) in the ideal region, the set $F_r(x_1^r, x_2^r) + F_s(x_1^s, x_2^s)$ does not necessarily include (y_1, y_2) for an arbitrary (x_1^r, x_2^r) and (x_1^s, x_2^s) such that $x_1^r + x_1^s = x_1$, $x_2^r + x_2^s = x_2$, $x_i^l \geqq 0$, $i = 1, 2$, and $l = r, s$. For example, when $F_r(1, 0) = 0$ and $F_s(0, 1) = 0$, it follows that $F_r(1, 0) + F_s(0, 1) = 0$. But, from (22), if we are given $(x_1, x_2) = (1, 1)$ in the ideal region, we can obtain any output combination in the set $F_r(\alpha, \beta) + F_s(1 - \alpha, 1 - \beta)$ for any $0 \leqq \alpha \leqq 1$, $0 \leqq \beta \leqq 1$ in this region. In particular, when $((x_1, x_2), (x_1 + y_1, y_2)) \in \Delta S$, the total new production (y_1, y_2) can be obtained in the space system only when x_1 and x_2 are distributed between the two regions in a very special way.

defined by

$$\bar{\lambda} = \max \{\lambda | \lambda X \in T(X), X \geq 0\},$$
$$\bar{X} \in \{X \geq 0 | \bar{\lambda} X \in T(X)\}. \tag{24}$$

The existence of $\bar{X}$ and $\bar{\lambda} > 0$ is guaranteed by Assumption A1.[9] To obtain them by geometrical means, let us first find the *maximum balanced growth point* $\bar{z}$ on the ideal technology set which is defined as follows:

$$\bar{\lambda} = \max \{\lambda | (x, \lambda x) \in S, x \geq 0\},$$
$$\bar{z} = (\bar{x}, \bar{\lambda}\bar{x}) \quad \text{for } \bar{x} \in \{x \geq 0 | (x, \bar{\lambda} x) \in S\}, \tag{25}$$

where we may call $\bar{x}$ and $\bar{\lambda}$, respectively, the *maximum balanced growth vector* and the *maximum growth factor in the ideal region*.

Given a point $z^1 = [(1, 0), (\lambda, 0)]$, which is on the right $y_1 + x_1'$ axis in Figure 5.5, and a point $z^2 = [(0, 1), (0, \lambda)]$, which is on the left y_2 axis, let us denote the line segment $\overline{z^1 z^2}$ by $L(\lambda)$, which is clearly a function of λ. Then, in Figure 5.5, the maximum balanced growth point for the ideal region is obtained as the common point of S and $L(\bar{\lambda})$, where $\bar{\lambda}$ is the maximum value of λ for which $S \cap L(\lambda) \neq \emptyset$.[10] In Figure 5.5, $\bar{z}$ is the maximum balanced growth point. By definition (21) this point can be represented as

$$\bar{z} = \bar{\alpha}\bar{z}_r + (1 - \bar{\alpha})\bar{z}_s \quad \text{for some } 0 \leq \bar{\alpha} \leq 1, \bar{z}_r \in S_r, \text{ and } \bar{z}_s \in S_s. \tag{26}$$

Let us denote $\bar{z}_r$ and $\bar{z}_s$ by

$$\bar{z}_l = [(a_1^l, a_2^l), (a_1^l + b_1^l, b_2^l)] \equiv [a_l, b_l], \quad l = r, s, \tag{27}$$

where

$$a_l = (a_1^l, a_2^l), \quad b_l = (a_1^l + b_1^l, b_2^l), \quad l = r, s.$$

Then, from (26) and since $\bar{z}$ is the maximum balanced growth point on S,

$$\bar{z} = [(\bar{\alpha}a_1^r + (1 - \bar{\alpha})a_1^s, \bar{\alpha}a_2^r + (1 - \bar{\alpha})a_2^s), (\bar{\alpha}(a_1^r + b_1^r)$$
$$+ (1 - \bar{\alpha})(a_1^s + b_1^s), \bar{\alpha}b_2^r + (1 - \bar{\alpha})b_2^s)]$$
$$= [(\bar{\alpha}a_1^r + (1 - \bar{\alpha})a_1^s, \bar{\alpha}a_2^r + (1 - \bar{\alpha})a_2^s), \bar{\lambda}(\bar{\alpha}a_1^r + (1 - \bar{\alpha})a_1^s,$$
$$\bar{\alpha}a_2^r + (1 - \bar{\alpha})a_2^s)].$$

[9]See, Theorem 1 in Gale (1956, p. 290). Also see Theorems 13.2 and 13.3 in Nikaido (1968).

[10]For the detailed procedure for this, see Koopmans (1964, pp. 366–70).

Next we define

$$a_i = \bar{\alpha} a'_i + (1 - \bar{\alpha}) a^s_i, \qquad b_i = \bar{\alpha} b'_i + (1 - \bar{\alpha}) b^s_i, \quad i = 1, 2. \tag{28}$$

Then, the above relation can be rewritten as follows:

$$\begin{aligned}
\bar{z} &= [(a_1, a_2), (a_1 + b_1, b_2)] \\
&= [(a_1, a_2), \bar{\lambda}(a_1, a_2)] (\equiv [a, \bar{\lambda} a]).
\end{aligned} \tag{29}$$

Hence, we have

$$\bar{\lambda} = 1 + \frac{b_1}{a_1} = \frac{b_2}{a_2}. \tag{30}$$

Therefore the vector $a = (a_1, a_2)$ represents the maximum balanced growth vector and $\bar{\lambda}$ gives the corresponding maximum balanced growth factor in the ideal region.

In Figure 5.5, the points $\bar{z}$, $\bar{z}_r$ and $\bar{z}_s$ are each unique. While in general this may not be so, let us assume for simplicity that:

A5. $\bar{z}$, $\bar{z}_r$ and $\bar{z}_s$ are unique.

According to the figure, we must have $0 < \bar{\alpha} < 1$. When $\bar{\alpha} = 0$ or 1, an activity of only one region is used in Definition (26). That is, the maximum balanced growth vector for the ideal region consists of an activity in only one region. Intuitively, this is not an interesting case for the space system analysis. Thus, hereafter, let us assume

A6. $0 < \bar{\alpha} < 1$.

This assumption means that $\bar{z} \in \Delta S$.

The remaining work is to show that the vector

$$\bar{X} = [\bar{\alpha}(a'_1, a'_2), (1 - \bar{\alpha})(a^s_1, a^s_2)] \tag{31}$$

is the unique (up to scalar multiplication) maximum balanced growth vector for the space system with growth factor $\bar{\lambda}$. This is not obvious since the above analysis only says that vector $a = (a_1, a_2)$ can grow in the ideal region with growth rate $\bar{\lambda}$ by using activity (a_r, b_r) in region r and activity (a_s, b_s) in region s with relative weights $\bar{\alpha}$ and $(1 - \bar{\alpha})$, respectively. But growth in the ideal region depends on the assumption that the technologies of both regions can be used in a single region, that is, that capital can be moved freely between the two regions. To show that vector $\bar{X}$ is the maximum balanced growth vector in the space system, we must first show that vector $\bar{X}$ can grow in the space system at growth rate $\bar{\lambda}$ only by moving the newly produced outputs between locations in each period. That is, we must first show that $(\bar{X}, \bar{\lambda}\bar{X})$ is a feasible activity under T (i.e., $\bar{\lambda}\bar{X} \in T(\bar{X})$).

From (26), (27) and Definition (7), $\bar{\alpha}(b_1^r, b_2^r) \in F_r[\bar{\alpha}(a_1^r, a_2^r)]$ and $(1 - \bar{\alpha})(b_1^s, b_2^s) \in F_s[(1 - \bar{\alpha})(a_1^s, a_2^s)]$. And by assumption, newly produced products $b_1 = \bar{\alpha}b_1^r + (1 - \bar{\alpha})b_1^s$ and $b_2 = \bar{\alpha}b_2^r + (1 - \bar{\alpha})b_2^s$ can be distributed freely between regions. Let us now set

$$(\bar{\theta}_1 \equiv)\bar{\theta}_1^r = \frac{\bar{\alpha}a_1^r}{a_1}, \qquad \bar{\theta}_1^s = \frac{(1 - \bar{\alpha})a_1^s}{a_1},$$

$$(\bar{\theta}_2 \equiv)\bar{\theta}_2^r = \frac{\bar{\alpha}a_2^r}{a_2}, \qquad \bar{\theta}_2^s = \frac{(1 - \bar{\alpha})a_2^s}{a_2}. \tag{32}$$

Obviously, $\bar{\theta}_i^r + \bar{\theta}_i^s = 1$ and $\bar{\theta}_i^r, \bar{\theta}_i^s \geq 0$, $i = 1, 2$. After distributing new products b_1 and b_2 according to (32), we see from (30) that the stock of each commodity in each region is, respectively, given by:

$$\bar{\alpha}a_1^r + \bar{\theta}_1^r b_1 = \bar{\lambda}(\bar{\alpha}a_1^r), \qquad \bar{\theta}_2^r b_2 = \bar{\lambda}(\bar{\alpha}a_2^r),$$

$$(1 - \bar{\alpha})a_1^s + \bar{\theta}_1^s b_1 = \bar{\lambda}[(1 - \bar{\alpha})a_1^s], \qquad \bar{\theta}_2^s b_2 = \bar{\lambda}[(1 - \bar{\alpha})a_2^s]. \tag{33}$$

Hence, $\bar{\lambda}[\bar{\alpha}(a_1^r, a_2^r), (1 - \bar{\alpha})(a_1^s, a_2^s)] \in T[\bar{\alpha}(a_1^r, a_2^r), (1 - \bar{\alpha})(a_1^s, a_2^s)]$, that is

$$\bar{\lambda}\bar{X} \in T(\bar{X}), \tag{34}$$

as was to be shown.

The fact that $\bar{\lambda}$ is the maximum growth factor for the space system, that is, $\bar{\lambda} = \max \{\lambda | \lambda X \in T(X) \; X \geq 0\}$, is easily seen from definitions (21) and (25). And, under A5 and A6, it follows that the vector $\bar{X}$ defined by (31) is the unique maximum balanced growth vector. (These proofs are omitted.)

Figure 5.6 (p. 179) shows the relation between a_r, a_s and a in a two-dimensional figure. When the space system moves on the *maximum balanced growth path* $\eta\bar{X}$, the paths for region r, region s and the ideal region move, respectively, along the rays $\eta(\bar{\alpha}a_r)$, $\eta((1 - \bar{\alpha})a_s)$ and ηa with common multiplier η.

Since in our model there are only two basic goods, let us assume

A7. $a_r \equiv (a_1^r, a_2^r) > 0$ and $a_s \equiv (a_1^s, a_2^s) > 0$.

And we assume

A8. $\bar{\lambda} > 1$.

Assumption A7 means that, on the maximum balanced growth path, each region uses both goods as inputs. From (30), A8 implies

$$b_1 > 0 \quad \text{and} \quad b_2 > a_2. \tag{35}$$

That is, on the maximum balanced growth path, net outputs of both goods are positive.

5.2.5. *von Neumann price vector and von Neumann facet*

We now turn to the investigation of the maximum balanced growth (or, shrinking) path for the price system.

Under Assumption A1, T is a non-singular m.t.cc. from $\mathbf{R}_+^4$ into $\mathbf{R}_+^4$. Thus, from Theorem 1 in Gale (1956), there exists a vector $\bar{P} = (\bar{p}_r, \bar{p}_s) = [(\bar{p}_1', \bar{p}_2'), (\bar{p}_1^s, \bar{p}_2^s)] \geq 0$ such that

$$\bar{P} \cdot X_{t+1} - \bar{\lambda}\bar{P} \cdot X_t \leq 0 \quad \text{for all } (X_t, X_{t+1}) \in \text{graph } T, \tag{36}$$

where $\bar{\lambda}$ is the maximum growth factor obtained in the previous section. Such a price vector is called a *von Neumann price vector*.

We now show that

$$\bar{p}_1' = \bar{p}_1^s (\equiv \bar{p}_1) \quad \text{and} \quad \bar{p}_2' = \bar{p}_2^s (\equiv \bar{p}_2). \tag{37}$$

First, from the reasoning described just above Equation (32), the following activity

$$[\bar{\alpha}(a_1', a_2'), (1 - \bar{\alpha})(a_1^s, a_2^s)] \rightarrow [(\bar{\alpha}a_1' + \theta_1 b_1, \theta_2 b_2),$$
$$((1 - \bar{\alpha})a_1^s + (1 - \theta_1)b_1, (1 - \theta_2)b_2)], \tag{38}$$

is a feasible activity in the space system for any $0 \leq \theta_i \leq 1$, $i = 1, 2$. And when $\theta_i = \bar{\theta}_i$ for $i = 1, 2$ in (38), this activity satisfies (36) with an equality because of (33). But, from assumptions A6, A7 and (32), it follows that $0 < \bar{\theta}_i < 1$, $i = 1, 2$. Hence, if $\bar{p}_i' \neq \bar{p}_i^s$ for $i = 1$ or 2, then activity (38) can obtain positive profits by appropriately choosing $\theta_i = 1$ or 0. This contradicts the assumption that $\bar{P}$ is a von Neumann price vector. Thus (37) is true.

Then, it follows from the property of constant returns to scale in transformation T and from the definition of the set S by (22) that $\bar{P} = [(\bar{p}_1, \bar{p}_2), (\bar{p}_1, \bar{p}_2)]$ is a von Neumann price vector if and only if

$(\bar{p}_1, \bar{p}_2)$ satisfies the following condition.

$$\bar{p}_1(x_1' + y_1) + \bar{p}_2 y_2 - \bar{\lambda}(\bar{p}_1 x_1 + \bar{p}_2 x_2) \leqq 0,$$
for all $[(x_1, x_2), (x_1' + y_1, y_2)] \in S.$ (39)

That is, $\bar{P} = [(\bar{p}_1, \bar{p}_2), (\bar{p}_1, \bar{p}_2)]$ is a von Neumann price vector for the space system if and only if $(\bar{p}_1, \bar{p}_2)$ is a von Neumann price vector for the technology set S.

One of the von Neumann price vectors for S is obtained as follows.[11] In Figure 5.5, we represent the plane which is spanned by line segments $L(\bar{\lambda})$ and $\bar{z}_r \bar{z}_s$ as follows:

$$c_1 x_1 + p_1^*(x_1' + y_1) + p_2^* y_2 = c_2.$$

Using the condition that this plane passes through points $[(x_1, x_2), (x_1' + y_1, y_2)] = [(1, 0), (\bar{\lambda}, 0)]$ and $[(0, 1), (0, \bar{\lambda})]$, and substituting $x_2 = 1 - x_1$, the equation of this plane can be rewritten as follows:

$$\pi(z) \equiv p_1^*(x_1' + y_1) + p_2^* y_2 - \bar{\lambda}(p_1^* x_1 + p_2^* y_2) = 0.$$ (40)

We choose the signs of p_1^* and p_2^* so that $\pi(z) \leqq 0$ for all $z \in S$. Then, since $0_1, 0_2 \in S$ and $\bar{\lambda} > 1$, we must have $p_1^*, p_2^* \geqq 0$, but not $p_1^* = p_2^* = 0$, for in that case (40) is not the equation of a plane.

From assumptions A5 and A6, and by using the condition that the plane (40) passes through $\bar{z}_r$ and $\bar{z}_s$ we can show that

$$p_1^*, p_2^* > 0.$$ (41)

For the proof, see Appendix D.3. It can also be shown that under the von Neumann price vector (p_1^*, p_2^*) only the set of activities

$$k^* = \{(x_t, x_{t+1}) | x_t = \beta[\alpha a_1^r + (1 - \alpha)a_1^s, \ \alpha a_2^r + (1 - \alpha)a_2^s],$$
$$x_{t+1} = \beta[\alpha(a_1^r + b_1^r) + (1 - \alpha)(a_1^s + b_1^s),$$
$$\alpha b_2^r + (1 - \alpha)b_2^s] \quad \text{for some } \beta \geqq 0 \text{ and } 1 \geqq \alpha \geqq 0\},$$ (42)

breaks even in the ideal region (for the proof, see Appendix D.4). Then, because of the equivalence of (36) and (39), one easily sees that under von Neumann price vector $[(p_1^*, p_2^*), (p_1^*, p_2^*)]$ only the set of

---

[11]For the details of this procedure, see Koopmans (1964, pp. 370–74).

activities

$$K^* = \{(X_t, X_{t+1}) | X_t = \beta[\alpha(a_1^r, a_2^r), (1-\alpha)(a_1^s, a_2^s)],$$
$$X_{t+1} = \beta[(\alpha a_1^r + \theta_1(\alpha b_1^r + (1-\alpha)b_1^s),$$
$$\theta_2(\alpha b_2^r + (1-\alpha)b_2^s)), ((1-\alpha)a_1^s + (1-\theta_1)(\alpha b_1^r + (1-\alpha)b_1^s),$$
$$(1-\theta_2)(b_2^r + (1-\alpha)b_2^s))] \quad \text{for some } \beta \geq 0,$$
$$1 \geq \alpha \geq 0 \text{ and } 1 \geq \theta_i \geq 0, i = 1, 2\}, \tag{43}$$

breaks even in the space system. Observe that K^* is a closed convex cone in $\mathbf{R}_+^{4 \times 2}$.

Hereafter we use $P^* = [(p_1^*, p_2^*), (p_1^*, p_2^*)]$ as the von Neumann price vector in the space system. We designate K^* as the *von Neumann facet in the space system*, and k^* as the *von Neumann facet in the ideal region*.

Notice from the definition of the adjoint transformation T^* that definition (36) simply says that $\bar{\lambda}\bar{P} \in T^*(\bar{P})$. Moreover, given (41) and assumption A7, we can prove that $\bar{\lambda} = \min\{\lambda | \lambda P \in T^*(P) \text{ for some } P \geq 0\}$.[12] Hence, considering (24), the maximum balanced growth vector $\bar{X}$ given by (31), the von Neumann price vector P^* and the corresponding growth factor $\bar{\lambda}$ satisfy the next relations:

$$\bar{\lambda}\bar{X} \in T(\bar{X}),$$
$$\bar{\lambda}P^* \in T^*(P^*),$$
$$\bar{\lambda} = \max\{\lambda | \lambda X \in T(X), X \geq 0\} = \min\{\lambda | \lambda P \in T^*(P), P \geq 0\}. \tag{44}$$

5.2.6. Long-run behavior of the optimal path

In this section, we study the long-run behavior of the optimal paths for Problem I(a). For this purpose, we employ the following assumption on the initial conditions for Problem I(a):

A9. The initial vector $\bar{X}(0)$ satisfies the condition that there exists a feasible path starting from $\bar{X}(0)$ which can achieve a point on the maximum balanced growth path $\eta\bar{X}(\eta > 0)$ in a finite number of periods.

For example, this condition is satisfied if $\bar{X}(0) > 0$.

If the von Neumann facet K^* were a point, we could directly apply

---

[12]See Theorem 2 in Part 6 and Theorem 2 in Part 7 in Rockafellar (1967).

the proof in Radner (1961) to obtain the turnpike behavior of our space system. But from (43) it follows that K^* is not a point. Nevertheless, we can apply the method of the proof of Radner's turnpike theorem to prove convergence of the optimal path to the von Neumann facet K^* instead of the convergence to the maximum balanced growth path. This was done in McKenzie (1963).

First, let the norm of a vector z be $\|z\| = \Sigma_i |z_i|$ and the angle between two vectors z and z' be $d(z, z') = \Sigma_i |z_i/\|z\| - z_i'/\|z'\||$, that is, the norm of the vector joining the normalized vectors. And if C is a nonempty set of vectors, let $d(z, C) = \inf \{d(z, z')|z' \in C\}$. Then, by applying Theorem 1 in McKenzie (1963), we get the following result for our space system.[13]

Lemma 5.1. Under assumptions A1 to A9, let $\{\hat{X}_t\}_0^N$ be an optimal path for Problem I(a). Then for any $\epsilon > 0$, there is a number N_ϵ such that the number of periods in which $d[(\hat{X}_t, \hat{X}_{t+1}), K^*] \geqq \epsilon$ cannot exceed N_ϵ.

This statement implies that if the plan period N is sufficiently long, then the optimal path should stay most of that time within a very small neighborhood of the von Neumann facet K^*. To see the geometrical meaning of this, let us define the corresponding state of the ideal region by

$$x_t = x_r^t + x_s^t \tag{45}$$

when the state of the space system is $X_t = (x_r^t, x_s^t)$. Then one easily sees by comparing (42) and (43) and taking into account McKenzie's definition of distance that the above result implies the following:

Lemma 5.2. Under assumptions A1 to A9, let $\{\hat{x}_t\}_0^N$ be the (optimal) path in the ideal region which, by relation (45), corresponds to the optimal path $\{\hat{X}_t\}_0^t$ for I(a). Then for any $\epsilon > 0$, there is a number N_ϵ such that the number of periods in which $d[(\hat{x}_t, \hat{x}_{t+1}), k^*] \geqq \epsilon$ cannot exceed N_ϵ.

[13]The proof of Theorem 1 in McKenzie (1963) is given for a generalized Leontief model. But it is easily seen that this theorem is also valid, for example, under following conditions: (i) T is a non-singular m.t.cc.; (ii) there is a unique positive maximum balanced growth path; (iii) there is a positive von Neumann price vector; and (iv) either T satisfies the Lipschitz condition, or $\bar{X}(0) > 0$. All of these conditions are satisfied in our model.

Hence, it is known that the normalized input–output pair in the ideal region which corresponds to each optimal input–output pair in the space system stay most of the time in a small neighborhood of the line segment $\overline{z_r z_s}$ in Figure 5.5.

For convenience, let us define the ϵ-neighborhood of sets K^* and k^*, respectively, by

$$N(\epsilon, K^*) = \{(X_t, X_{t+1}) \mid d[(X_t, X_{t+1}), K^*] < \epsilon\}. \tag{46}$$

$$N(\epsilon, k^*) = \{(x_t, x_{t+1}) \mid d[(x_t, x_{t+1}), k^*] < \epsilon\}. \tag{47}$$

From Lemma 5.2, an optimal path $\{\hat{X}_t\}_0^N$ should stay most of the time in a small neighborhood $N(\epsilon, K^*)$ of the von Neumann facet K^*. And when we take the plan length N sufficiently large, ϵ can be taken as small as we like. Hence the movement of the optimal path in the set $N(\epsilon, K^*)$ can be approximated by an appropriate path which moves exactly on the set K^*. The movement of a path $\{X_t\}$ which stays exactly on the set K^* is obtained as follows.

Since $(X_t, X_{t+1}) \in K^*$ only when X_t is equal to $\beta_t[\alpha(a_1^r, a_2^r), (1 - \alpha)(a_1^s, a_2^s)]_t$ for some $\beta_t \geq 0$ and $1 \geq \alpha_t \geq 0$ (see (43)), it follows that $(X_{t+1}, X_{t+2}) \in K^*$ only when X_{t+1} can be represented in the same form. Thus the movement of a path on the set K^* is given by

$$\beta_t[(\alpha a_1^r + \theta_1(\alpha b_1^r + (1-\alpha)b_1^s), \theta_2(\alpha b_2^r + (1-\alpha)b_2^s)), ((1-\alpha)a_1^s$$
$$+ (1-\theta_1)(\alpha b_1^r + (1-\alpha)b_1^s), (1-\theta_2)(\alpha b_2^r + (1-\alpha)b_2^s))]_t$$
$$= \beta_{t+1}[\alpha(a_1^r, a_2^r), (1-\alpha)(a_1^s, a_2^s)]_{t+1} \tag{48}$$

where $\beta_t, \beta_{t+1} \geq 0$, $1 \geq \alpha_t \geq 0$, $1 \geq \alpha_{t+1} \geq 0$ and $1 \geq \theta_i^t \geq 0$, $i = 1, 2$.[14] Let

$$\beta_t \alpha_t = q_r^t, \beta_t(1 - \alpha_t) = q_s^t, \beta_{t+1} \alpha_{t+1} = q_r^{t+1} \quad \text{and}$$
$$\beta_{t+1}(1 - \alpha_{t+1}) = q_s^{t+1} \tag{49}$$

Then the period-to-period relation of the movement of a feasible path

[14]As known from this, the allowable range of K^* in which a feasible path can move during a certain time interval is restricted by two conditions. The first restriction, $1 \geq \alpha_t \geq 0$ for each t, is the condition under which the corresponding input–output pair moves consecutively on the set k^*, and this condition does not depend on whether capital can be moved freely or not. The second restriction, $1 \geq \theta_i^t \geq 0$ is the additional condition which is necessitated by the immobility of capital. We may call this allowable range of K^* the *restricted von Neumann facet*, and denote it by ΔK^*. It can be shown under A1 to A8 that ΔK^* is a closed convex cone in $\mathbf{R}_+^{4 \times 2}$ including the ray generated by the maximum balanced input–output pair $(\bar{X}, \lambda \bar{X})$ in its relative interior. For detailed analyses of the restricted von Neumann facet ΔK^*, see Fujita (1972, pp. 264–268).

on the set K^* is given by

$$q_r^t a_1^r + \theta_1^t(q_r^t b_1^r + q_s^t b_1^s) = q_r^{t+1} a_1^r,$$
$$\theta_2^t(q_r^t b_2^r + q_s^t b_2^s) = q_r^{t+1} a_2^r,$$
$$q_s^t a_1^s + (1 - \theta_1^t)(q_r^t b_1^r + q_s^t b_1^s) = q_s^{t+1} a_1^s,$$
$$(1 - \theta_2^t)(q_r^t b_2^r + q_s^t b_2^s) = q_s^{t+1} a_2^s. \tag{50}$$

Adding the first and the third equations and the second and the last equations, respectively, we get

$$\begin{bmatrix} a_1^r + b_1^r \\ b_2^r \end{bmatrix} q_r^t + \begin{bmatrix} a_1^s + b_1^s \\ b_2^s \end{bmatrix} q_s^t = \begin{bmatrix} a_1^r \\ a_2^r \end{bmatrix} q_r^{t+1} + \begin{bmatrix} a_1^s \\ a_2^s \end{bmatrix} q_s^{t+1}. \tag{51}$$

Hence the solution of (50) should satisfy (51). But (51) can be solved independently of the variables θ_i^t, $i = 1, 2$. In addition, by using (33), we obtain the following relations from (49)

$$\theta_1^t - \bar{\theta}_1 = (\alpha_t - \bar{\alpha}) \frac{(a_2^r - a_2^s)a_1^r a_1^s}{b_1(a_1^r a_2^s - a_1^s a_2^r)},$$
$$\theta_2^t - \bar{\theta}_2 = (\alpha_t - \bar{\alpha}) \frac{(a_1^r - a_1^s)a_2^r a_2^s}{b_2(a_1^r a_2^s - a_1^s a_2^r)}, \tag{52}$$

where $\bar{\alpha}$ and $\bar{\theta}_i$ are given in (26) and (32). Hence, observing from (50) that $\alpha_t = q_r^t/(q_r^t + q_s^t)$, we see that the path on K^* can be determined by examining (51) first, and then using (52). The solution of (51) is obtained by finding the roots of the following equation:[15]

$$\begin{vmatrix} a_1^r + b_1^r - \lambda a_1^r & a_1^s + b_1^s - \lambda a_1^s \\ b_2^r - \lambda a_2^r & b_2^s - \lambda a_2^s \end{vmatrix} = 0. \tag{53}$$

It is easily seen from (28) and (29) that the maximum growth factor $\bar{\lambda}$ is one of the solutions of this equation and $(\bar{\alpha}, 1 - \bar{\alpha})$ is the corresponding characteristic vector. Since (53) has at most two roots and one is given by $\bar{\lambda}$ which is real, it follows that the other root is also real. Let us denote the other root by $\bar{\lambda}'$ and the corresponding characteristic vector by (h_1, h_2). Then the general solution of (51) is given by[16]

---

[15]This is equivalent to the characteristic equation of the matrix

$$\begin{bmatrix} a_1^r & a_1^s \\ a_2^r & a_2^s \end{bmatrix}^{-1} \begin{bmatrix} a_1^r + b_1^r & a_1^s + b_1^s \\ b_2^r & b_2^s \end{bmatrix}$$

[16]Let us suppose, for simplicity, that (53) has two distinct roots.

$$\begin{bmatrix} q_r^t \\ q_s^t \end{bmatrix} = \mu_1 \begin{bmatrix} \bar{\alpha} \\ 1 - \bar{\alpha} \end{bmatrix} (\bar{\lambda})^t + \mu_2 \begin{bmatrix} h_1 \\ h_2 \end{bmatrix} (\bar{\lambda}')^t \tag{54}$$

where μ_1 and μ_2 are determined by the starting point on K^*.

From the definition of the maximum balanced growth factor, we know that $\bar{\lambda} > \bar{\lambda}'$ if $(h_1, h_2) > 0$. Hence, we have four possibilities:

(i) $|\bar{\lambda}'| < \bar{\lambda}$,

(ii) $\bar{\lambda}' < -\bar{\lambda}$,

(iii) $\bar{\lambda}' > \bar{\lambda}$ and $(h_1, h_2) \not> 0$,

(iv) $\bar{\lambda}' = -\bar{\lambda}$.

Let us first examine case (i). Suppose $\bar{\lambda}' > 0$ and $(h_1, h_2) > 0$. Then it could happen that $\mu_1 = 0$ and $\mu_2 > 0$, depending on the location of the starting point on K^*. But it can be shown that this does not happen. For, if it were true, then we would have $(q_r^t, q_s^t) = \mu_2(h_1, h_2)(\bar{\lambda}')^t$. Hence the equation of the path would be $X_t = \mu_2(h_1(a_1^r, a_2^r), h_2(a_1^s, a_2^s))(\bar{\lambda}')^t$. But the input–output pair (X_t, X_{t+1}) on this path does not satisfy (36) with an equality, since $\bar{\lambda}' < \bar{\lambda}$. Hence $(X_t, X_{t+1}) \notin K^*$ and this contradicts the presumption that (54) gives the movement of the path on K^*. Hence, in case (i), we must have

$$(q_r^t, q_s^t) \to \mu_1(\bar{\alpha}, 1 - \bar{\alpha})(\bar{\lambda})^t \quad \text{as } t \to \infty. \tag{55}$$

Of course, for each input–output pair (X_t, X_{t+1}) to be in the set K^* for an extended period, μ_1 must be positive; that is, the starting location of a path on K^* should be in the region which yields $\mu_1 > 0$.

Therefore, when N is very large, from (55) the path on K^* spends most of its time in a very small neighborhood of the maximum balanced growth ray generated by (31). This, in turn, implies that the paths for regions r and s spend most of their time in small neighborhoods of the rays generated by vectors a_r and a_s in Figure 5.6, respectively. And the corresponding path in the ideal region is near the ray generated by the vector a.

Next, let us examine case (ii). When μ_2 is sufficiently large, the second term dominates the first term in (54). Thus q_r^t and q_s^t oscillate between positive values and negative values. But this clearly violates the conditions $q_l^t \geq 0$, $l = r, s$. Hence, for the solution of (51) to satisfy the nonnegativity of q_l^t, $l = r, s$, for a very long time, the starting position of the path on K^* should be such that it gives μ_2 a negligibly small value compared to μ_1. That is, the starting location of the path on K^* should be very close to the maximum balanced growth ray. But

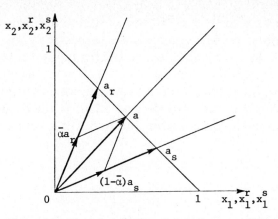

Figure 5.6. Maximum balanced growth vector in the space system.

the movement of such a path, in terms of angular distance (i.e., McKenzie's distance defined at the beginning of this section) from the maximum balanced growth ray, is very slow when the path is near that ray compared to when it is far from the ray. Hence, in this case also, the path spends most of its time in a small neighborhood of the maximum balanced growth ray.

In case (iii), we can also get a similar conclusion. Hence, in cases (i) to (iii), a path which moves on the set K^* for a long time should spend most of its time in a very small neighborhood of the maximum balanced growth ray.

However, case (iv) is an exception. When the starting location of the path on ΔK^* is chosen appropriately, the path determined by (54) can make undamped oscillations around the maximum balanced growth ray.

Hence, considering Lemmas 5.1 and 5.2, we obtain the following conclusion.

Theorem 5.1 (turnpike theorem). Suppose we have assumptions A1 to A9. Then, if the plan period N is sufficiently long, the optimal path $\{\hat{X}_t\}_0^N$ for Problem I(a) stays most of the time in a very small neighborhood of the von Neumann facet K^*.[17] And, except for case (iv), it spends most of its time in a very small neighborhood of the maximum

---

[17]We can replace "the von Neumann facet K^*" with "the restricted von Neumann facet ΔK^*". For this point, see footnote 14.

balanced growth ray generated by (31). In this case, the optimal paths for the ideal region, region r and region s stay, respectively, in very small neighborhoods of the rays generated, respectively, by vectors a, a_r and a_s (refer to Figure 5.6).[18]

Thus, the optimal path for Problem I(a) may be depicted as in Figure 5.7(a) and (b). Figure 5.7(a) is for case (i), and 5.7(b) is for cases (ii) and (iii); in 5.7(a) the optimal path is stable, but in 5.7(b) it is unstable.

5.2.7. Price equalization

In Section 5.2.3, it was shown that for each optimal path in Problem I(a) there was an associated optimal path in Problem I(b). Hence, let us examine the properties of the dual optimal path which enable the commodity path to move most of the time in a small neighborhood of the von Neumann facet and the maximum balanced growth ray.

Suppose at the beginning of period t, a part of the optimal path $\{\hat{X}_t\}_0^N$ which lies within the set $N(\epsilon, K^*)$ for some small ϵ is in a very small neighborhood of the maximum balanced growth ray, that is, $\hat{X}_t \in \tilde{N}(\epsilon', \bar{X}) \equiv \{X | d(X, \eta \bar{X}) < \epsilon'$ for some $\eta > 0\}$ for small ϵ'. Then we may ask: what are the prices which will enable the optimal path to remain inside of this neighborhood in the next period? Let $\hat{P}_{t+1} = [(\hat{p}_1^r, \hat{p}_2^r), (\hat{p}_1^s, \hat{p}_2^s)]_{t+1}$ be the optimal price vector at the beginning of period $t + 1$, and let $(\hat{y}_1^r, \hat{y}_2^r)_t$ and $(\hat{y}_1^s, \hat{y}_2^s)_t$ be the optimal net outputs of the commodities in each region at the end of period t. Then, the optimal allocation plan θ_1^i and θ_2^i should satisfy (20). Hence, no matter how small the difference in prices of a good between regions is, the total output of the investment good as well as the material good is allocated entirely to the region with the higher price. But, when $\hat{X}_t \in \tilde{N}(\epsilon', \bar{X})$ for a sufficiently small ϵ', it is possible that $\hat{X}_{t+1} \in$

---

[18]We got this conclusion by assuming that the optimal path stays in the set $N(\epsilon, K^*)$ continuously except for certain periods at the beginning and the end of the plan period. But this statement holds true even if the part of the optimal path goes out of $N(\epsilon, K^*)$ for a finite time (which is surely finite) in the intermediate periods. For the strict proof of this, see Theorem 2 in McKenzie (1963). The same reasoning of this theorem can be directly applied to our problem.

Also, we have not proved that the optimal path stays continuously in a small neighborhood of the maximum balanced growth path except for certain periods at the beginning and the end. This can be proved, but the proof is highly complicated. For example, see Nikaido (1968) and Tsukui (1966).

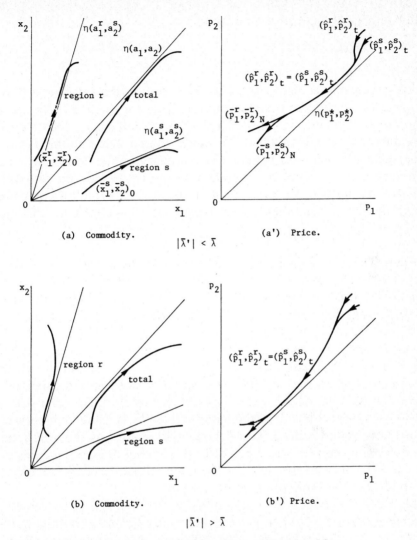

(a) Commodity.

$|\bar{\lambda}'| < \bar{\lambda}$

(a') Price.

(b) Commodity.

$|\bar{\lambda}'| > \bar{\lambda}$

(b') Price.

Figure 5.7. Optimal paths for Problem I(a) and Problem I(b).

$\tilde{N}(\epsilon', \bar{X})$ and $(\hat{X}_t, \hat{X}_{t+1}) \in N(\epsilon, K^*)$ for small ϵ only when $\hat{\theta}_1^t$ and $\hat{\theta}_2^t$ have, respectively, approximately the same values as $\bar{\theta}_1$ and $\bar{\theta}_2$ given by (32). And, from A6 and A7, $1 > \bar{\theta}_i > 0$, $i = 1, 2$. Hence, it is impossible for $\hat{X}_{t+1}$ to be in the set $\tilde{N}(\epsilon', X)$ with very small ϵ' and for $(\hat{X}_t, \hat{X}_{t+1})$ to be in the set $N(\epsilon, K^*)$ with small ϵ if $\hat{p}_i^r(t+1) \neq \hat{p}_i^s(t+1)$ for $i = 1$, or 2. Therefore, the price of each good should be exactly the same in each region when the commodity path moves near the maximum balanced growth ray. That is

$$\hat{p}_i^r(t) = \hat{p}_i^s(t), \quad i = 1, 2 \tag{56}$$

when $\hat{X}_t$ lies in the set $\tilde{N}(\epsilon', \bar{X})$.

This conclusion is also true when $\hat{X}_t$ moves in the set $N(\epsilon, K^*)$ but not necessarily close to the maximum balanced growth ray. As shown above, when $\hat{X}_t$ moves in the set $N(\epsilon, K^*)$ for small ϵ, the allocation rule of outputs is approximated by (52). Hence $\hat{\theta}_1^i$ and $\hat{\theta}_2^i$ are neither 0 nor 1, except possibly when $\hat{X}_t$ is near the boundary of the allowable range of the set K^* (i.e., near the boundary of the restricted von Neumann facet defined in footnote 14), which implies that the dual prices which sustain this movement should also satisfy (56). Thus we may conclude that:

Theorem 5.2 (price equalization theorem).   Suppose we have assumptions A1 to A9. Then if the plan period N is sufficiently long on the optimal path for I(b) the price of each good should be exactly the same in both regions for most of the time, except possibly for case (iv).

Even for case (iv), the above statement is true as long as the optimal path does not oscillate between points near the boundary of the allowable range of the set K^*. Hence, *price equalization* may hold even when the commodity path does not exhibit turnpike behavior.

Moreover, since the material good $(y_2^r + y_2^s)$ is allocated between regions according to rule (20), it follows that as long as the production in each region requires the input of both commodities, price equalization will hold most of the time for the material good even when capital does not have the same price in each region.

The relation between the commodity vector and the price vector is depicted in Figure 5.7 when price equalization holds for both goods in periods $t + 1$ and t. When $\hat{P}_{t+1} = ((\hat{p}_1, \hat{p}_2), (\hat{p}_1, \hat{p}_2))_{t+1}$, the point $(\hat{p}_1, \hat{p}_2)_t$ should lie in the intersection of the set, $(\hat{p}_1, 0)_{t+1} + F_r^*(\hat{p}_1, \hat{p}_2)_{t+1}$, and the set, $(\hat{p}_1, 0)_{t+1} + F_s^*(\hat{p}_1, \hat{p}_2)_{t+1}$, for price equalization to hold in both goods in period t. Moreover, since $\hat{P}_t$ is determined according to rule (17), the point $(\hat{p}_1, \hat{p}_2)_t$ should be a boundary point of the above two sets, that is, it should lie in the intersection of the boundaries of the above two sets as depicted in Figure 5.8.

It may be noted that $\hat{p}_1^r(t) = \hat{p}_1^s(t)$ implies that the immobility of capital does not cause any burden to the system in this period. This follows from (22) in Chapter 4 (refer also to (40) in Chapter 2), that is,

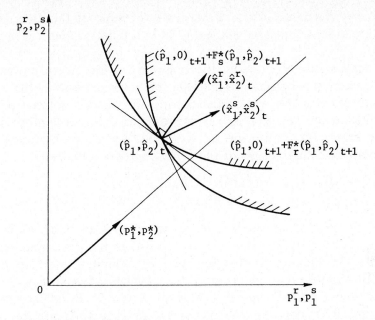

Figure 5.8. Relation between the optimal commodity vector and the optimal price vector under price equalization.

from the fact that

$$\frac{\partial \mu_t^-(\hat{X}_t)}{\partial x_i^l(t)} \geq \hat{p}_i^l(t) \geq \frac{\partial \mu_t^+(\hat{X}_t)}{\partial x_i^l(t)}, \quad i = 1, 2, \quad l = r, s$$

where $\mu_t(\hat{X}_t) = \max \{\bar{P}_N \cdot X_N | X_N \in T^{N-t}(\hat{X}_t)\}$ (refer to (38b) in Chapter 2). Thus, when $\hat{p}_i^r(t) = \hat{p}_i^s(t)$, it follows that if we subtract one unit of good i from, say $\hat{x}_i^r(t)$ and add one unit of i to $\hat{x}_i^s(t)$, the result is at best unchanged with respect to the maximization of the objective function. Thus, when $\hat{p}_i^r(t) = \hat{p}_i^s(t)$, we do not need to reallocate any of commodity i between the regions – even if we can move it freely. This in particular implies that no capital need be reallocated between the regions. This point is also a direct consequence of the pricing mechanism. As stated above, when $\hat{p}_i^r(t) = \hat{p}_i^s(t)$ $(= \hat{p}_i(t))$ for $i = 1$, and 2, the point $(\hat{p}_1, \hat{p}_2)_t$ should lie in the intersection of the set, $(\hat{p}_1, 0)_{t+1} + F_r^*(\hat{p}_1, \hat{p}_2)_{t+1}$ and the set, $(\hat{p}_1, 0)_{t+1} + F_s^*(\hat{p}_1, \hat{p}_2)_{t+1}$, that is, the prices of the stock of goods are determined as if the technology of both regions could be used in any one region.

Therefore, since we have seen that the price of the stock of each good is exactly the same in the two regions for most of the time, our space system attains the same efficiency as in a point system for most of the time.

Finally, let us briefly examine the behavior of the corresponding price path when the optimal commodity path moves within the set $N(\epsilon, K^*)$ for small ϵ. As was stated before, the part of optimal path $\{\hat{X}_t\}_0^N$ moving in the set $N(\epsilon, K^*)$ is approximated by an appropriate path which moves exactly in the set K^*. And, when a path moves in K^*, only convex combinations of the following two activities,

$$[(a_1^r, a_2^r), (0, 0)] \rightarrow [(a_1^r + b_1^r, b_2^r), (0, 0)],$$
$$[(0, 0), (a_1^s, a_2^s)] \rightarrow [(0, 0), (a_1^s + b_1^s, b_2^s)],$$

are used in the production process in the space system. In addition, it can be seen from (i) of Theorem 2.1′ (refer also to the reasoning used in footnote 7 in Chapter 4) that any activity which is used on the optimal path should break even under the corresponding prices. And when the commodity path moves in the set $N(\epsilon, K^*)$, price equalization holds for both goods except possibly for case (iv), and hence, $\hat{P}_t = [(p_1, p_2), (p_1, p_2)]_t$ for appropriate p_1^l and p_2^l. Thus the following equality should be satisfied approximately:

$$p_1^t a_1^l + p_2^t a_2^l = p_1^{t+1}(a_1^l + b_1^l) + p_2^{t+1} b_2^l, \quad l = r, s.$$

That is, the behavior of the price vectors is approximated by the difference equation

$$\begin{bmatrix} a_1^r \\ a_1^s \end{bmatrix} p_1^t + \begin{bmatrix} a_2^r \\ a_2^s \end{bmatrix} p_2^t = \begin{bmatrix} a_1^r + b_1^r \\ a_1^s + b_1^s \end{bmatrix} p_1^{t+1} + \begin{bmatrix} b_2^r \\ b_2^s \end{bmatrix} p_2^{t+1} \tag{57}$$

Comparing (51) to the above equation, it can be shown that they are mutually dual. Hence, when the time superscripts are reversed in (57), the behavior of (p_1^l, p_2^l) parallels that of (q_r^l, q_s^l) determined by (51). Since plane (40) was determined so as to pass through the points $\bar{z}_r$ and $\bar{z}_s$, it is easily seen that (p_1^*, p_2^*) is on the balanced growth path of Equation (57) with growth factor $1/\lambda$. Moreover, it can be shown as in the previous section that the solution of (57) exhibits the same turnpike behavior around the balanced growth ray generated by (p_1^*, p_2^*), except for case (iv). But since the solutions of (51) and (57) exhibit the same turnpike behavior when time is reversed in (57), they have the relation of *dual instability* under ordinary time (i.e., when time is not reversed in (57)). Hence we obtain Figures 5.7(a′) and 5.7(b′) corresponding to 5.7(a) and 5.7(b).

5.3. Optimum growth in a space system with transport inputs

5.3.1. *Problem formulation*

In the previous section, the properties of optimum growth paths in a two-region two-good monotone space system without transport inputs were studied. In particular, it was shown that when the plan period is sufficiently long, the investment good is allocated between regions in a "balanced manner" for most of the time, and the optimal path exhibits turnpike behavior around the maximum balanced growth path in the space system. Moreover, the prices of each good which sustain this optimal path behavior are exactly the same in both regions for most of the plan period.

The purpose of this section is to show that the above characteristics of the optimum growth path essentially do not change when transport inputs are required for commodity movements. The analysis in this section parallels that of Section 5.2, and is presented in a simplified form.

Assumptions of the model in this section are exactly the same as those of the model in Section 5.2 except for the introduction of transport inputs for the movement of investment goods. That is, we assume that the movement of one unit of investment good from location r to location s (location s to location r) requires v_{rs} units (v_{sr} units) of material good. And it is assumed that the material good can be moved freely between the two locations. Though this form of transport inputs is very simple, the essential role of transport inputs can be studied adequately under this model.

According to these assumptions, the primal problem in this section is formulated as follows:

Problem II(a)

$$\text{Maximize} \sum_{l=r,s} \sum_{i=1,2} \bar{p}_i^l(N)x_i^l(N),$$

$$\text{subject to } 0 \leq x_1^l(t+1) \leq x_1^l(t) + y_1^{rl}(t) + y_1^{sl}(t),$$

$$0 \leq x_2^l(t+1) \leq \theta_2^l(t)(y_2^r + y_2^s - v_{rs}y_1^{rs} - v_{sr}y_1^{sr})_t, \quad l = r, s,$$

$$\text{for some}$$

$$y_1^l(t) \geq y_1^{lr}(t) + y_1^{ls}(t), \ y_1^{lr}(t), \ y_1^{ls}(t) \geq 0, \quad l = r, s,$$

$$(y_1^l, y_2^l)_t \in f_l(x_1^l, x_2^l)_t, \quad l = r, s,$$

$$\theta_2^r(t) + \theta_2^s(t) = 1, \ \theta_2^r(t), \ \theta_2^s(t) \geq 0, \quad t = 0, 1, \ldots, N-1,$$

$$(1')$$

given

$$(x_1^l, x_2^l)_0 = (\bar{x}_1^l(0), \bar{x}_2^l(0)), \quad l = r, s.$$

And to make this problem different from Problem I(a), let us assume explicitly that

A10. $v_{rs} > 0, v_{sr} > 0.$ (58)

In a compact form this problem can be represented by (2) as before, but the space transformation T is defined by the following equation instead of (3):

$$T(X_t) = \{X_{t+1} = ((x_1^r, x_2^r), (x_1^s, x_2^s))_{t+1} | \text{subject to } (1')\}.$$ (3′)

We retain assumptions A1 and A2. Hence, because of A1, the transformation T defined by (3′) is a non-singular m.t.cc. from $\mathbf{R}_+^4$ into $\mathbf{R}_+^4$. As before this space transformation T can be decomposed into production activities and transportation activities. For production activities, there is no change. That is, when T is represented by a composite mapping of T_1 and T_2 such as (6), T_1 is exactly the same as (4). But T_2 is modified as follows:

$$T_2(Y_t) = T_2((x_1^{\prime r}, y_1^r, y_2^r), (x_1^{\prime s}, y_1^s, y_2^s))_t$$
$$= \{X_{t+1} = ((x_1^r, x_2^r), (x_1^s, x_2^s))_{t+1} | \text{for some}$$
$$0 \leqq x_1^l(t+1) \leqq x_1^{\prime l}(t) + y_1^{rl}(t) + y_1^{sl}(t),$$
$$0 \leqq x_2^l(t+1) \leqq \theta_2^l(t)(y_1^r + y_1^s - v_{rs}y_1^{rs} - v_{sr}y_1^{sr})_t,$$
$$y_1^l(t) \geqq y_1^{lr}(t) + y_1^{ls}(t), y_1^{lr}(t), y_1^{ls}(t) \geqq 0, \quad l = r, s,$$
$$\theta_2^r(t) + \theta_2^s(t) = 1, \theta_2^r(t), \theta_2^s(t) \geqq 0\}.$$ (5′)

Thus T_2 is a non-singular m.t.cc. from $\mathbf{R}_+^6$ into $\mathbf{R}_+^4$, and as before, assumption A1 implies that T_1 is a non-singular m.t.cc. from $\mathbf{R}_+^4$ into $\mathbf{R}_+^6$.

5.3.2. Dual problem

The dual system for Problem II(a) can be obtained in essentially the same way as before, that is, by the simple rule that no feasible activity in the space system can obtain positive profits.

Since there is no change for T_1, the adjoint transformation of T_1 is defined by (8) and (10). The adjoint transformation of T_2 is defined as follows:

$$T_2^*(P_{t+1}) = T_2^*(p_1^r, p_2^r), (p_1^s, p_2^s))_{t+1}$$
$$= \{\Phi_t = ((p_1^{\prime r}, \phi_1^r, \phi_2^r), (p_1^{\prime s}, \phi_1^s, \phi_2^s))_t | \text{for some}$$
$$p_1^{\prime l}(t) \geq p_1^l(t+1), \phi_1^l(t) \geq p_1^l(t+1), \quad l = r, s,$$
$$\phi_1^r(t) + v_{rs}\phi_2^r(t) \geq p_1^s(t+1), \phi_1^s(t) + v_{sr}\phi_2^s(t) \geq p_1^r(t+1),$$
$$\phi_1^r(t), \phi_2^s(t) \geq p_2(t+1)\}, \tag{9'}$$

where

$$p_2(\tau) = \max(p_2^r, p_2^s)_\tau, \quad \tau = 0, 1, \ldots, N. \tag{59}$$

The economic meaning of each condition in the above definition of T_2^* is clear. For example, the condition, $\phi_1^r(t) + v_{rs}\phi_2^r(t) \geq p_1^s(t+1)$ says that (price of the investment good in r) + (transport cost for one unit of the investment good from r to s) ≥ (price of capital in region s).

From (10), (11) and (9′), the adjoint transformation of the space transformation T (defined by (3′)) is given as follows:

$$T^*(P_{t+1}) = \{P_t = ((p_1^r, p_2^r), (p_1^s, p_2^s))_t | \text{for some}$$
$$(p_1^r, p_2^r)_t \in (p_1^r, 0)_{t+1} + F_r^*(\max(p_1^r, p_1^s - v_{rs}p_2), p_2)_{t+1},$$
$$(p_1^s, p_2^s)_t \in (p_1^s, 0)_{t+1} + F_s^*(\max(p_1^s, p_1^r - v_{sr}p_2), p_2)_{t+1}\}, \tag{12'}$$

where F_l^*, $l = r, s$, are defined by (8).

Hence, the dual problem for II(a) is given as follows:

Problem II(b)

Minimize $\sum\limits_{l=r,s} \sum\limits_{i=1,2} p_i^l(0)\bar{x}_i^l(0)$

subject to $(p_1^r, p_2^r)_t \in (p_1^r, 0)_{t+1} + F_r^*(\max(p_1^r, p_1^s - v_{rs}p_2), p_2)_{t+1}$,
$(p_1^s, p_2^s)_t \in (p_1^s, 0)_{t+1} + F_s^*(\max(p_1^s, p_1^r - v_{sr}p_2), p_2)_{t+1}$,
$t = 0, 1, \ldots, N - 1$,

given

$$(p_1^l, p_2^l) = (\bar{p}_1^l(N), \bar{p}_2^l(N)), \quad l = r, s,$$

where

$$p_2(\tau) = \max(p_2^r, p_2^s)_\tau, \quad \tau = 0, 1, \ldots, N.$$

In a compact form, this is represented by (13).

5.3.3. Stepwise procedures for the optimal path

We also retain Assumption A3. Then, as before, it follows that by using optimality conditions (14) and (15) (where the transformations T_1, T_2, T_1^* and T_2^* are defined by (4), (5'), (10) and (9') respectively), it is easily seen that the optimal variables should satisfy the following conditions:

$$
\begin{aligned}
&\hat{p}_1^{l'}(t) = \hat{p}_1^l(t+1), \quad l = r, s, \\
&\hat{\phi}_2^r(t) = \hat{\phi}_2^s(t)(\equiv \hat{\phi}_2(t)) = \max \, (\hat{p}_2^r, \hat{p}_2^s)_{t+1}(\equiv \hat{p}_2(t+1)), \\
&\hat{\phi}_1^r(t) = \max \, (\hat{p}_1^r, \hat{p}_1^s - v_{rs}\hat{p}_2)_{t+1}, \\
&\hat{\phi}_1^s(t) = \max \, (\hat{p}_1^s, \hat{p}_1^r - v_{sr}\hat{p}_2)_{t+1}.
\end{aligned} \tag{16'}
$$

Hence, from (16') and the right equality in (15), the optimal value of $(p_1^l, p_2^l)_t$, $l = r, s$, is determined for given vectors $(\hat{p}_1^l, \hat{p}_2^l)_{t+1}$ and $(\hat{x}_1^l, \hat{x}_2^l)_t$, $l = r, s$, so as to:

Minimize $p_1^r(t)\hat{x}_1^r(t) + p_2^r(t)\hat{x}_2^r(t)$,

subject to $(p_1^r, p_2^r)_t \in (\hat{p}_1^r, 0)_{t+1} + F_r^*(\max \, (\hat{p}_1^r, \hat{p}_1^s - v_{rs}\hat{p}_2), \hat{p}_2)_{t+1}$,

$$\tag{17'.1}$$

and,

minimize $p_1^s(t)\hat{x}_1^s(t) + p_2^s(t)\hat{x}_2^s(t)$,

subject to $(p_1^s, p_2^s)_t \in (\hat{p}_1^s, 0)_{t+1} + F_s^*(\max \, (\hat{p}_1^s, \hat{p}_1^r - v_{sr}\hat{p}_2), \hat{p}_2)_{t+1}$.

$$\tag{17'.2}$$

Dually, from (4), (16') and the left equality in (14), the optimal production schedule is determined in each region, for a given $(\hat{x}_1^l, \hat{x}_2^l)_t$ and $(\hat{p}_1^l, \hat{p}_2^l)_{t+1}$, $l = r, s$, so as to:

Maximize $\max \, (\hat{p}_1^r, \hat{p}_1^s - v_{rs}\hat{p}_2)_{t+1}y_1^r(t) + \hat{p}_2(t+1)y_2^r(t)$,

subject to $(y_1^r, y_2^r)_t \in F_r(\hat{x}_1^r, \hat{x}_2^r)_t$,

$$\tag{18'.1}$$

and,

Maximize $\max \, (\hat{p}_1^s, \hat{p}_1^r - v_{sr}\hat{p}_2)_{t+1}y_1^s(t) + \hat{p}_2(t+1)y_2^s(t)$,

subject to $(y_1^s, y_2^s)_t \in F_s(\hat{x}_1^s, \hat{x}_2^s)_t$.

$$\tag{18'.2}$$

Condition (19) is retained as it is. Finally, from (5'), (16') and the left equality of (15), the optimal transportation plan is determined, for a given $(\hat{y}_1^l, \hat{y}_2^l)_t$ and $(\hat{p}_1^l, \hat{p}_2^l)_{t+1}$, $l = r, s$, so as to:

Maximize $\hat{p}_1^r(t+1)y_1^{rr}(t) + (\hat{p}_1^s - v_{rs}\hat{p}_2)_{t+1}y_1^{rs}(t)$,

subject to $\hat{y}_1^r(t) \geq (y_1^{rr} + y_1^{rs})_t$, $y_1^{rr}(t)$, $y_1^{rs}(t) \geq 0$,

$$\tag{20'.1}$$

and,

$$\text{maximize } \hat{p}_1^s(t+1)y_1^{ss}(t) + (\hat{p}_1^r - v_{sr}\hat{p}_2)_{t+1}y_1^{sr}(t),$$
$$\text{subject to } \hat{y}_1^s(t) \geq (y_1^{ss} + y_1^{sr})_t, \; y_1^{ss}(t), \; y_1^{sr}(t) \geq 0, \tag{20'.2}$$

where, of course we must have,

$$(\hat{y}_2^r + \hat{y}_2^s)_t \geq (v_{rs}y_1^{rs} + v_{sr}y_1^{sr})_t. \tag{20'.3}$$

And, the material good is allocated so as to:

$$\text{Maximize } [\hat{p}_2^s + \theta_2^t(\hat{p}_2^r - \hat{p}_2^s)]_{t+1}(\hat{y}_2^r + \hat{y}_2^s - v_{rs}\hat{y}_1^{rs} - v_{sr}\hat{y}_1^{sr})_t$$
$$\text{subject to } 1 \geq \theta_2^t \; (\equiv \theta_s^r(t)) \geq 0. \tag{20'.4}$$

Assuming that constraint (20'.3) is automatically satisfied on the optimal path (i.e., enough transport inputs are produced on the optimal path),[19] an extreme allocation of investment good is avoided only in the following three cases:

$$\hat{p}_1^r(t+1) = \hat{p}_1^s(t+1) - v_{rs}\hat{p}_2(t+1), \quad \text{or} \tag{60}$$
$$\hat{p}_1^s(t+1) = \hat{p}_1^r(t+1) - v_{sr}\hat{p}_2(t+1), \quad \text{or} \tag{61}$$
$$\left.\begin{array}{l} \hat{p}_1^r(t+1) > p_1^s(t+1) - v_{rs}\hat{p}_2(t+1), \quad \text{and} \\ \hat{p}_1^s(t+1) > \hat{p}_1^r(t+1) - v_{sr}\hat{p}_2(t+1) \end{array}\right\} \tag{62}$$

In the case of (62), each region invests all of the investment good produced in that location in its own region. When the investment good is not produced in one of the two regions, (62) also gives an extreme allocation of the investment good.

The extreme allocation of the material good is avoided only when $\hat{p}_2^r = \hat{p}_2^s$.

5.3.4. Maximum balanced growth path

The maximum balanced growth vector $\bar{X}$ is defined as before by (24) and the existence of such vector is guaranteed by the fact that the space transformation T defined by (4), (5') and (6) is a non-singular m.t.cc. from $\mathbf{R}_+^4$ into $\mathbf{R}_+^4$ under assumption A1.[20] Let us first examine

---

[19]This is true when the production function in each location has smooth output substitution and the optimal output point is not a corner point. In our special model, this may be satisfied by assuming that a part of the material good produced in a period is used as the payment to labor in the next period and thus that not all of the material good is used for transport inputs.

[20]See the references in footnote 9.

how the maximum balanced growth path is described by activities.

The maximum balanced growth path is a combination of production activities in the two regions and transportation activities. Thus, let us first suppose that the following two production activities are used on the maximum balanced growth path:

$$\bar{\alpha}[(a_1^r, a_2^r), (0, 0)] \rightarrow \bar{\alpha}[(a_1^r + b_1^r, b_2^r), (0, 0)], (1 - \bar{\alpha})[(0, 0), (a_1^s, a_2^s)]$$
$$\rightarrow (1 - \bar{\alpha})[(0, 0), (a_1^s + b_1^s, b_2^s)], \quad 0 \leq \bar{\alpha} \leq 1, \tag{63}$$

where $[(a_1^l, a_2^l), (a_1^l + b_1^l, b_2^l)] \in S_l$, $l = r, s$, and $\bar{\alpha}$ is an intensity level of the normalized production activity in region r. Thus when the production activities in each region are combined, the production activity in the space system on that maximum balanced growth path is represented as follows:

$$[\bar{\alpha}(a_1^r, a_2^r), (1 - \bar{\alpha})(a_1^s, a_2^s)] \rightarrow [\bar{\alpha}(a_1^r + b_1^r, b_2^r), (1 - \bar{\alpha})(a_1^s + b_1^s, b_2^s)]. \tag{64}$$

Next, let us suppose, for example, that a part of the investment good produced in region r is transported to region s and a part of the material good produced in region s is transported to region r on this maximum balanced growth path. Thus the transportation activity on this maximum balanced growth path is represented as follows:

$$[\bar{\alpha}(a_1^r + b_1^r, b_2^r), (1 - \bar{\alpha})(a_1^s + b_1^s, b_2^s)]$$
$$\rightarrow [(\bar{\alpha}(a_1^r + b_1^r) - c_1^{rs}, \bar{\alpha}b_2^r + c_2^{sr} - v_1^{rs}c_1^{rs}),$$
$$((1 - \bar{\alpha})(a_1^s + b_1^s) + c_1^{rs}, (1 - \bar{\alpha})b_2^s - c_2^{sr})], \tag{65}$$

where c_1^{rs} is the amount of the investment good transported from r to s and c_2^{sr} is the amount of the material good transported from s to r on the maximum balanced growth path. We assume that $c_1^{rs}, c_2^{sr} = 0$, $\bar{\alpha}b_2^r + c_2^{sr} - v_1^{rs}c_1^{rs} \geq 0$ and $(1 - \bar{\alpha})b_2^s - c_2^{sr} \geq 0$. Then, by assumption, we must have

$$[(\bar{\alpha}(a_1^r + b_1^r) - c_1^{rs}, \bar{\alpha}b_2^r + c_2^{sr} - v_1^{rs}c_1^{rs}), ((1 - \bar{\alpha})(a_1^s + b_1^s)$$
$$+ c_1^{rs}, (1 - \bar{\alpha})b_2^s - c_2^{sr})] = \bar{\lambda}[\bar{\alpha}(a_1^r, a_2^r), (1 - \bar{\alpha})(a_1^s, a_2^s)], \tag{66}$$

where $\bar{\lambda}$ is the maximum balanced growth factor defined by (24).

Expressions (63) to (66) describe one possible case of the maximum balanced growth path. But it can be shown that any maximum balanced growth path in the space system defined by (4), (5') and (6) can be described by a set of activities similar to (63) to (66), with minor variations. For example, depending on the technology in the space system, transport inputs for the movement of investment goods may be supplied from both locations on a balanced growth path. For another example, the investment good may be transported from region s to region r. But we do not need two different sets of activities similar to (63) to (65) to describe the maximum balanced growth path, i.e., it does not happen that the maximum balanced growth path is obtained only as a convex combination of two different sets of activities such as (63) to (65). It is easily seen in the present model that if the maximum balanced growth path is obtained by a convex combination of two such sets of activities, then that maximum balanced growth path can be represented by a single set of activities similar to (63) to (65).

Thus, as a typical example, we assume that (63) to (65) describes the maximum balanced growth path in the present space system. Hence, if we set

$$\bar{X} = [\bar{\alpha}(a_1^r, a_2^r), (1 - \bar{\alpha})(a_1^s, a_2^s)], \tag{31'}$$

$\bar{X}$ is the maximum balanced growth vector in the present space system. Next, we explicitly assume that:

A5′. The maximum balanced growth vector is unique up to scalar multiplication and is described by (63) to (66), and $\bar{\alpha}b_1^r > c_1^{rs} > 0$, $c_2^{sr} \geqq 0$.

A6′. $0 < \bar{\alpha} < 1$.

A7′. $a_r \equiv (a_1^r, a_2^r) > 0$ and $a_s \equiv (a_1^s, a_2^s) > 0$.

A8′. $\bar{\lambda} > 1$.

A5′ to A8′ have the same interpretation as A5 to A8, respectively.

The maximum balanced growth path described by (63) to (66) may be depicted as in Figure 5.9. Since the material good can be moved freely between the regions, the specific location of material good is irrelevant and thus they can be added together as in this figure. The coordinates of each point in Figure 5.9 are given as follows:

$$a_r = [a_1^r, 0, a_2^r], \ a_s = [0, a_1^s, a_2^s],$$
$$a = \bar{\alpha}a_r + (1 - \bar{\alpha})a_s = [\bar{\alpha}a_1^r, (1 - \bar{\alpha})a_1^s, \bar{\alpha}a_2^r + (1 - \bar{\alpha})a_2^s],$$
$$b_r = [a_1^r + b_1^r, 0, b_2^r], \ b_s = [0, a_1^s + b_1^s, b_2^s],$$
$$e_1 = \bar{\alpha}b_r + (1 - \bar{\alpha})b_s$$
$$\quad = [\bar{\alpha}(a_1^r + b_1^r), (1 - \bar{\alpha})(a_1^s + b_1^s), \bar{\alpha}b_2^r + (1 - \bar{\alpha})b_2^s],$$

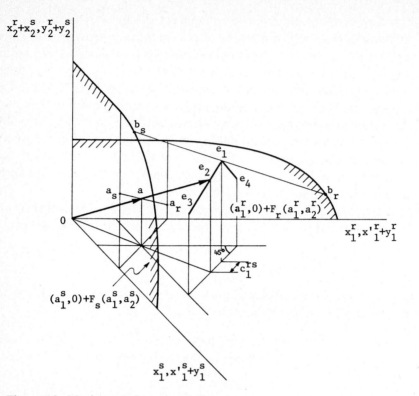

Figure 5.9. Maximum balanced growth vector.

$$e_2 = [\bar{\alpha}(a_1^r + b_1^r) - c_1^{rs}, (1 - \bar{\alpha})(a_1^s + b_1^s) + c_1^{rs},$$
$$\bar{\alpha}b_2^r + (1 - \bar{\alpha})b_2^s - v_1^{rs}c_1^{rs}] = \bar{\lambda}a,$$
$$e_3 = [\bar{\alpha}a_1^r, (1 - \bar{\alpha})(a_1^s + b_1^s) + \bar{\alpha}b_1^r, \bar{\alpha}b_2^r + (1 - \bar{\alpha})b_2^s - v_1^{rs}\bar{\alpha}b_1^r],$$
$$e_4 = [\bar{\alpha}(a_1^r + b_1^r) + (1 - \bar{\alpha})b_1^s, (1 - \bar{\alpha})a_1^s,$$
$$\bar{\alpha}b_2^r + (1 - \bar{\alpha})b_2^s - v_1^{sr}(1 - \bar{\alpha})b_1^s],$$

where the first, the second and the third coordinates represent, respectively, x_1^r (or, $x_1^{rr} + y_1^r$), x_1^s (or, $x_1^{rs} + y_1^s$) and $x_2^r + x_2^s$ (or, $y_2^r + y_2^s$).

Point a_l ($l = r, s$) represents the basic production input vector in each region used on the maximum balanced growth path. Point b_l ($l = r, s$) is one of output points obtainable from input a_l and is chosen to lie on the maximum balanced growth path. And, as described by (64), point $\bar{\alpha}b_r + (1 - \bar{\alpha})b_s$ is obtained from point $\bar{\alpha}a_r + (1 - \bar{\alpha})a_s$, that is, point e_1 is obtained from point a as the result of the production activity on the maximum balanced growth path. Thus the amount, c_1^{rs}, of the investment good is transported from r to s by consuming the amount, $v_1^{rs}c_1^{rs}$, of the material good, that is, point e_2 is obtained from e_1 as the result of the transportation activity on the maximum balanced growth path. By assumption, point e_2 must be on the ray Oa and satisfy $e_2 = \bar{\lambda}a$, as is depicted in the figure. Point e_3 is obtained when all the newly produced investment good in region r is transported to region s; and e_4 denotes the opposite case. Line segments $\overline{e_1 e_3}$ and $\overline{e_1 e_4}$ constitute the efficient transport output set obtained from the transport input point e_1.

5.3.5. Brief discussion of turnpike behavior and the spatial price relation

As shown by previous analyses, there are few essential differences between cases in which transport inputs are required and those in which they are not. It is also true that essential results in the analyses of the turnpike behavior of the optimal path and the spatial price relations are independent of transport inputs. Thus let us summarize them briefly.

First, the von Neumann price vectors in the present problem are also defined by (36). And when $\bar{P} = [(\bar{p}_1^r, \bar{p}_2^r), (\bar{p}_1^s, \bar{p}_2^s)]$ is a von Neumann price vector, it can be shown under assumptions A5' to A7' that

$$\bar{p}_1^r = \bar{p}_1^s - v_{rs}\bar{p}_2, \bar{p}_2^r = \bar{p}_2^s (= \bar{p}_2). \tag{37'}$$

This is easily seen in Figure 5.9 from the fact that the normal vector of any hyperplane which sustains point e_2 should pass through points e_1 and e_3. Such a price vector may not be unique, and the set of von Neumann price vectors is in general a closed convex cone in $\mathbf{R}_+^{*4}$. It can be shown using A5' that one of the von Neumann price vectors is obtained as the normal vector of the hyperplane passing through points b_r, b_s and e_3 in Figure 5.9. That is, when the normal vector of that hyperplane is $(p_1^* - v_{rs}p_2^*, p_1^*, p_2^*)$, the vector

$$P^* = [(p_1^* - v_{rs}p_2^*, p_2^*), (p_1^*, p_2^*)]$$

is a von Neumann price vector. Hereafter we use P^* as the von Neumann price vector, and we assume that:

A11. $p_1^* > 0, p_2^* > 0, p_1^* - v_{rs}p_1^* > 0.$

As seen from Figure 5.9, not only e_1 does break even under P^* but also any convex combination of points b_r, b_s and e_3 breaks even under P^*. That is, any activity in the set

$$
\begin{aligned}
K^* = \{(X_t, X_{t+1}) | X_t &= \beta [\alpha(a_1^r, a_2^r), (1 - \bar{\alpha})(a_1^s, a_2^s)], \\
X_{t+1} &= \beta [(\alpha(a_1^r + b_1^r) - y_1^{rs}, \alpha b_2^r + y_2^{sr} - v_1^{rs} y_1^{rs}), \\
&\quad ((1 - \alpha)(a_1^s + b_1^s) + y_1^{rs}, (1 - \alpha)b_2^s - y_2^{sr})], \\
1 &\geqq \alpha \geqq 0, \beta \geqq 0, \alpha b_1^r \geqq y_1^{rs} \geqq 0, \\
\alpha b_2^r &+ y_2^{sr} \geqq v_1^{rs} y_1^{rs}, (1 - \alpha)b_2^s \geqq y_2^{sr}\},
\end{aligned} \tag{43'}
$$

breaks even under P^*. Moreover, it can be shown that only the activities in K^* break even under P^*. As before, we call K^* the von Neumann facet and retain assumption A9. Then, by applying Theorem 1 in McKenzie (1963), we obtain the following:

Lemma 5.3.   Under assumptions A1 to A4, A5' to A8' and A9 to A11, let $\{\hat{X}_t\}_0^N$ be an optimal path for Problem II(a). Then for any $\epsilon > 0$, there is a number N_ϵ such that the number of periods in which $d[(\hat{X}_t, \hat{X}_{t+1}), K^*] \geqq \epsilon$ cannot exceed N_ϵ.

Thus, when the plan period is sufficiently long, the optimal path should stay most of the time in a small neighborhood of the von Neumann facet K^*.[21] And, as before, by examining the movement of a feasible path on this set, it can be shown that:

Theorem 5.3 (turnpike theorem).   Suppose we have assumptions A1 to A4, A5' to A8' and A9 to A11. Then, if the plan period is sufficiently long, the optimal path $\{\hat{X}_t\}_0^N$ for Problem II(a) stays most of the time in a very small neighborhood of the von Neumann facet K^* defined by (43'). And except for the oscillation case, it spends most of its time in a very small neighborhood of the maximum balanced growth ray generated by (31').

Then, as in Section 5.2.7, by examining the character of the dual optimal path which enables the commodity path to move most of the

[21]By the condition that the optimal path stays on the facet K^* in successive periods, this set is further restricted to a subset of K^* which may be called the restricted von Neumann facet (refer to footnote 14). It can be shown that the restricted von Neumann facet is a closed convex cone in $\mathbf{R}_+^{4 \times 2}$ including the ray generated by the maximum balanced growth input–output pair $(\bar{X}, \bar{\lambda}\bar{X})$ in its relative interior under these assumptions. Thus, the optimal path spends most of its time in a small neighborhood of this set.

time in a small neighborhood of the (restricted) von Neumann facet and the maximum balanced growth ray, the following theorem can be derived.

Theorem 5.4. Suppose we have A1 to A4, A5′ to A8′ and A9 to A11. Then, if the plan period is sufficiently long, on the optimal path for II(b) (except possibly for the oscillation case), the price of the stock of each good should satisfy the following relations for most of the time:

$$\hat{p}_2^r(t) = \hat{p}_2^s(t)(= \hat{p}(t)), \hat{p}_1^r(t) = \hat{p}_1^s(t) - v_{rs}\hat{p}(t). \tag{67}$$

Considering these statements together with A5′, it is easily seen that each newly produced commodity is allocated between the regions in a "balanced manner" during most of the plan period.

Under the special assumption A5′, the latter condition in (67) (that is, (60)) defines the spatial price relation of capital which must be satisfied most of the time by the optimal price path. But, in general, one of the relations (60) to (62) defines a possible set of spatial price relations, depending on the technology in the system. Relation (61) is simply the opposite case of (60). In case (62), each region is entirely self-sufficient with respect to the investment good. This situation can occur when the amount of transport inputs necessary for the movement of the investment goods is relatively large.

In Section 4.4.3, it was pointed out that though the dual variables corresponding to immobile commodities can theoretically assume any values (depending on the initial conditions and the objective function) when the plan period is short, there will be some kind of "balanced" spatial price relation of immobile commodities if the plan period is sufficiently long. The two examples in this chapter reinforce the validity of this conjecture.

PART III

DEVELOPMENT IN SPACE SYSTEMS WITH NON-CONVEX STRUCTURES

Chapter 6

REGIONAL ALLOCATION OF INVESTMENT UNDER CONDITIONS OF VARIABLE RETURNS TO SCALE

6.1. Introduction

Since the analysis in Part II was limited to monotone space systems, it is of interest to explore the restrictions due to the convexity assumption inherent in monotone space systems. In this chapter (Part III), we attempt to examine these restrictions by employing a specific model. We develop a generalization of the Rahman model by introducing scale economies and diseconomies into the production process and, within the context of this model, examine the optimum growth implications of the following alternative assumptions of production: decreasing, constant, increasing and variable returns to scale.

In Section 6.2, we formulate an aggregate problem of the optimal allocation of investment among regions. This problem is obtained by introducing *scale economies and diseconomies* in the production process into the original Rahman model.[1] But, unlike all the previous studies, we assume that all the regions in the economy have the same

[1]An aggregate investment allocation model was first analyzed by Rahman (1963). The problem examined was how to allocate the total investment funds, which are the sum of savings in two regions, between two regions at each time so as to maximize the total income in the final period. This problem was later reformulated in a continuous-time framework by Intriligator (1964), Rahman (1966) and Takayama (1967), and generalized by Sakashita (1967) and Ohtsuki (1971). The central concern in all of these studies has been to analyze how the *regional differences* in technology and/or propensity to save would affect the optimal growth pattern of the spatial economy. To examine this, they employed constant-coefficient or constant-returns-to-scale production functions, and hence the problem of scale economies and diseconomies could not be investigated in these studies.
Note that the terms, scale economies and agglomeration economies, are used interchangeably in this chapter.

production function. Hence all the regions have exactly the same economic characteristics except for development stages (which are represented by the amount of capital stock in each region). This enables us to concentrate on the pure effects that scale economies and diseconomies have on growth processes of the spatial economy.

In Section 6.3, we summarize the optimality conditions for the problem obtained in the previous section. Then, in Section 6.4, we study the optimal growth path in the two-region economy. First, we investigate the characteristics of the optimal growth path under the conditions of variable returns to scale. The main concern here is to investigate the nature of the *switching function* which determines the time for switching investment from the more developed region to the less developed region. Then, by applying the results of this analysis, the optimal growth path will be obtained for each of the other cases: decreasing returns to scale, constant returns to scale, and increasing returns to scale.

In Section 6.5, we generalize the results from Section 6.4 to the case of the n-region economy.

Finally, to make clear the meanings of optimal growth paths obtained in the previous sections, in Section 6.6, we compare them with other types of growth paths. We first describe the competitive economy in disaggregated terms. And then, we compare the three kinds of growth paths obtained under the three allocation rules: the *Optimum rule*, the *Private Marginal Productivity* (*PMP*) *rule*, and the *Social Marginal Productivity* (*SMP*) *rule*. It is shown that the PMP rule and the SMP rule tend to concentrate capital in the more developed region in excess of the optimal amount when the conditions of variable returns to scale prevail in production. This will not take place in the other three cases.

In this chapter, we only study the characteristics of optimal growth paths for final state problems. However, it is shown in Appendix F that the main results obtained in this chapter also hold to be true for consumption stream problems.

The problem investigated in this chapter is quite simple in its formulation, and the continuous-time maximum principle is found to be more convenient for use than the discrete-time one. Therefore, unlike in the preceeding chapters, we adopt here the assumption of continuous time.

For simplicity, only intuitive explanations of Lemmas and Theorems are given in the text, and most of the proofs are given in Appendix E.

6.2. Problem formulation

Suppose we have a closed economy consisting of n regions. The relation between the amount of income, $Y_l(t)$, produced in region l at time t and the amount of capital stock, $K_l(t)$, in that region at that time is described by

$$Y_l(t) = F(K_l(t)), \quad l = 1, 2, \ldots, n. \tag{1}$$

Let us call the function, F, the *regional production function*[2] and assume this function is the same for every region in the economy. We also assume that the outputs $Y_l(t)$, which are aggregatively called *income*, can be moved among the regions without transport costs. But once they are invested in one region, they become a part of the capital stock in that region which cannot be moved from that region to any other region. That is, we assume that capital stock is interregionally immobile.

Next, suppose that the saving ratio, s, for the whole economy is always positive and, for simplicity of notation, is constant over time.[3] Then, the total available investment fund of the economy at time t will be equal to $s \sum_{l=1}^{n} Y_l(t)$.

Therefore, if $\theta_l(t)$, termed the "investment ratio," is the proportion of investment allocated to region l at time t, we have

$$\dot{K}_l(t) = \theta_l(t) s \sum_{j=1}^{n} Y_j(t) \quad \text{for } l = 1, 2, \ldots, n,$$

where $\dot{K}_l(t) = dK_l(t)/dt$. Here we neglect, for simplicity, the depreciation of capital. The role of the central authority is to control the investment ratios, $\theta_l(t)$, $l = 1, 2, \ldots, n$, at each time so as to achieve its planning objective. We here assume that its objective is to maximize the total income, $\sum_{l=1}^{n} Y_l(t) = \sum_{l=1}^{n} F(K_l(t))$, at the end of the plan period, $t = T$.[4] Hence the problem is summarized as follows.

---

[2] See Section 6.6 for the derivation of this aggregated form of regional production function from the individual production functions $Y_i = g(K_i, L_i, \bar{D}_i) f(K_i, L_i, D_i)$.

[3] We see later that there is no essential difference in characteristics of the solution for our problem regardless of whether s is constant or not over time.

That is, the reader can use $s(t)$ instead of s in the following analysis and see that all the results hold to be true when s is replaced by $s(t)$.

It is not necessary to consider that ratio s reflects the "natural propensity to save" in the economy. s can be the target saving ratio planned by the central authority, and this level can be enforced, for example, by taxation. We briefly analyze, in Appendix F, the case where the saving ratio is also a control variable in the problem.

[4] For analyses under alternative objective functions, see Appendix F.

Problem A. Choose values of investment ratios $\theta_l(t)$, $l = 1, 2, \ldots, n$, at each time so as to maximize the total income in the final period,

$$\sum_{l=1}^{n} F(K_l(T)), \tag{2}$$

subject to

$$\dot{K}_l(t) = \theta_l(t) \, s \sum_{j=1}^{n} F(K_j(t)), \quad l = 1, 2, \ldots, n, \tag{3}$$

$$\sum_{l=1}^{n} \theta_l(t) = 1, \, \theta_l(t) \geqq 0, \quad l = 1, 2, \ldots, n, \, 0 \leqq t \leqq T, \quad \text{and} \tag{4}$$

$$K_l(0) = K_l^0, \quad l = 1, 2, \ldots, n, \tag{5}$$

where K_l^0 is the amount of capital in region l at the initial time, $t = 0$, and it is assumed that $\Sigma_{l=1}^{n} K_l^0 > 0$ and $s > 0$.

In the following sections we study the solution of Problem A under each of the following alternative assumptions on the form of regional production function F:

(i) decreasing returns to scale, namely,

$\quad$ $F'(K) > 0$ and $F''(K) < 0$ for all $K \geqq 0$. $\hfill$ (6)

(ii) constant returns to scale, namely,

$\quad$ $F(K) = aK$ for all K where a is a positive constant. $\hfill$ (7)

(iii) increasing returns to scale, namely,

$\quad$ $F'(K) > 0$ and $F''(K) > 0$ for all $K \geqq 0$. $\hfill$ (8)

(iv) variable returns to scale, namely,

$\quad$ $F'(K) > 0$ for all $K \geqq 0$, and $F''(K) > 0$ for $0 \leqq K < K^*$,

$\quad$ $F''(K) = 0$ for $K = K^*$, $F''(K) < 0$ for $K > K^*$. $\hfill$ (9)

Here $F'(K) = dF(K)/dK$ and $F''(K) = dF'(K)/dK$. In addition, we assume in all cases that

$$F(0) \geqq 0, \tag{10}$$

and $F(K)$ is twice continuously differentiable at each $K > 0$.

6.3. Optimality conditions

In this section, we obtain the optimality conditions for Problem A, and conduct some preliminary analyses. For this purpose, first we introduce auxiliary variables p_l, $l = 1, 2, \ldots, n$, and define the Hamiltonian function H by

$$H((p_l)_1^n, (K_l)_1^n, (\theta_l)_1^n) \equiv \sum_{l=1}^n p_l \dot{K}_l = \left(\sum_{l=1}^n \theta_l p_l \right) \left(s \sum_{l=1}^n F(K_l) \right). \tag{11}$$

Then, by applying the maximum principle from optimal control theory, we have

Optimality condition A.[5] Suppose that $\{(K_l(t))_1^n\}_0^T$ and $\{(\theta_l(t))_1^n\}_0^T$ are, respectively, an optimal growth path and an optimal allocation path for Problem A. Then there exits a "price path" $\{(p_l(t))_1^n\}_0^T$ which satisfies the next three conditions.

(a) $H((p_l(t))_1^n, (K_l(t))_1^n, (\theta_l(t))_1^n) = \max \Big\{ H((p_l(t))_1^n, (K_l(t))_1^n,$

$$(\theta_l)_1^n) \Big| \sum_{l=1}^n \theta_l = 1, \theta_l \geq 0, l = 1, 2, \ldots, n \Big\} \quad \text{for } 0 \leq t \leq T, \tag{12}$$

(b) $\dot{p}_l(t) = - \partial H((p_l(t))_1^n, (K_l(t))_1^n, (\theta_l(t))_1^n) / \partial K_l$

$$\text{for } l = 1, 2, \ldots, n, \text{ and } 0 \leq t \leq T, \tag{13}$$

(c) $p_l(T) = F'(K_l(T)) \quad \text{for } l = 1, 2, \ldots, n,$ (14)

where $\dot{p}_l(t) = dp(t)/dt$.

From programming theory we know that the auxiliary variable, $p_l(t)$, represents the incremental amount of the final income which is obtained by exogenously increasing the stock of capital in region l by one unit at time t on the optimal path.[6] Namely, $p_l(t)$ represents the accounting price (shadow price) of a unit of capital in region l at time t in terms of the final income. We simply call $p_l(t)$ the *price of capital* in region l at time t.

We define $p(t)$ by

$$p(t) = \max_l p_l(t) \tag{15}$$

---

[5] This is the continuous-time version of Theorem 2.8. For the details of the maximum principle in continuous time, see, for example, Pontryagin et al. (1962). For a convenient summary of it, see, for example, Arrow and Kurz (1970) or Takayama (1974).

[6] Refer to Sections 2.3.6 and 4.4. There is no difference in the economic meanings of $p_l(t)$ whether we use the continuous representation or the discrete representation of time. For the case of the continuous representation of time, see, for example, Arrow and Kurz (1970).

for each t, $0 \leqq t \leqq T$. Then, according to (12), the *optimal rule* for investment allocation is given by

$$\sum_{l \in \mathcal{L}(t)} \theta_l(t) = 1 \quad \text{where } \mathcal{L}(t) = \{l | p_l(t) = p(t)\}. \tag{16}$$

That is, at each time, all of the investment fund of the economy should be allocated among regions which have the highest price of capital at that time. Since we call $p_l(t)$ the price of capital in region l, it is appropriate to term $p(t)$ the *price of investment* at time t.[7]

Next, from (13), we have

$$\dot{p}_l(t) = -\left(\sum_{j=1}^{n} \theta_j(t) p_j(t)\right) s F'(K_l(t)), \quad l = 1, 2, \ldots, n, \tag{17}$$

at each t, $0 \leqq t \leqq T$, where $F'(K_l(t)) = dF(K_l(t))/dK_l$. Thus, from (15) to (17),

$$\dot{p}_l(t) = -p(t) s F'(K_l(t)), \quad l = 1, 2, \ldots, n. \tag{18}$$

Therefore, using (14) and (18), we observe that

Lemma 6.1. In any of the cases (i) to (iv), $p_l(t) > 0$ and $\dot{p}_l(t) < 0$ for each l, $l = 1, 2, \ldots, n$, and for all $t \geqq 0$.

Since the marginal productivity of capital, $F'(K)$, is assumed to be positive at any $K \geqq 0$, the price of capital, $p_l(t)$, should naturally be positive in any region at any time. Then, from (18), $\dot{p}_l(t)$ should be negative.

6.4. Optimal growth in the case of two regions ($n = 2$)

In this section, we examine the characteristics of the optimal growth path for Problem A assuming that the economy consists of two

---

[7]Recall that we have assumed that the capital stock cannot be moved from one region to another region. Hence, the "price of capital" $p_l(t)$ may differ among regions. On the other hand, we have assumed that the investment good (i.e., the part of outputs which are saved) is free to move among regions, and hence, the investment good should have the same price over all of the regions. Further, since Equation (3) says that one unit of investment to a region increases one unit of capital in that region, the "price of investment" should be equal to the price of capital in the regions to which the investment good is allocated. Of course, on the optimal path, all of the investment good should be allocated to the regions which have the highest price of capital among regions as expressed by (16).

regions, $l = r, s$. In this two-region case, Equations (3), (16) and (18) are, respectively, rewritten as follows:

$$\dot{K}_l(t) = \theta_l(t)s(F(K_r(t)) + F(K_s(t))), \quad l = r, s. \tag{19}$$

$$\left.\begin{array}{l} \text{if } p_r(t) > p_s(t), \text{ then } \theta_r(t) = 1 \text{ and } \theta_s(t) = 0, \\ \text{if } p_r(t) < p_s(t), \text{ then } \theta_r(t) = 0 \text{ and } \theta_s(t) = 1, \\ \text{if } p_r(t) = p_s(t), \text{ then } \theta_l(t) \in [0, 1], \, l = r, \, s, \text{ and} \\ \qquad \theta_r(t) + \theta_s(t) = 1. \end{array}\right\} \tag{20}$$

$$\dot{p}_l(t) = - \max(p_r(t), p_s(t))sF'(K_l(t)), \quad l = r, s. \tag{21}$$

We first examine variable returns to scale since the rest of the cases can be considered as its special cases. Under assumption (9), the production function, $F(K)$, and the marginal productivity curve, $F'(K)$, can be depicted, respectively, as in Figure 6.1(a) and (b). We say that a region is in the *increasing phase* or in the *decreasing phase* depending on whether the amount of its capital is less than K^* or more than K^*, where K^* is the inflection point defined in (9).

When both regions are in the decreasing phase, the marginal productivity of capital is always higher in the *less developed region* (i.e., in the region with the smaller amount of capital stock); moreover, the marginal productivity continues to decrease in both regions as the capital increases in the future. Hence, the price of capital should be higher in the less developed region. That is,

Lemma 6.2. In the case of variable returns to scale,

(a) if $K_r(t) > K_s(t) \geqq K^*$, then $p_r(t) < p_s(t)$,
(b) if $K_r(t) = K_s(t) \geqq K^*$, then $p_r(t) = p_s(t)$,

where K^* is the inflection point defined in (9).

For the proof, see Appendix E. From this lemma, we immediately have

Lemma 6.3. In the case of variable returns to scale,

(a) if $K_r(t) > K_s(t) \geqq K^*$, then $\theta_r(t) = 0$ and $\theta_s(t) = 1$,
(b) if $K_r(t) = K_s(t) \geqq K^*$, then $\theta_r(t) = \theta_s(t) = \frac{1}{2}$.

Namely, when both regions are in the decreasing phase, all of the investment should be allocated to the less developed region until both

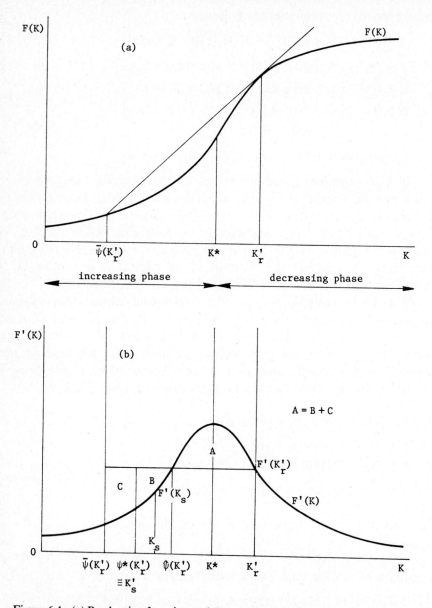

Figure 6.1. (a) Production function and (b) marginal productivity curve for the case of variable returns to scale.

regions come to have the same amount of capital. After that, the investment should be equally split between the two regions.

On the other hand, when both regions stay in the increasing phase, the marginal productivity of capital is greater in the *more developed region* (i.e., the region with the larger amount of capital stock). Hence the price of capital should be higher in the more developed region. That is,

Lemma 6.4.   In the case of variable returns to scale,

(a) if $K^* > K_r(t) > K_s(t)$, then $p_r(t) > p_s(t)$,
(b) if $K^* > K_r(t) = K_s(t)$, then $p_r(t) > p_s(t)$ or $p_r(t) < p_s(t)$, and the choice of one relation among the two is arbitrary.

The proof of this property is not simple, and see Appendix E for it. From this property we have

Lemma 6.5.   In the case of variable returns to scale,

(a) if $K^* > K_r^0 > K_s^0$, the whole investment should be allocated to region r (i.e., to the more developed region) at least until it goes out of the increasing phase.
(b) if $K^* > K_r^0 = K_s^0$, we must arbitrarily choose one of the two regions at the initial time, and all of the investment should be allocated to that region at least until it goes out of the increasing phase.

This property implies that when both regions are in the increasing phase at the outset, one region must stay at the initial point at least until the other region goes out of its increasing phase. Part (b) says that when both regions have the same amount of capital at the initial time, we must choose one of them as our favourite region. To enable us to make this "asymmetric" choice, initial values of capital prices should have the "asymmetric" relation as described in Lemma 6.4(b). In a sense, Lemma 6.4(b) is a surprising property, since it says that capital prices should have different values in the two regions even though the two regions are indistinguishable in terms of economic stage. But this is a reasonable result if we consider that the price of capital reflects the contributions of a marginal unit of capital for the rest of the plan period. That is, the region which we choose as our favourite region should have a *much* higher capital price than the other region since the marginal productivity of capital continuously increases in the former region and it stays the same in the latter region for a long period in the future.

Let us now turn to the investigation of relations between the capital prices when the two regions locate on opposite sides of the inflection point K^*. For this purpose we define function $\hat{\psi}(K)$ by

$$F'(\hat{\psi}(K)) = F'(K) \quad \text{and} \quad K \geq K^* \geq \hat{\psi}(K). \tag{22}$$

The meaning of function $\hat{\psi}(K)$ is clear from Figure 6.1(b). From the assumption of variable returns to scale, (9), we have

$$\mathrm{d}\hat{\psi}(K)/\mathrm{d}K < 0 \quad \text{where} \quad K > K^*. \tag{23}$$

Suppose first we have $K_r(t) > K^*$ and $K_r(t) > K_s(t) \geq \hat{\psi}(K_r(t))$. Then, the marginal productivity of capital is always greater in region s until it catches up to region r. Even if $K_s(t) = \hat{\psi}(K_r(t))$, that is, even if both regions have the same marginal productivity of capital at present (i.e., at time t), the marginal productivity of capital is going to increase in region s and it is going to decrease in region r. Hence we should have

Lemma 6.6 In the case of variable returns to scale,

(a) if $K_r(t) > K^*$ and $K_r(t) > K_s(t) > \hat{\psi}(K_r(t))$, then $p_r(t) < p_s(t)$,
(b) if $K_r(t) > K^*$ and $K_r(t) > K_s(t) = \hat{\psi}(K_r(t))$, then $p_r(t) < p_s(t)$ when $0 \leq t < T$, and $p_r(t) = p_s(t)$ when $t = T$.

For the proof, see Appendix E.

Hence, when $K_r(t) > K^*$ and $K_r(t) > K_s(t) > \hat{\psi}(K_r(t))$, all of the investment should be allocated to region s until it catches up to region r. Even if $K_s(t) = \hat{\psi}(K_r(t))$, that is, even if the marginal productivity of capital is the same in the two regions at present, we "prefer" region s (i.e., $p_r(t) < p_s(t)$ and hence $\theta_s(t) = 1$) because of its future. Further, (b) implies that when the rest of planning period, $T - t$, is long enough, $p_r(t)$ is possibly equal to $p_s(t)$ only when $K_s(t)$ is sufficiently smaller than $\hat{\psi}(K_r(t))$. That is, even if the marginal productivity of capital in region s is slightly less than that in region r at the present time, we should prefer region s because of its future.

Then a question of interest is: what is the lower limit of such a range? That is, given $K_r(t)$ $(K_r(t) \geq K^*)$, what is the value of $K_s(t)$ under which $p_r(t)$ becomes equal to $p_s(t)$? We denote such a value of

$K_s(t)$ by $\psi^*(K_r(t))$. Namely, function ψ^* is defined by the condition:

$$p_r(t) = p_s(t) \quad \text{when } K_s(t) = \psi^*(K_r(t)), \text{ where } K_r(t) \geqq K^* \geqq K_s(t).$$
(24)

We call function ψ^* the *switching function* since, as we see later, it represents when the allocation of investment should be switched from the more developed region to the less developed region.

To obtain function ψ^*, we first suppose that given a time, $t = t'$, the rest of plan period $T - t'$ is sufficiently long (the minimum length of $T - t'$ under which the following analysis is true is obtained later). And suppose we have that

at $t = t'$, $K_r(t') = K_r'$, $K_s(t') = K_s'$ and $p_r(t') = p_s(t')$
where $K_r' \geqq K^* \geqq K_s'$.
(25)

Then, from Lemma 6.6, we have

$$K_s' < \hat{\psi}(K_r') \quad \text{and hence} \quad F'(K_r') > F'(K_s').$$
(26)

In addition, from (21)

$$\dot{p}_r(t) - \dot{p}_s(t) = - \max(p_r(t), p_s(t)) s (F'(K_r(t)) - F'(K_s(t))).$$
(27)

Hence $\dot{p}_r(t) < \dot{p}_s(t)$ for $t \in [t', t' + \epsilon)$ for some $\epsilon > 0$. Then, considering (25), $p_r(t) < p_s(t)$ for $t \in [t', t' + \epsilon)$ for some $\epsilon > 0$. Therefore, all investment should be allocated to region s at least for a while. But as long as $K_s(t) < \psi(K_r')$, we have $\dot{p}_r(t) < \dot{p}_s(t)$ and hence $p_r(t) < p_s(t)$. In addition, from Lemma 6.6, $p_r(t) < p_s(t)$ as long as $K_r' > K_s(t) \geqq \hat{\psi}(K_r')$. Hence all of the investment should be allocated to region s until it catches up to region r.

That is, when $T - t'$ is sufficiently large, there is time t'' ($t'' \leqq T$) such that

$$p_r(t) < p_s(t) \quad \text{for } t \in (t', t''), \text{ and}$$
(28)
$$p_r(t'') = p_s(t'') \text{ and } K_r(t'') = K_s(t'') = K_r',$$
(29)

and

$$\left. \begin{array}{l} \dot{K}_r(t) = 0 \\ \dot{K}_s(t) = s(F(K_r') + F(K_s(t))) \end{array} \right\} \quad \text{for } t \in [t', t''),$$
(30)

$$\left. \begin{array}{l} \dot{p}_r(t) = - p_s(t) s F'(K_r') \\ \dot{p}_s(t) = - p_s(t) s F'(K_s(t)) \end{array} \right\} \quad \text{for } t \in [t', t'').$$
(31)

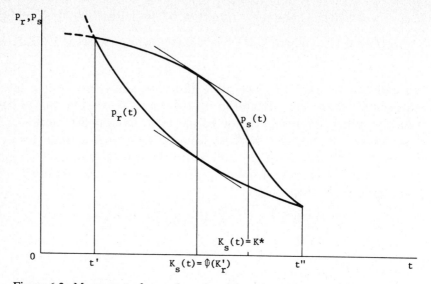

Figure 6.2. Movement of p_r and p_s along Equations (30) and (31).

The movement of $p_r(t)$ and $p_s(t)$ along Equations (30) and (31) is depicted in Figure 6.2.

Next, from (30) and (31), we have

$$\left.\begin{aligned}\frac{dp_r}{dK_s} &= \frac{-p_s F'(K'_r)}{F(K'_r)+F(K_s)} \\ \frac{dp_s}{dK_s} &= \frac{-p_s F'(K_s)}{F(K'_r)+F(K_s)}\end{aligned}\right\} \quad \text{for } K_s \in [K'_s, K'_r). \tag{32}$$

Denote by $p_r(K_s; K'_r)$ and $p_s(K_s; K'_r)$ the solution of (32) under the initial condition (25), where $K'_s \leqq K_s \leqq K'_r$. Then, from (25), (28) and (29),

$$\left.\begin{aligned}p_r(K_s = K'_s; K'_r) &= p_s(K_s = K'_s; K'_r), \\ p_r(K_s = K'_r; K'_r) &= p_s(K_s = K'_r; K'_r),\end{aligned}\right\} \tag{33}$$

$$p_r(K_s; K'_r) < p_s(K_s; K'_r) \text{ for } K_s \in (K'_s, K'_r). \tag{34}$$

Figure 6.3 depicts the curve, $p_s(K_s; K'_r) - p_r(K_s; K'_r)$, for $K'_s \leqq K_s \leqq K'_r$.

Hence, from (32) and (33), we get

$$\int_{K'_s}^{K'_r} \frac{F'(K_s) - F'(K'_r)}{F(K'_r)+F(K_s)} p_s(K_s; K'_r) \, dK_s = 0, \tag{35}$$

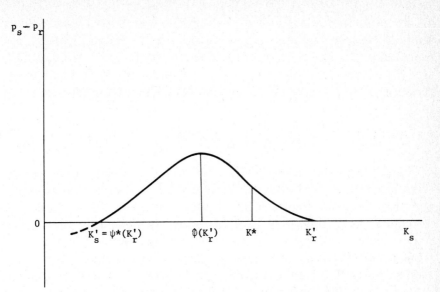

Figure 6.3. Change of $(p_s(K_s; K'_r) - p_r(K_s; K'_r))$ along Equation (32).

where, from the second equation of (32),

$$p_s(K_s; K'_r) = p_s(K'_r; K'_r) \exp\left(\int_{K_s}^{K'_r} \frac{F'(K)}{F(K'_r) + F(K)} \, dK \right)$$

$$= p_s(K'_r; K'_r) \frac{2F(K'_r)}{F(K'_r) + F(K_s)}. \tag{36}$$

Finally, taking into account the facts $p_s(K'_r; K'_r) = p_s(t'') > 0$ (from Lemma 6.1) and $F(K'_r) > 0$ (from (9) and (10)), (35) and (36) are summarized as follows:

$$\int_{K'_s}^{K'_r} \frac{F'(K_s) - F'(K'_r)}{(F(K'_r) + F(K_s))^2} \, dK_s = 0. \tag{37}$$

Solving the above equation for K'_s under each value of $K'_r (K'_r \geqq K^*)$, we get a function, $K'_s = \psi^*(K'_r)$, which describes the switching function defined by (24) under the condition that the rest of plan period $T - t'$ is sufficiently long. Here, t' is defined by (25) and it is the time when $(K_r(t), K_s(t))$ reaches (i.e., starts from) a point on the switching line $K'_s = \psi^*(K'_r)$.

The minimum length of $T - t'$ under which the above analysis holds to be true is equal to $t'' - t'$, where t'' is defined by (29), and it represents the time period required for $K_s(t)$ to move from $\psi^*(K'_r)$ to K'_r. Since the length, $t'' - t'$, is a function of K'_r (equivalently, a function of K'_s), we denote it by $T_{\min}(K'_r)$. Then, from (30) we get

$$T_{\min}(K'_r) = \int\limits_{\psi^*(K'_r)}^{K'_r} \frac{1}{s(F(K'_r) + F(K_s))} \, dK_s. \tag{38}$$

Though we have analytically obtained the equation of switching function (37), it is not difficult to obtain this equation from an intuitive argument by using Figure 6.1(b). Suppose the amount of capital stock in region r at a time, say $t = t'$, is given by the point, K'_r, in Figure 6.1(b). Then, if $K_s(t')$ locates at a point between $\hat{\psi}(K'_r)$ and K'_r, the marginal productivity of capital is always higher in region s until K_s catches up to K'_r. Hence, recalling that $p_l(t)$, $l = r$ or s, measures the value of the total contribution of an additional unit of capital in region l from time t to T, to the final income in the economy, it should be true that $p_s(t') > p_r(t')$. Even if $K_s(t')$ is slightly less than $\hat{\psi}(K'_r)$, it should be true that $p_s(t') > p_r(t')$ since the marginal productivity of capital in region s can soon exceed that of region r. Hence, for it to be true that $p_s(t') = p_r(t')$, $K_s(t')$ must be much less than $\hat{\psi}(K'_r)$; and we denote this $K_s(t')$ by $K'_s (\equiv \psi^*(K'_r))$ in Figure 6.1(b).

Next suppose that, at a time t $(t \geq t')$, region r locates at K'_r (in terms of the amount of capital) in Figure 6.1(b), and region s at K_s where $K'_s < K_s < K'_r$. Then the value of the contribution of one unit of capital in region l per unit of time at time t (to the increase of the final income of the economy) is given by

$p_s sF'(K_l) =$ (the value of one unit of the investment fund in the economy) $\times$ (the saving ratio of the economy) $\times$ (the marginal product of one unit of capital in region l),

where $l = r$, s. Note, in the above calculation, the value of one unit of the investment fund in the economy should be measured by the price of capital in region s at that time since it is higher in region s as long as $K'_s < K_s < K'_r$. Hence, the difference of the contributions of one unit of capital in the two regions per unit of time is given by

$p_s sF'(K_s) - p_s sF'(K'_r)$.

Hence, for the time interval dt, the difference is

$[p_s sF'(K_s) - p_s sF'(K'_r)] \, dt$.

Then, since all investment is allocated to region s when $K_s' < K_s < K_r'$, we have $dK_s/dt = s(F(K_r') + F(K_s))$, and hence $dt = dK_s/s(F'(K_r') + F'(K_s))$. Therefore, the above difference can be rewritten as

$$\frac{F'(K_s) - F'(K_r')}{F'(K_r') + F'(K_s)} p_s \, dK_s.$$

This difference is negative when $K_s' \le K_s < \hat{\psi}(K_r')$, and positive when $\hat{\psi}(K_r') < K_s < K_r'$. Hence, for "the losses" and "the gains" to cancel out each other while K_s moves from K_s' to K_r', the next relation should hold to be true.

$$\int_{K_s'}^{K_r'} \frac{F'(K_s) - F'(K_r')}{F'(K_r') + F'(K_s)} p_s(K_s; K_r') \, dK_s = 0,$$

which is the same with (35). And, from this equation we can obtain the equation of switching function (37) as before.

Next, if we extend the solution of (30) and (31) to the interval $[0, t')$, we get $p_r(t) > p_s(t)$ for $t \in [0, t')$ as shown in Figure 6.2. Hence, taking (34) into account, the solution of (32) can be depicted as in Figure 6.3. Hence, for each value of K_r' ($K_r' \ge K^*$), Equation (37) determines a unique value (if it exists) for K_s' ($K_s' \le K^*$). That is, Equation (37) uniquely defines switching function $K_s' = \psi^*(K_r')$.

Finally, observe that

$$\begin{aligned} &\text{if } p_t(t') = p_s(t') \text{ and } K_r(t') > K^* > K_s(t') \text{ at a time } t', \\ &\text{then } p_r(t) > p_s(t) \text{ for all } t \in [0, t'). \end{aligned} \tag{39}$$

This is not difficult to see by repeating the same procedures given just after Equation (26). In this case, we must take time backward from t' to 0.

In sum, we get

Theorem 6.1. Define function $\psi^*(K)$ by the solution of the following equation.

$$\int_{\psi^*(K)}^{K} \frac{F'(K_s) - F'(K)}{(F(K) + F(K_s))^2} \, dK_s = 0 \quad \text{where } K \ge K^* \ge \psi^*(K). \tag{40}$$

And define function $T_{\min}(K)$ by

$$T_{\min}(K) = \int_{\psi^*(K)}^{K} \frac{1}{s(F(K) + F(K_s))} \, dK_s. \tag{41}$$

Then, on the optimal path for Problem A, if

$$T - t \geqq T_{min}(K_r(t)),$$

then, the relation

$$p_r(t) = p_s(t) \quad \text{and} \quad K_r(t) \geqq K^* \geqq K_s(t),$$

holds to be true if and only if $K_s(t) = \psi^*(K_r(t))$. Function $\psi^*(K)$ is a single valued function of K (on its domain), and we call it the (*optimum*) *switching function* of investment allocation.

We next summarize properties of switching function $\psi^*(K)$. First, the following corollary immediately follows from Equation (40).

Corollary 6.1. Switching function $\psi^*(K)$ defined by Equation (40) has the following properties.

(a) It is independent of time t.
(b) It is independent of saving ratio s (or, $s(t)$).
(c) It is symmetrical with respect to K_r and K_s; namely, if $K_s = \psi^*(K_r)$ is the switching function under condition $K_r \geqq K^* \geqq K_s$, then $K_r = \psi^*(K_s)$ is the switching function under condition $K_s \geqq K^* \geqq K_r$.

Next, to examine more detailed properties of function $\psi^*(K)$, it is convenient to generalize Equation (40) as follows:

$$\int_{\psi^*(K, M)}^{K} \frac{F'(K_s) - F'(K)}{(F(K) + F(K_s) + M)^2} dK_s = 0 \quad \text{and}$$

$$K \geqq K^* \geqq \psi^*(K, M), \quad \text{where } M \geqq 0. \tag{42}$$

We may call the function $\psi^*(K, M)$ defined by the solution of the above equation the *generalized switching function*. The meaning of this function becomes clear if we observe that Equation (42) is obtained by using, instead of (30), the following equation

$$\left.\begin{aligned}\dot{K}_r(t) &= 0 \\ \dot{K}_s(t) &= s(F(K_r') + F(K_s(t)) + M)\end{aligned}\right\} \quad \text{for } t \in [t', t''), \tag{43}$$

in the previous analysis. That is, function $\psi^*(K, M)$ defines the "switching function" under the condition that a constant amount of income, M, exogenously flows into the two-region system at each time.

By definition, the two-function, $\psi^*(K)$ and $\psi^*(K, M)$, have the following relation.

$$\psi^*(K) = \psi^*(K, 0). \tag{44}$$

Finally, we define another function, $\bar{\psi}(K)$, by the solution of the following equation.

$$F'(K) = \frac{F(K) - F(\bar{\psi}(K))}{K - \bar{\psi}(K)} \quad \text{and} \quad K \geqq K^* \geqq \bar{\psi}(K). \tag{45}$$

The meaning of function $\bar{\psi}(K)$ is clear from Figure 6.1(a) and (b). Observe that Equation (45) is equivalent to the following equation.

$$\int_{\bar{\psi}(K)}^{K} (F'(K_s) - F'(K)) \, dK_s = 0. \tag{46}$$

Hence, comparing (42) and (46), we see that the two functions, $\bar{\psi}(K)$ and $\psi^*(K, M)$, have the following relation:

$$\bar{\psi}(K) = \lim_{M \to \infty} \psi^*(K, M). \tag{47}$$

The following theorem summarizes the properties of function $\psi^*(K)$ and the relations between functions $\psi^*(K)$, $\psi^*(K, M)$, $\hat{\psi}(K)$ and $\bar{\psi}(K)$ (for the proof, see Appendix E).

Theorem 6.2. Define functions $\psi^*(K)$, $\psi^*(K, M)$, $\hat{\psi}(K)$ and $\bar{\psi}(K)$, respectively, by (41), (42), (22) and (45). Then, we have:

(a) $\dfrac{\partial \psi^*(K, M)}{\partial K} < 0, \quad \dfrac{\partial \psi^*(K, M)}{\partial M} < 0 \quad \text{and} \quad \dfrac{d\psi^*(K)}{dK} < 0.$

Hence,

$$\frac{d\psi_M^{*-1}(K_s)}{dK_s} < 0, \quad \frac{d\psi_K^{*-1}(K_s)}{dK_s} < 0 \quad \text{and} \quad \frac{d\psi^{*-1}(K_s)}{dK_s} < 0,$$

where $\psi_M^{*-1}(K_s)$ and $\psi_K^{*-1}(K_s)$ are, respectively, the inverse functions of $\psi^*(K, M)$ under each fixed value of M and K, respectively; $\psi^{*-1}(K_s)$ is the inverse function of $\psi^*(K)$.

(b) $\dfrac{dK}{dM}\bigg|_{\psi^*(K, M) = \text{constant}} < 0.$

(c) The domains of functions $\psi^*(K, M)$ and $\psi^*(K)$ with respect to K are, respectively, $[K^*, \psi_M^{*-1}(0)]$ and $[K^*, \psi^{*-1}(0)]$. Hence, the domains of functions $\psi_M^{*-1}(K_s)$ and $\psi^{*-1}(K_s)$ with respect to K_s are, both, $[0, K^*]$.

(d) $\hat{\psi}(K) > \psi^*(K) > \psi^*(K, M) > \bar{\psi}(K)$ whenever these values are defined, where $K \geqq K^*$ and $M > 0$. Hence, $\hat{\psi}^{-1}(K_s) > \psi^{*-1}(K_s) > \psi^{*-1}(K_s, M) > \bar{\psi}^{-1}(K_s)$ for each $K_s \in [0, K^*)$, where $M > 0$.

Using Corollary 6.1 and Theorem 6.2, the *optimal switching line*, $K_s = \psi^*(K_r)(K_r \geqq K^* \geqq K_s)$ and $K_r = \psi^*(K_s)(K_s \geqq K^* \geqq K_r)$, can be depicted as in Figure 6.4(a). From (a) of Theorem 6.2, $\psi^*(K_r)$ is a downward sloping curve of K_r. And, from (d) of Theorem 6.2, the curve $\psi^*(K_r)$ lies between the two curves, $\hat{\psi}(K_r)$ and $\bar{\psi}(K_r)$.

Now suppose the plan period, T, is sufficiently long. Then, by using Lemmas 6.2 to 6.6 and taking (34) and (39) into account, optimal growth paths in the case of variable returns to scale can be depicted as in Figure 6.4. The upper half of Figure 6.4 depicts optimal capital-stock paths starting from various initial points, and the lower half depicts the corresponding capital-price paths. For example, when the initial capital stock (K_r^0, K_s^0) is located at point b in Figure 6.4 (a), the optimal capital-stock path is given by a curve, $b \rightarrow c \rightarrow d \rightarrow e \rightarrow f$, in Figure 6.4(a), and the corresponding capital-price path is given by a curve, $b \rightarrow c \rightarrow d \rightarrow e \rightarrow f$ in Figure 6.4(b). On the other hand, when the initial capital stock (K_r^0, K_s^0) is located, for example, at point a in Figure 6.4(a), we must choose one region as our favourite region. If we choose region r, the optimal capital-stock path is given by a curve, $a \rightarrow b \rightarrow c \rightarrow d \rightarrow e \rightarrow f$, in Figure 6.4(a), and the corresponding capital-price path is given by a curve, $a \rightarrow b \rightarrow c \rightarrow d \rightarrow e \rightarrow f$, in Figure 6.4(b). But if we choose region s, the capital-stock path is given by $a \rightarrow b' \rightarrow c' \rightarrow d' \rightarrow e \rightarrow f$ in Figure 6.4(a), and the capital-price path is given by $a' \rightarrow b' \rightarrow c \rightarrow d' \rightarrow e \rightarrow f$ in Figure 6.4(b).

Distances between the three curves, $\hat{\psi}(K)$, $\psi^*(K)$ and $\bar{\psi}(K)$, depend on the form of the production function F and the initial point (K_r^0, K_s^0). Figure 6.5 gives an example. Suppose the initial capital stock in region s is given by K_s^0 in Figure 6.5, and suppose $K_r^0 > K_s^0$. Then, first, region r moves to the point $\psi^{*-1}(K_s^0)$, while region r stays there. From Equation (40), the point $K_r' \equiv \psi^{*-1}(K_s^0)$ is obtained by solving

$$\int_{K_s^0}^{K_r'} \frac{F'(K_s) - F'(K_r')}{(F(K_r') + F(K_s))^2} \, dK_s = 0 \quad \text{where } K_r' \geqq K^*, \tag{48}$$

for the unknown K_r'. Then, if $F(\bar{\psi}^{-1}(K_s^0))$ is much larger than $F(K_s^0)$ (i.e., $F(\bar{\psi}^{-1}(K_s^0)) \gg F(K_s^0)$), we can suggest, by the following intuitive argument, that $\psi^{*-1}(K_s^0)$ is very close to neither $\bar{\psi}^{-1}(K_s^0)$ nor $\hat{\psi}^{-1}(K_s^0)$.

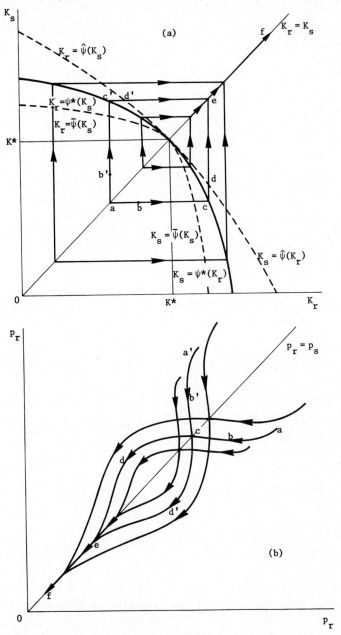

Figure 6.4. Optimal paths in the case of variable returns to scale ($n = 2$): (a) Capital-stock path, (b) Capital-price path.

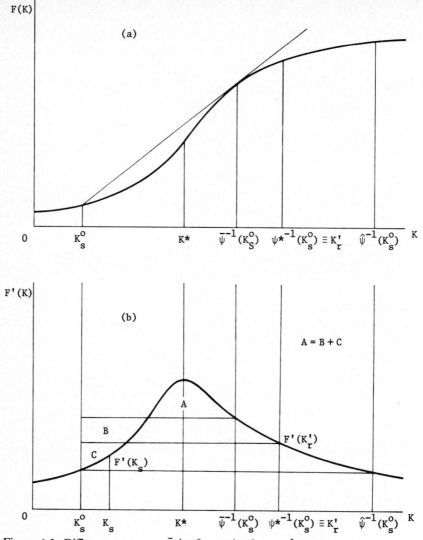

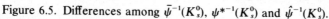

Figure 6.5. Differences among $\bar{\psi}^{-1}(K_s^0)$, $\psi*^{-1}(K_s^0)$ and $\hat{\psi}^{-1}(K_s^0)$.

If $F(K_r') + F(K_s)$ were constant when K_s moved from K_s^0 to $\bar{\psi}^{-1}(K_s^0)$ in Figure 6.5, then $\psi*^{-1}(K_s^0)$ would be equal to $\bar{\psi}(K_s^0)$ (recall the equivalence of (45) and (46)). But, if $F(\bar{\psi}^{-1}(K_s^0)) \gg F(K_s^0)$, then the "value of discount factor," $(F(K_r') + F(K_s))^2$, in (48) is close to $F(K_r')^2$ when K_s is close to K_s^0, and it is close to $4(F(K_r'))^2$ when K_s is close to K_r' ($\equiv \psi*^{-1}(K_s^0)$). Hence, the difference in the values of discount factor

$(F(K'_r) + F(K_s))^2$ when K_s moves from K_s^0 to K'_r is four-fold. Thus, $\psi^{*-1}(K_s^0)(\equiv K'_r)$ cannot be close to $\bar{\psi}^{-1}(K_s^0)$.

On the other hand, from (48), $\psi^{*-1}(K_s^0)$ could be very close to $\hat{\psi}^{-1}(K_s^0)$ only when there was an extremely large difference in the value of the discount factor $(F(K'_r) + F(K_s))^2$ when K_s moved from K_s^0 to K'_r (refer to (48) and Figure 6.5(b)). But, by the same argument as above, the maximum difference is always less than four-fold. Hence $\psi^{*-1}(K_s^0)$ cannot be very close to $\hat{\psi}^{-1}(K_s^0)$.

Hence, we can suggest that the optimal switching point is generally not very close to $\hat{\psi}^{-1}(K_s^0)$ or $\bar{\psi}^{-1}(K_s^0)$. This implies that (i) the investment allocation should be switched from the more developed region to the less developed region *much before* the marginal productivity of capital become the same in the two regions, and (ii) when $F(K_s^0)$ is much smaller than $F(\bar{\psi}^{-1}(K_s^0))$, the investment allocation should be switched from the more developed region to the less developed region *considerably after* the average productivity of the total investment in the more developed region becomes maximum.

In Figure 6.4 it is assumed that the plan period T is so long that when each capital-stock path reaches the switching line $K_s = \psi^*(K_r)$, the rest of the plan period is no less than $T_{\min}(K_r)$ which is given by (41). Under this condition, we see from Corollary 6.1(a) that optimal growth paths are time-invariant and independent of the length of plan period T. We also see from Corollary 6.1(b) that optimal growth paths are also independent of the saving ratio s (or, $s(t)$). Namely, though the movement of the capital-stock path in Figure 6.4(a) is faster as the saving ratio is larger, the locus of a capital-stock path depends only on its initial position in Figure 6.4(a). This point is also true for capital-price paths in Figure 6.4(b).

On the other hand, when the plan period T is not sufficiently long, the switching time (and hence, switching position) of capital from the more developed region to the less developed region depends on the plan length T. But, since in this chapter we are mainly interested in the case where the plan period is sufficiently long, we only state the following conclusion. When the plan period T is not sufficiently long, the (short-run) optimum switching curve lies between the two curves, $K_s = \hat{\psi}(K_r)(K_r = \hat{\psi}(K_s))$ and $K^* = K_r$ $(K^* = K_s)$, and it converges to the (long-run) optimum switching curve $K_s = \psi^*(K_r)$ $(K_r = \psi^*(K_s))$ as T becomes longer.

Finally, let us examine the optimal growth paths for the rest of the cases defined at the end of Section 6.2. By applying Lemmas 6.2 and

6.4 to the cases of decreasing returns to scale and increasing returns to scale, we get

Lemma 6.7.   In the case of decreasing returns to scale,

(a) if $K_r(t) > K_s(t)$, then $p_r(t) < p_s(t)$,
(b) if $K_r(t) = K_s(t)$, then $p_r(t) = P_s(t)$.

Lemma 6.8.   In the case of increasing returns to scale,

(a) if $K_r(t) > K_s(t)$, then $p_r(t) > p_s(t)$,
(b) if $K_r(t) = K_s(t)$, then $p_r(t) > p_s(t)$ or $p_r(t) < p_s(t)$, and the choice of one relation between the two is arbitrary.

Next, in the case of constant returns to scale, we see from Definition (7) and from relations (20) and (21) that $\dot{p}_l(t) = -sa \cdot \max(p_r(t), p_s(t))$. In addition, from terminal condition (14), $p_r(T) = P_s(T) = a$. Hence we get

Lemma 6.9.   In the case of constant returns to scale,

  $p_r(t) = p_s(t)$ for all $t \in [0, T]$.

Therefore, the optimal capital-stock paths and their corresponding capital-price paths can be depicted for each case as in Figures 6.6, 6.7 and 6.8. Figure 6.6(a) corresponds to the part of Figure 6.4(a) which is above the switching line, and Figure 6.8(a) corresponds to the lower part of it. Hence, decreasing returns to scale in production is a basic

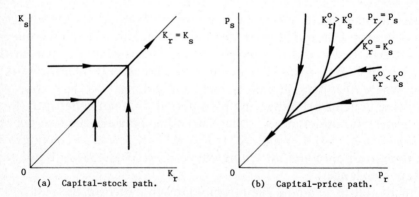

Figure 6.6. Optimal paths in the case of decreasing returns to scale.

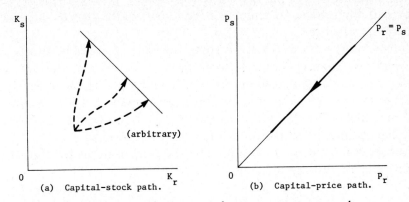

Figure 6.7. Optimal paths in the case of constant returns to scale.

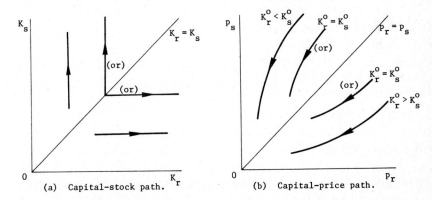

Figure 6.8. Optimal paths in the case of increasing returns to scale.

force for spatial dispersion, and increasing returns to scale in production is a basic force for spatial concentration. On the other hand, in the case of constant returns to scale, the relative size of the capital stock between the two regions has no influence on the total output of the economy, and hence the investment allocation can be arbitrary at each time.

6.5. Optimal growth in the case of many regions ($n \geqq 3$)

In this section, we examine the properties of optimal growth paths in the case of many regions. As in the previous section, we first study the case of variable returns to scale, and then analyze the rest of the cases as special cases of it.

In Subsections 6.5.1 to 6.5.3, we study the case of variable returns to scale. First, in Subsection 6.5.1, some basic properties of the optimal growth paths are obtained under general conditions. Then, in Subsection 6.5.2, we closely examine the optimal growth path of the four-region economy. Next, in Subsection 6.5.3, we attempt to further clarify the optimal growth path in the n-region economy.

In Subsection 6.5.4, we obtain the optimal growth-paths for the other specifications of returns to scale.

6.5.1. Some basic properties

First, we must confirm that

Theorem 6.3. In the case of variable returns to scale, there exists an optimal path for Problem A (assuming both n and T to be finite).

For the proof, see Appendix E.

Next, we see that the following lemma is the direct extension of Lemmas 6.2, 6.4(a) and 6.6.

Lemma 6.10. In the case of variable returns to scale, we have, for any two regions, r and s, that:

(a) if $K_r(t) = K_s(t) \geqq K^*$, then $p_r(t) = p_s(t)$,
(b) if $K_r(t) > K^*$ and $K_r(t) > K_s(t) > \hat{\psi}(K_r(t))$, then $p_r(t) < p_s(t)$,
(c) if $K_r(t) > K^*$ and $K_r(t) > K_s(t) = \hat{\psi}(K_r(t))$, then $p_r(t) < p_s(t)$ when $0 \leqq t < T$, and $p_r(t) = p_s(t)$ when $t = T$,
(d) if $K^* > K_r(t) > K_s(t)$, then $p_r(t) > p_s(t)$.

Here, function $\hat{\psi}$ is defined by (22).

From (a) in the above lemma, we have the following result which is the generalization of Lemma 6.3(b).

Lemma 6.11. In the case of variable returns to scale, if $K_1(t) = K_2(t) = \cdots = K_{n'}(t) \geqq K^*$ and $p_1(t) > p_j(t)$ for all $j = n' + 1, \ldots, n$, then $\theta_l(t) = 1/n'$ for all $l = 1, 2, \ldots, n'$. Here $n \geqq n' \geqq 1$. (Note, in this case, $p_1(t) = p_2(t) = \cdots = p_{n'}(t)$.)

Next, the following lemma is the generalization of Lemma 6.4(b).

Lemma 6.12. In the case of variable returns to scale, suppose we have (on the optimal path) that $K^* > K_1(t) = K_2(t) = \cdots = K_{n'}(t)$ at some time $t \in [0, T]$ and that $K_l(T) > K_l(t)$ for $l = 1, 2, \ldots, n''$ and $K_l(T) = K_l(t)$ for $l = n'' + 1, \ldots, n'$, where $n \geq n' \geq n'' \geq 0$. Then, one of the following alternative relations holds to be true, and the choice of one relation among them can be arbitrary. Alternative relations are:

$$p_1(t) > p_2(t) > \cdots > p_{n''}(t) > p_{n''+1}(t) = \cdots = p_{n'}(t),$$

and all of its variations in terms of the ordering of the subscripts $1, 2, \ldots, n''$.[8]

In the following two Subsections, 6.5.2 and 6.5.3, our main concern is to study the properties of the optimal growth path under the next initial condition (which seems to be the most basic case):

$$K^* > K_1(0) = K_2(0) = \cdots = K_n(0) \equiv K^0 > 0. \tag{49}$$

Hence, we next summarize some basic properties of the optimal growth paths under assumption (49).

Lemma 6.13. In the case of variable returns to scale, if the initial condition is given by (49), the amount of capital in any region cannot become greater than $\hat{\psi}^{-1}(K^0)$ when some (or, one) regions remain at K^0.

This can be proved as follows. Suppose, on the contrary, that there is a time t $(T > t > 0)$ such that $K_r(t) \geq \hat{\psi}^{-1}(K^0)$ and $K_s(t) = K^0$ for any two regions, r and s. Since $K_r(0) = K^0$, the fact $K_r(t) \geq \hat{\psi}^{-1}(K^0)$ implies that there was a time t' such that $K_r(t') \geq \hat{\psi}^{-1}(K^0)$, $p_r(t') = \max_l p_l(t)$ (recall the optimal allocation rule (16)) and $K_s(t') = K^0$. Hence, recalling (23), we have $K_r(t') > K_s(t') = K^0 \geq \hat{\psi}(K_r(t'))$ and $p_r(t') \geq p_s(t')$, which is a contradiction of (a) and (b) in Lemma 6.10.

From Lemma 6.13, the next lemma immediately follows.

Lemma 6.14. In the case of variable returns to scale, suppose that the initial condition is given by (49). Then, if the plan period is "sufficiently long," we have $K_l(T) > K^0$ for all $l = 1, 2, \ldots, n$.

---

[8]Examples are:

$$p_2(t) > p_1(t) > p_3(t) > \cdots > p_{n''}(t) > p_{n''+1}(t) = \cdots = p_{n'}(t),$$

and

$$p_{n''}(t) > p_2(t) > p_3(t) > \cdots > p_1(t) > p_{n''+1}(t) = \cdots = p_{n'}(t).$$

The minimum length of the plan period under which the above lemma holds to be true cannot be known before obtaining the optimal growth path. But, it is shown after Lemma 6.18 that an upper bound for that minimum length can be easily obtained a priori. Next, from Lemmas 6.12 and 14, we immediately have

Lemma 6.15. In the case of variable returns to scale, suppose that the initial condition is given by (49) and that the plan period is sufficiently long. And suppose that $n - n'$ regions, $l = n' + 1, n' + 2, \ldots, n$, remains at K^0 (i.e., have K^0 units of capital) at a time t, where $n \geqq n' \geqq 0$ and $t \in [0, T)$. Then the following relation is one of the alternative relations which can be chosen arbitrarily.

$$p_{n'+1}(t) > p_{n'+2}(t) > \cdots > p_n(t).$$

And, the following lemma is always true from Lemma 6.12 and the optimal allocation rule (16).

Lemma 6.16. In the case of variable returns to scale, if the initial condition is given by (49), then no two regions can start to grow simultaneously.

When the initial condition is given by (49), the naming of the regions can be arbitrary. Therefore, considering Lemma 6.16, we adopt the following rule for the rest of Section 6.5:

Rule 6.1. When the initial condition is given by (49), region 1 is the first region which starts to grow (if there is any), region 2 is the second (if any), . . . , and region n is the last region to start to grow (if all the regions eventually move from K^0).

Next, we have

Lemma 6.17. In the case of variable returns to scale, suppose we have initial condition (49). Then, if a region once starts to grow, the allocation of investment cannot be switched to other regions before that region goes out of its increasing phase. In other words, if the rest of the plan period is sufficiently long, that region must go out of its increasing phase without switching the allocation of investment to other regions.

The proof is as follows. Suppose region s starts to grow at time t'. Then, from Lemma 6.16, no other region starts to grow at that time. Hence, from Lemma 6.10(d), the allocation of investment cannot be switched to the rest of those regions which were at K^0 at time t'. Next, the fact that region s starts to grow at time t' implies from allocation rule (16) that $p_s(t') = \max_l p_l(t')$. Hence, from Lemma 6.10(d), no region exists with an amount of capital in the range (K^0, K^*) at that time. It can be proved by using the same method which was used just after Equation (27) that the allocation of the investment cannot be switched to those regions which were in the range $[K^*, \hat{\psi}^{-1}(K^0))$ at time t' before region s reaches (one of) them. Finally, from Lemma 6.13, there exists no region in the range $[\hat{\psi}^{-1}(K^0), \infty)$ at time t'. Consequently the lemma holds to be true.

Therefore, from Lemmas 6.13, 6.17 and 6.10(b), we conclude that

Lemma 6.18. In the case of variable returns to scale, suppose that the plan period is sufficiently long. Then, all the regions move into the decreasing phase before any region goes beyond $\hat{\psi}^{-1}(K^0)$. Hence, if the rest of the plan period is sufficiently long, all the regions eventually pass the point $\hat{\psi}^{-1}(K^0)$ together.

Returning to the meaning of Lemma 6.14, an upper bound for the minimum length of the plan period under which that lemma holds to be true is given by the time period which is required for the completion of the following (non-optimal) processes: region 1 first starts to grow and its amount of capital reaches $\hat{\psi}^{-1}(K^0)$, then region 2 starts to grow and reaches $\hat{\psi}^{-1}(K^0)$, ..., and finally region n starts to grow and reaches $\hat{\psi}^{-1}(K^0)$. The time period which is required for the completion of these processes can easily be calculated. Hence, in this chapter, when the phrase, "the plan period is sufficiently long," is used, we may interpret it to mean "the plan period is not shorter than the time period which is required for each of the n-regions in turn to reach $\hat{\psi}^{-1}(K^0)$."

Next, from Lemmas 6.11 and 6.17, we immediately get

Lemma 6.19. In the case of variable returns to scale, suppose that the initial condition is given by (49) and we adopt Rule 6.1. Then, the following relation holds to be true at any time $t \in [0, T]$.

$$K_1(t) \geq K_2(t) \geq \cdots \geq K_n(t).$$

That is, any region cannot pass any other region which started to grow before it.

Recall from Section 6.4 that, in the case of the two-region economy, if $K^* > K_r(0) = K_s(0) \equiv K^0 > 0$ and if we denote by K'_r the point at which the allocation of investment is switched from region r to region s, then we have $\hat{\psi}^{-1}(K^0) > K'_r > \bar{\psi}^{-1}(K^0)$. This is true in general, and we have

Lemma 6.20. In the case of variable returns to scale, suppose that the initial condition is given by (49) and we adopt Rule 6.1. And suppose it happens (on the optimal path) that at time t' region l starts to grow and it reaches region $l - 1$ without switching the allocation of investment to other regions (i.e., it directly catches up to region $l - 1$). Then, it is true that

$$\hat{\psi}^{-1}(K^0) > K_{l-1}(t') > \bar{\psi}^{-1}(K^0).$$

Here $\hat{\psi}$ and $\bar{\psi}$ are defined, respectively, by (22) and (45). Note, by definition, $K_{l-1}(t')$ is the amount of capital in region $l - 1$ when the allocation of investment is switched from region $l - 1$ to region l and region l starts to grow, and it is also the position at which region l catches up to region $l - 1$.

To see the above lemma, observe that the movements of the capital-stock paths and capital-price paths of regions l and $l - 1$ during the corresponding time interval are given by:

$$\left.\begin{array}{l} \dot{K}_{l-1}(t) = 0, \\ \dot{K}_l(t) = s[K_{l-1}(t') + F(K_l(t)) \\ \qquad + \displaystyle\sum_{j=1}^{l-2} (F(K_j(t')) + (n - l)F(K^0)], \end{array}\right\} \text{ for } K_l(t) \in [K^0, K_{l-1}(t')].$$

$$\left.\begin{array}{l} \dot{p}_{l-1}(t) = -p_l(t)sF'(K_{l-1}(t')), \\ \dot{p}_l(t) = -p_l(t)sF'(K_l(t)), \end{array}\right\} \text{ for } K_l(t) \in [K^0, K_{l-1}(t')].$$

Hence, by employing the same method as was used in obtaining (37), we have

$$\int_{K^0}^{K_{l-1}(t')} \frac{F'(K_l) - F'(K_{l-1}(t'))}{(F(K_{l-1}(t')) + F(K_l) + M)^2} \, dK_l = 0 \tag{50}$$

where

$$M = \sum_{j=1}^{l-2} F(K_j(t')) + (n-l)F(K^0).$$

Therefore, from Definition (42) and Theorem 6.2(d), we get Lemma 6.20.

6.5.2. Case of four regions (n = 4); variable returns to scale

Next, let us study in detail the optimal growth path for the four-region economy. In doing so, we will discover some clues for the further characterization of the optimal growth path in the n-region economy.

In this section, we only study the optimal growth path under the following initial condition assuming that the plan period is sufficiently long.

$$K^* > K_1^0 = K_2^0 = K_3^0 = K_4^0 \equiv K^0 > 0. \tag{51}$$

We adopt Rule 6.1. Then, setting $n' = 0$ and $t = 0$ in Lemma 6.15, we have

$$p_1(0) > p_2(0) > p_3(0) > p_4(0).$$

and, region 1 starts to grow first. Then, from Lemma 6.17, region 1 must go out of its increasing phase; but, from Lemma 6.13, it should stop before reaching $\hat{\psi}^{-1}(K^0)$. Suppose it stops at point K_1' at time t_1 and the allocation of investment is switched to region 2. Then, considering Lemma 6.15, we have

$$p_1(t_1) = p_2(t_1) > p_3(t_1) > p_4(t_1), \ \hat{\psi}^{-1}(K^0) > K_1(t_1) \equiv K_1' \geqq K^* \text{ and}$$
$$K_l(t_1) = K^0, \quad l = 1, 2, 3. \tag{52}$$

The equality, $p_1(t_1) = p_2(t_1)$, is obtained from the continuity of price path $p_l(t)$ (for every l) with respect to time t.

Once region 2 starts to grow, from Lemma 6.17, it must go out of its increasing phase. Then, we have the following two alternative possibilities:

(α) Region 2 reaches region 1 before the allocation of investment is switched to region 3.

(β) Region 2 stops growing before reaching region 1, and the allocation of investment is switched to region 3.

In the following, we show by counterexample that (α) cannot happen.

Suppose (α) is true. Then, we have for some $t_2(>t_1)$ that

$$\left.\begin{aligned}\dot{K}_2(t) &= s(F(K_1') + F(K_2(t)) + 2F(K^0)),\\ \dot{K}_l(t) &= 0, \quad l = 1, 3, 4,\end{aligned}\right\} \quad \text{for } t \in [t_1, t_2). \tag{53}$$

$$\left.\begin{aligned}\dot{p}_1(t) &= -p_2(t)sF'(K_1'),\\ \dot{p}_2(t) &= -p_2(t)sF'(K_2(t)),\\ \dot{p}_l(t) &= -p_2(t)sF'(K^0), \quad l = 3, 4,\end{aligned}\right\} \quad \text{for } t \in [t_1, t_2). \tag{54}$$

and

$$p_1(t_2) = p_2(t_2) \geqq p_3(t_2) > p_4(t_2),$$
$$\hat{\psi}^{-1}(K^0) > K_1(t_2) = K_2(t_2) = K_1' \geqq K^*. \tag{55}$$

From Equations (52) to (55), the following relation is derived.

$$\int_{K^0}^{K_1'} \frac{F'(K_2) - F'(K_1')}{(F(K_1') + F(K_2) + 2F(K^0))^2} \, dK_2 = 0, \tag{56}$$

that is,

$$K^0 = \psi^*(K_1', 2F(K^0)), \tag{57}$$

where ψ^* is defined by (42).

Next, when region 2 reaches region 1, the allocation of investment may be switched to region 3, or regions 1 and 2 may grow together for a while. In the latter case, from Lemma 6.13, they cannot go beyond $\hat{\psi}^{-1}(K^0)$. Hence, in any case, we can consider that the allocation of investment is switched to region 3 at some point K' $(\hat{\psi}^{-1}(K^0) > K' \geqq K_1')$ at some time $t'(\geqq t_2)$, and we have

$$p_1(t') = p_2(t') = p_3(t') > p_4(t'),$$
$$\hat{\psi}^{-1}(K^0) > K_1(t') = K_2(t') \equiv K' \geqq K_1' \geqq K^*, \quad \text{where } t' \geqq t_2. \tag{58}$$

After region 3 starts to grow, from Lemma 6.17, it must go out of its increasing phase. And, again, we have the following two alternative possibilities:

(α-1) Region 3 reaches regions 1 and 2 before the allocation of investment is switched to region 4.

(α-2) Region 3 stops growing before reaching regions 1 and 2, and the allocation of investment is switched to region 4.

But, we can easily see by the following reason that (α-1) cannot happen.

Suppose (α-1) is true. Then, by the same method as was used in obtaining (56), we get

$$\int_{K^0}^{K'} \frac{F'(K_3) - F'(K')}{(F(K') + F(K_3) + F(K') + F(K^0))^2} \, dK_3 = 0,$$

that is,

$$K^0 = \psi^*(K', F(K') + F(K^0))$$

where ψ^* is defined by (42). However, (57) and the above relation is a contradiction of Theorem 6.2(b) since $F(K') + F(K^0) > 2F(K^0)$ and $K' \geq K_1'$ from (58).

Therefore, if (α) is true, then only (α-2) remains as a possibility. That is, region 3 must stop growing at some point $K_3'(< K')$ at some time $t_3(> t')$, and the allocation of investment must be switched to region 4. Hence we have

$$\left.\begin{aligned}
\dot{K}_3(t) &= s(2F(K') + F(K_3(t)) + F(K^0)), \\
\dot{K}_l(t) &= 0, \quad l = 1, 2, 4,
\end{aligned}\right\} \quad \text{for } t \in [t', t_3). \tag{59}$$

$$\left.\begin{aligned}
\dot{p}_l(t) &= -p_3(t)sF'(K'), \quad l = 1, 2, \\
\dot{p}_3(t) &= -p_3(t)sF'(K_3(t)), \\
\dot{p}_4(t) &= -p_3(t)sF'(K^0),
\end{aligned}\right\} \quad \text{for } t \in [t', t_3). \tag{60}$$

and

$$\begin{aligned}
p_3(t_3) &= p_4(t_3) > p_1(t_3) = p_2(t_2), \\
\hat{\psi}^{-1}(K^0) &> K_1(t_3) = K_2(t_3) = K' > K_3(t_3) \equiv K_3' \geq K^*.
\end{aligned} \tag{61}$$

Once region 4 starts to grow, from Lemma 6.17 it must go out of its increasing phase. Then, since no region is left in the increasing phase, from Lemma 6.10(b), region 4 continues to grow until it reaches region 3. Hence, there exists a time $t_4(> t_3)$ such that

$$\left.\begin{aligned}
\dot{K}_4(t) &= s(2F(K') + F(K_3') + F(K_4(t))), \\
\dot{K}_l(t) &= 0, \quad l = 1, 2, 3,
\end{aligned}\right\} \quad \text{for } t \in [t_3, t_4). \tag{62}$$

$$\left.\begin{aligned}
\dot{p}_l(t) &= -p_4(t)sF'(K'), \quad l = 1, 2, \\
\dot{p}_3(t) &= -p_4(t)sF'(K_3'), \\
\dot{p}_4(t) &= -p_4(t)sF'(K_4(t)),
\end{aligned}\right\} \quad \text{for } t \in [t_3, t_4). \tag{63}$$

and

$$p_3(t_4) = p_4(t_4) > p_1(t_4) = p_2(t_4),$$
$$\hat{\psi}^{-1}(K^0) > K_1(t_4) = K_2(t_4) = K' > K_3(t_4) = K_4(t_4) = K_3' > K^*. \tag{64}$$

Here, the relation $K_3' > K^*$ is obtained from Lemma 6.20.

Next, from Lemma 6.10(b) and Lemma 6.11, regions 2 and 3 continue to grow together until they catch up to regions 1 and 2. Therefore, there exists a time $t_5(> t_4)$ such that

$$\left. \begin{array}{l} \dot{K}_l(t) = s(F(K') + F(K_l(t))), \quad l = 3, 4, \\ K_3(t) = K_4(t), \\ \dot{K}_l(t) = 0, \quad l = 1, 2, \end{array} \right\} \quad \text{for } t \in [t_4, t_5). \tag{65}$$

$$\left. \begin{array}{l} \dot{p}_l(t) = -p_l(t)sF'(K_l(t)), \quad l = 3, 4, \\ p_3(t) = p_4(t), \\ \dot{p}_l(t) = -p_3(t)sF'(K'), \quad l = 1, 2, \end{array} \right\} \quad \text{for } t \in [t_4, t_5). \tag{66}$$

and

$$p_1(t_5) = p_2(t_5) = p_3(t_5) = p_4(t_5),$$
$$\hat{\psi}^{-1}(K^0) > K_l(t_5) = K' > K_3' > K^*, \quad l = 1, 2, 3, 4. \tag{67}$$

We now show that there is an inconsistency among relations (56) to (67).

According to (58) and (67), $p_2(t)$-path and $p_3(t)$-path intersect at times t' and t_5, respectively. Hence, it should be true that

$$0 = \int_{t'}^{t_5} (\dot{p}_2(t) - \dot{p}_3(t)) \, dt$$

$$= \int_{t'}^{t_3} (\dot{p}_2(t) - \dot{p}_3(t)) \, dt + \int_{t_3}^{t_4} (\dot{p}_2(t) - \dot{p}_3(t)) \, dt + \int_{t_4}^{t_5} (\dot{p}_2(t) - \dot{p}_3(t)) \, dt.$$

Hence, if we set

$$A = \int_{t'}^{t_3} (\dot{p}_2(t) - \dot{p}_3(t)) \, dt, \qquad B = \int_{t_3}^{t_4} (\dot{p}_2(t) - \dot{p}_3(t)) \, dt,$$

$$C = \int_{t_4}^{t_5} (\dot{p}_2(t) - \dot{p}_3(t)) \, dt,$$

then, it must be true that

$$A + B + C = 0. \tag{68}$$

On the other hand, according to (59) to (61), (62) to (64), and (65) to (69), respectively, we have (refer to the procedure used in obtaining (35) and (36)):

$$A = \int_{K^0}^{K_3'} \left(\frac{F'(K_3) - F'(K')}{2F(K') + F(K_3) + F(K^0)} p_3(t_3) \right.$$

$$\times \exp\left(\int_{K_3}^{K_3'} \frac{F'(K)}{2F(K') + F(K) + F(K^0)} dK \right) \right) dK_3$$

$$= p_3(t_3)(2F(K') + F(K_3'))$$

$$+ F(K^0)) \int_{K^3}^{K_3'} \frac{F'(K_3) - F'(K')}{(2F(K') + F(K_3) + F(K^0))^2} dK_3,$$

$$B = \int_{K^0}^{K_3'} \left(\frac{F'(K_3') - F'(K')}{2F(K') + F(K_3') + F(K_4)} p_3(t_3) \right.$$

$$\times \exp\left(-\int_{K^0}^{K_4} \frac{F'(K)}{2F(K') + F(K_3') + F(K)} dK \right) \right) dK_4,$$

$$C = \int_{K_3'}^{K'} \left(\frac{F'(K_3) - F'(K')}{F(K') + F(K_3)} p_3(t_3) \right.$$

$$\times \exp\left(-\int_{K^0}^{K_3'} \frac{F'(K)}{2F(K') + F(K_3') + F(K)} dK \right.$$

$$- \int_{K_3'}^{K_3} \frac{F'(K)}{F(K') + F(K_3)} dK \right) \right) dK_3$$

$$= p_3(t_3)(2F(K') + F(K_3') + F(K^0)) \int_{K_3'}^{K'} \frac{F'(K_3) - F'(K')}{2(F(K') + F(K_3))^2} dK_3.$$

From Lemma 6.1, $p_3(t_3) > 0$; from (64), $K' > K_3' > K^*$. Hence, it must be true that $B > 0$. Therefore, if (66) were true, we would have

$$\int_{K^0}^{K_3'} a(K_3)\, dK_3 + \int_{K_3'}^{K'} c(K_3)\, dK_3 < 0, \tag{69}$$

where

$$a(K_3) = (F'(K_3) - F'(K'))/(2F(K') + F(K_3) + F(K^0))^2,$$
$$c(K_3) = (F'(K_3) - F'(K'))/2(F(K') + F(K_3))^2.$$

However, (69) cannot be true for the following reason.

Since $F(K') > F(K^0)$, from (56) (i.e., (57)) and (a) (i.e., $\partial\psi^*(K, M)/\partial K < 0$) and (b) of Theorem 6.2 we have

$$0 < \int_{K^0}^{K'} a(K_3)\, dK_3 = \int_{K^0}^{K_3'} a(K_3)\, dK_3 + \int_{K_3'}^{K'} a(K_3)\, dK_3,$$

hence,

$$\int_{K^0}^{K_3'} a(K_3)\, dK_3 > - \int_{K_3'}^{K'} a(K_3)\, dK_3.$$

Thus,

$$\int_{K^0}^{K_3'} a(K_3)\, dK_3 + \int_{K_3'}^{K'} c(K_3)\, dK_3 > \int_{K_3'}^{K'} (c(K_3) - a(K_3))\, dK_3.$$

And, by direct calculations, we see that $a(K_3)/c(K_3) < 1$ for $K_3 \in [K^0, K']$.[9] Therefore, since $a(K_3) > 0$ and $c(K_3) > 0$ for $K_3 \in [K_3', K')$, we have

$$\int_{K_3'}^{K'} (c(K_3) - a(K_3))\, dK_3 > 0,$$

---

[9] $a(K_3)/c(K_3) = (F(K') + F(K_3))^2/2(F(K') + [(F(K_3) + F(K_0))/2])^2$
$\quad\quad\quad\quad < (F(K') + F(K_3))^2/2(F(K') + [F(K_3)/2])^2$
$\quad\quad\quad\quad = \frac{1}{2}\left(\dfrac{1 + [F(K_3)/F(K')]}{1 + [K(K_3)/2F(K')]}\right)^2$
$\quad\quad\quad\quad \leq \frac{1}{2}\left(\dfrac{1+1}{1+(\frac{1}{2})}\right)^2 \quad$ (since $K' \geq K_3$ and $F(K') \geq F(K_3)$)
$\quad\quad\quad\quad \leq 8/9.$

and hence

$$\int_{K^0}^{K_3^s} a(K_3) \, dK_3 + \int_{K_3^s}^{K'} c(K_3) \, dK_3 > 0,$$

which is a contradiction of (69).

Therefore, we conclude that the first assumption, (α), must be wrong. Hence, recalling Theorem 6.3, we see that assumption (β) must be true.

According to (β), region 2 stops growing at some point $K_2'(< K_1')$ at some time t_2, and the allocation of investment is switched to region 3. Hence, we have

$$p_2(t_2) = p_3(t_2) > \max(p_1(t_2), p_4(t_2)),$$
$$\hat{\psi}^{-1}(K^0) > K_1(t_2) = K_1' > K_2(t_2) \equiv K_2' \geqq K^*, \text{ and} \tag{70}$$
$$K_3(t_2) = K_4(t_2) = K^0.$$

Once region 3 starts to grow, from Lemma 6.17, it must go out of its increasing phase. Then, again, we have the following two alternative possibilities:

(β-1) Region 3 reaches region 2 before the allocation of investment is switched to region 4.
(β-2) Region 3 stops growing before reaching region 1, and the allocation of investment is switched to region 4.

We can easily show, as follows, that $(\beta - 1)$ cannot occur.

Suppose $(\beta$-1) is true. Then, by the same method as was used in obtaining (56), we see that the following relation must hold to be true.

$$\int_{K^0}^{K_2'} \frac{F'(K_3) - F'(K_2')}{(F(K_2') + F(K_3) + F(K_1') + F(K^0))^2} \, dK_3 = 0. \tag{71}$$

And, the allocation of investment must eventually be switched to region 4 (recall Lemma 6.13). Assume region 4 starts to grow at time t'. Then, from Lemmas 6.11 and 6.19, $K_1(t') \geqq K_2(t') = K_3(t') \geqq K_2'$, and $K_4(t') = K^0$. And, from Lemma 6.17, region 4 must go out of its increasing phase, and then, from Lemma 6.10 it must reach regions 2 and 3 (recall, there remains no region at K^0). Hence, as above, we have

$$\int_{K^0}^{K_3(t')} \frac{F'(K_4) - F'(K_3(t'))}{[F(K_3(t')) + F(K_4) + F(K_1(t')) + F(K_2(t'))]^2} \, dK_4 = 0. \tag{72}$$

However, (71) and (72) is a contradiction of Theorem 6.2(b) since $F(K_1(t')) + F(K_2(t')) > F(K_1') + F(K^0)$ and $K_3(t') \geqq K_2'$ by assumption.

Consequently, recalling Theorem 6.3, we conclude that $(\beta\text{-}2)$ must be true. That is, there exist a point K_3' and a time t_3 such that

$$p_3(t_3) = p_4(t_3) > p_2(t_3) > p_1(t_3),$$

$$\hat{\psi}^{-1}(K^0) > K_1(t_3) = K_1' > K_2(t_3) = K_2' > K_3(t_3) \equiv K_3' \geqq K^*, \quad \text{and}$$

$$t_3 > t_2. \tag{73}$$

Once region 4 starts to grow, from Lemmas 6.17 and 6.10, region 4 must directly catch up to region 3. Hence, there exists a time t_4 such that

$$p_3(t_4) = p_4(t_4) > p_2(t_4) > p_1(t_4),$$

$$\hat{\psi}^{-1}(K^0) > K_1(t_4) = K_1' > K_2(t_4) = K_2' > K_3(t_4) = K_3'$$

$$= K_4(t_4) > \hat{\psi}^{-1}(K^0), \quad \text{and } t_4 > t_3. \tag{74}$$

Here, the relation $K_4(t_4) > \hat{\psi}^{-1}(K^0)$ is derived from 6.20.

The rest of the optimal growth processes are clear from Lemma 6.10(b) and Lemma 6.11.

In sum, when the plan period is sufficiently long, the optimal growth path of the four-region economy under initial condition (51) passes through the seven stages as depicted in Figure 6.9. The capital-stock path is depicted in 6.9(a), and the capital-price path in 6.9(b).

6.5.3. Case of n regions: variable returns to scale

The most important characteristic of the optimal growth processes of the four-region economy, which are depicted in Figure 6.9, is that, except for the last region, no region directly catches up to another region. This characteristic also seems to be true for the case of the n-region economy, and we would have

Conjecture 6.1. In the case of variable returns to scale, suppose that the initial condition is given by (49) and that the plan period is sufficiently long. And we adopt Rule 6.1. Then, before region n reaches region $n-1$, any two regions l and $l+1$ cannot come together in the decreasing phase, where

$$1 \leqq l \leqq n-2.$$

Here it is assumed that $n \geqq 3$.

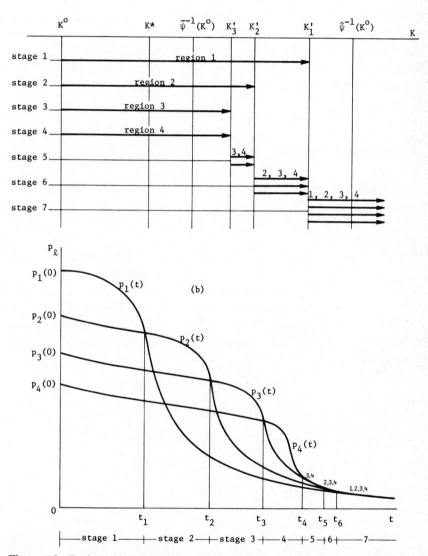

Figure 6.9. Optimal growth path in the case of four regions ($K^* > K_1^0 = K_2^0 = K_3^0 = K_4^0 \equiv K^0 > 0$); (a) capital-stock path, (b) capital-price path.

Since the above statement has not been completely proven yet, it is called a conjecture. However, the following lemma is proved (See Appendix E).

Lemma 6.21.   In the case of variable returns to scale, suppose that the initial condition is given by (49) and that the plan period is sufficiently long. And we adopt Rule 6.1. Then, before region n catches up to region $n-1$, any two regions l and $l+1$ cannot come together in the decreasing phase, where,

$$\frac{-3+\sqrt{1+4n}}{2} \leqq l \leqq n-2. \tag{75}$$

Here we assume that $n \geqq 3$.

Let us denote by l^* the smallest integer which satisfies relation (75). The relation between l^* and n is given in Table 6.1. For example, when n is between 7 and 12, any number l no less than 2 (nor greater than $n-2$) satisfies relation (75).

Hence, when the number of regions, n, is not greater than 6, Lemma 6.21 is equivalent to Conjecture 6.1. But, when n is greater than 6, Lemma 6.21 cannot guarantee the validity of Conjecture 6.1 for the first several regions (note that $l^*/n \to 0$ as $n \to \infty$).[10] However, this does not mean that Conjecture 6.1 would not always be true. The proof of Lemma 6.21 in Appendix E suggests that it should be possible to prove Conjecture 6.1.[11]

Table 6.1. The number of regions n and the smallest integer l^* which satisfies relation (75).

n	3–6	7–12	13–20	21–30	. . .	100	. . .	1000
l^*	1	2	3	4	. . .	9	. . .	31

[10]For example when $n = 21$, Lemma 6.21 cannot deny the possibility that regions 1 and 2, or regions 2 and 3, come together in the decreasing phase before region n catches up to region $n-1$.

[11]That is, the proof of Lemma 6.21 in Appendix E is obtained without effectively using the term, B. Hence, if we would effectively take into account this term, we could obtain the proof of Conjecture 6.1. It is hoped that some one will give the complete proof of Conjecture 6.1.

Let us assume that Conjecture 6.1 is true. And suppose that the initial condition is given by (49) and that the plan period is sufficiently long. And we adopt Rule 6.1. Then, the optimal growth in the n-region economy passes through the following stages:

Stage 1.   Region 1 starts to grow, goes out of its increasing phase, and stops at some K_1', where $\hat{\psi}^{-1}(K^0) > K_1' > \check{\psi}^{-1}(K^0)$.

Stage 2.   Region 2 starts to grow, goes out of its increasing phase, and stops at some K_2', where $\psi^{-1}(K^0) > K_1' > K_2' > \bar{\psi}^{-1}(K^0)$.

Stage $n-1$.   Region $n-1$ starts to grow, goes out of its increasing phase, and stops at some point K_{n-1}', where $\psi^{-1}(K^0) > K_1' > K_2' > \cdots > K_{n-1}' > \bar{\psi}^{-1}(K^0)$.

Stage n.   Region n starts to grow, and reaches region $n-1$ at K_{n-1}'.

Stage $n+1$.   Region $n-1$ and n catches up to region $n-2$ at K_{n-2}'.

Stage $2n-2$.   Regions 2 to $n-1$ catches up to region 1 at K_1'.

Stage $2n-1$.   All the regions grow together for the rest of the plan period.

In the above, the relation, $\psi^{-1}(K^0) > K_1'$, is obtained from Lemma 6.13. And, taking into account Conjecture 6.1, the relation $K_l' > \bar{\psi}^{-1}(K^0)$, $l = 1, 2, \ldots, n-1$, can be derived from Lemma 6.20 and the behavior in Stage n.

It is important to observe in the above optimal growth process that, before region n reaches region $n-1$, no two regions have the same marginal productivity of capital. That is, in the case of variable returns to scale, the "equality of the marginal productivity of capital" is not an effective decision rule for the optimal growth path (until all the regions move into the decreasing phase).

Besides condition (49), there are many other types of initial conditions. And, by employing procedures similar to the above, we can explore the characteristics of the optimal growth path under each type of initial condition. For example, let us take the case of the three-region economy, and examine briefly the processes of optimal growth under the following initial condition assuming that the period is sufficiently long.

$$K_1^* > K_1^0 > K_2^0 > K_3^0 \geqq 0. \tag{76}$$

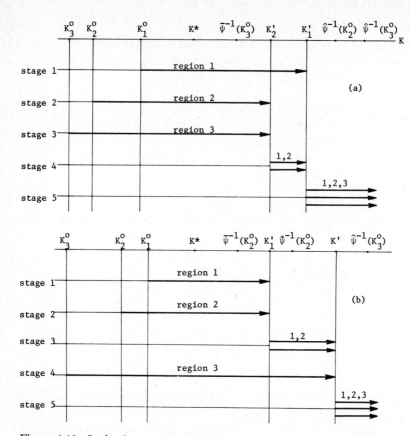

Figure 6.10. Optimal growth paths in the case of three regions ($K^* > K_1^0 > K_2^0 > K_3^0 \geq 0$); K_2^0 and K_3^0 are close in (a), and they are far apart in (b).

Under this condition, there are essentially two different patterns in the optimal growth processes as depicted in (a) and (b) of Figure 6.10.[12] Pattern (a) occurs when K_2^0 and K_3^0 are close, and Pattern (b) happens when they are far apart.

From Lemma 6.10(d), we have

$$p_1(0) > p_2(0) > p_3(0).$$

Hence, region 1 starts to grow first, and from Lemma 6.10(d), it goes out of its increasing phase. But, from (b) and (c) of Lemma 6.10, region

[12]Note that, in Figure 6.10(b), the relative positions of K' and $\hat{\psi}^{-1}(K_2^0)$ depend on the relative positions of the initial capital stock in the three regions.

1 must stop growing before its amount of capital reaches $\hat{\psi}^{-1}(K_2^0)$. Thus, the first stage is essentially the same in both patterns.

Next, region 2 starts to grow (recall Lemma 6.10(d)). By the same reasoning as used in the proof of Lemma 6.17, we see that region 2 must go out of its increasing phase before the allocation of investment is switched to other regions. And, we now have two alternative pattern of growth. In one pattern (Figure 6.10(a)), region 2 stops growing before it reaches region 1, and the allocation of investment is switched to region 3. In the other pattern (Figure 6.10(b)), region 2 directly reaches region 1, and then the allocation of investment is switched to region 3.

Stages 3 and 4 in each pattern are depicted in (a) and (b) of Figure 6.10. Stage 5 is essentially the same in both patterns.

6.5.4. Decreasing returns to scale, constant returns to scale and increasing returns to scale: n regions

First, by neglecting the information about the increasing phase in Lemma 6.10, we get

Lemma 6.22. In the case of decreasing returns to scale, we have, for any two regions, r and s, that:

(a) if $K_r(t) = K_s(t)$, then $p_r(t) = p_s(t)$,
(b) if $K_r(t) > K_s(t)$, then $p_r(t) < p_s(t)$.

Hence, in the case of decreasing returns to scale, we must allocate all of the investment to the least developed region; and if there is more than one least developed region, we must equally split the investment among them.

Next, similarly to Lemma 6.10(d) and Lemma 6.12, we obtain

Lemma 6.23. In the case of increasing returns to scale, we have:

(a) For any two regions, r and s, if $K_r(t) > K_s(t)$, then $p_r(t) > p_s(t)$.
(b) Suppose $K_1(t) = K_2(t) = \cdots = K_{n'}(t) > K_l(t)$ for all $l = n' + 1, \ldots, n$, where $1 \le n' \le n$. Then, one of the following alternative relations holds to be true, and the choice of one relation among them can be arbitrary. Alternative relations are: $p_1(t) > p_2(t) = p_3(t) = \cdots = p_{n'}(t) > p_l(t)$ for all $l = n' + 1, \ldots, n$, and all of its variations in terms of the ordering of the subscripts $1, 2, \ldots, n'$.

Hence, in the case of increasing returns to scale, all of the investment must be allocated to the most developed region throughout the plan period.

Finally, in the case of constant returns to scale, we can easily show, as in the two-region economy, that the allocation of investment is arbitrary at each time.

6.6. Comparison of growth paths under three alternative allocation rules

Meanings of optimal growth paths obtained in the previous sections will become clearer when they are compared with growth paths which would be realized under a competitive market mechanism. In this section, by employing the two-region economy (for simplicity), we perform the comparison of the growth paths obtained under the three allocation rules: the Optimum rule, the Private Marginal Productivity rule, and the Social Marginal Productivity rule, which are, respectively, defined below.

For this purpose, we must first make clear the competitive market economy which underlies the optimal planning problem described in Section 6.2. Although increasing returns to scale and variable returns to scale are generally not consistent with competition, they are consistent in the following special case.

Suppose the production function of firm i in region l ($l = r$ or s) is described as follows:

$$Y_i = g(K_l, L_l, \bar{D}_l)f(K_i, L_i, D_i)^{13} \tag{77}$$

where Y_i is the rate of output by firm i, K_i is the amount of capital, L_i is the amount of labor, and D_i is the amount of land used by firm i, respectively. And, K_l and L_l represent, respectively, the amount of total capital and the amount of total labor employed by all firms in region l, and $\bar{D}_l$ represents the amount of total land *existing* in region l. In this equation, $g(K_l, L_l, \bar{D}_l)$ represents the agglomeration economies (or diseconomies) which are enjoyed by each firm in region l as the result of the increased size of the total production activities in that region. Hence function g represents the so-called Marshallian external economies which are external to each firm but internal to the region.[14]

---

[13]Instead of (77), we may assume alternatively that $Y_i = g(Y_l)f(K_i, L_i, D_i)$ where Y_l represents the total output in region l. But under assumptions given later, Y_l becomes a function of $(K_l, L_l, \bar{D}_l)$, and hence the two representations are equivalent.

[14]For Marshallian external economies, for example, see Negishi (1972), Chapter 5.

Although different firms might produce different commodities, we make a simplifying assumption that every firm has the same production function irrespective of its output-commodity and its region. That is, functions g and f are common for all firms in the two regions. In addition it is assumed that

f is a linearly homogeneous function of K_i, L_i and D_i
        and concave with respect to K_i, L_i and D_i,   (78)

and

$L_i/K_i = \alpha$ (a constant) for each i,   (79)

namely, there is no input substitutability between capital and labor.

Next, we assume that the behavior of each firm is described as follows in each region at each time.

$$\max_{\{Y_i, K_i, L_i, D_i\}} Y_i - R_K^l(t)K_i - R_L^l(t)L_i - R_D^l(t)D_i, \text{ subject to}$$

$$Y_i = g(K_l, L_l, \bar{D}_l)f(K_i, L_i, D_i), \tag{80}$$

where $R_K^l(t)$, $R_L^l(t)$ and $R_D^l(t)$, respectively, represent rental price of capital, wage rate and land rent in region l at time t. That is, each firm takes the size of the regional economy as given and chooses the best combination of input factors.

Since the production function is common for all commodities, the production frontier curve in the output-commodity-space is flat under assumption (78), and its slope (equal to one) is independent of region and time. Consequently, recalling that each output can be moved without transport-inputs between the two regions, we see that all of the output-commodities in the economy have the same price (at each time). And, we define factor prices in terms of the output price for each time period.

As resource constraints in each region we have

$$K_l\left(\equiv \sum_i K_i \text{ in region } l \right) \leq K_l(t),$$

$$L_l\left(\equiv \sum_i L_i \text{ in region } l \right) \leq L_l(t), \tag{81}$$

$$D_l\left(\equiv \sum_i D_i \text{ in region } l \right) \leq \bar{D}_l,$$

where $K_l(t)$ and $L_l(t)$ are, respectively, the amount of capital and the amount of labor in region l at time t. For simplicity, we assume that the two regions have the same amount of land, namely

$$\bar{D}_r = \bar{D}_s = \bar{D},$$ (82)

and that the labor constraint does not become binding for any region throughout the plan period.[15]

Under these assumptions, the equilibrium conditions in factor markets are given by

$$g(K_l(t), \alpha K_l(t), \bar{D}) \frac{\partial f(K_l(t), \alpha K_l(t), \bar{D})}{\partial K_l} = R^l_K(t) + \alpha R^l_L(t),$$ (83)

$$g(K_l(t), \alpha K_l(t), \bar{D}) \frac{\partial f(K_l(t), \alpha K_l(t), \bar{D})}{\partial D_l} = R^l_D(t),$$ (84)

and we get the following relation.

$$Y_l(t) = g(K_l(t), \alpha K_l(t), \bar{D}) f(K_l(t), \alpha K_l(t), \bar{D}),$$ (85)

where $Y_l(t)$ is the sum of outputs (= the sum of output values at output price $p_i = 1$ for each i) in region l at time t, and it is the regional income at that time.

Next, to simplify notation, we represent (83) and (85) by

$$g(K_l(t)) f'(K_l(t)) = R^l_K(t) + \alpha R^l_L(t),$$ (86)

$$Y_l(t) = g(K_l(t)) f(K_l(t)) \equiv F(K_l(t)),$$ (87)

where

$$g(K_l(t)) = g(K_l(t), \alpha K_l(t), \bar{D}),$$
$$f(K_l(t)) = f(K_l(t), \alpha K_l(t), \bar{D}),$$ (88)
$$f'(K_l(t)) = \partial f(K_l(t), \alpha K_l(t), \bar{D}) / \partial K_l.$$

Equation (87) defines the regional production function previously described in equation (1). We for convenience call $g(K_l) f'(K_l)$ the *private marginal productivity of capital* in region l, though it is not literally true.

Since conditions (83) and (84) are not sufficient to determine the values of $R^l_K(t)$ and $R^l_L(t)$, we must consider that they will be determined through some institutional mechanisms. But, instead of specifying these mechanisms, we simply assume here that values of $R^l_K(t)$ determined through these mechanisms always satisfy the following

---

[15]For a justification of this assumption, we may suppose that the economy as a whole has surplus labor at each time and that labor has sufficient mobility between the two regions.

property;

$$\text{if } g(K_r(t))f'(K_r(t)) > g(K_s(t))f'(K_s(t)), \quad \text{then } R_K^r(t) > R_K^s(t). \quad (89)$$

Next, we assume as in Section 6.2 that the average propensity to save for the whole economy is known and is a function of time t, which we present by $s(t)$, or simply by s.[16] Hence the total investment of the whole economy at time t is given by $s(t)\Sigma_l Y_l(t)$ (or, $s\Sigma_l Y_l(t)$). We suppose that investment allocation ratios, $\tilde{\theta}_r(t)$ and $\tilde{\theta}_s(t)$, determined by the market have the following property:

if $g(K_r(t))f'(K_r(t)) > g(K_s(t))f'(K_s(t))$, then

  $\tilde{\theta}_r(t) = 1$ and $\tilde{\theta}_s(t) = 0$,

if $g(K_r(t))f'(K_r(t)) < g(K_s(t))f'(K_s(t))$, then

  $\tilde{\theta}_r(t) = 0$ and $\tilde{\theta}_s(t) = 1$,   (90)

if $g(K_r(t))f'(K_r(t)) = g(K_s(t))f'(K_s(t))$, then

  $\tilde{\theta}_l(t) \in [0, 1]$ for $l = r$ and s, and $\tilde{\theta}_r(t) + \tilde{\theta}_s(t) = 1$.

Namely we assume that all investors are myopic in their investment decisions and that they simply follow the signal of the rental price of capital at each time.

In addition to (20) (i.e., (16)) and (90), let us consider the following investment allocation rule as the third alternative.

if $F'(K_r(t)) > F'(K_s(t))$,   then $\hat{\theta}_r(t) = 1$ and $\hat{\theta}_s(t) = 0$,

if $F'(K_r(t)) < F'(K_s(t))$,   then $\hat{\theta}_r(t) = 0$ and $\hat{\theta}_s(t) = 1$,   (91)

if $F'(K_r(t)) = F'(K_s(t))$,   then $\hat{\theta}_l(t) \in [0, 1]$ for

  $l = r, s$ and $\hat{\theta}_r(t) + \hat{\theta}_s(t) = 1$.

This investment allocation rule has an advantage in its simplicity. Namely, though allocation rule (20) is very difficult to apply for the planning authority since values of $p_l(t)$ can be known only after obtaining the solution of Problem A, the above allocation rule can be used by decentralized planning authorities before the solution to Problem A is determined. In contrast to $g(K_l)f'(K_l)$, we call $F'(K_r)$ the *social marginal productivity of capital* in region l.

We call investment allocation rules described by (20), (90) and (91), respectively, the *Optimum rule*, the *PMP rule* (private marginal productivity rule), the *SMP rule* (social marginal productivity rule).

---

[16]Footnote 3 is valid also in this case.

The nature of production function F defined by (87) depends on functions g and f. Therefore, corresponding to each of (6) to (9), we specify functions g and f as follows, respectively.

(i) decreasing returns to scale,

$$g(K) = 1 \text{ for all } K, \text{ and } f(0) = 0, f'(K) > 0 \text{ and}$$
$$f''(K) < 0 \text{ for all } K. \tag{92}$$

(ii) constant returns to scale,

$$g(K) = 1 \text{ for all } K, \text{ and } f(K) = aK \text{ for some positive } a, \tag{93}$$

(iii) increasing returns to scale,

$$g(0) = 1, g'(K) > 0 \text{ and } g''(K) > 0 \text{ for all } K, \text{ and}$$
$$f(K) = aK \text{ for some positive } a, \tag{94}$$

(iv) variable returns to scale,

$$g(0) = 1, f(K) = aK \text{ for some positive } a, \text{ and}$$
$$F'(K) > 0 \text{ for all } K, F''(K) > 0 \text{ for } K < K^*,$$
$$F''(K) < 0 \text{ for } K > K^*, \tag{95}$$

where $g'(K) = dg(K)/dK$, $g''(K) = dg'(K)/dK$ and so on. Since $f'(K)$ represents $\partial f(K, \alpha K, \bar{D})/\partial K$ as noted in (88), case (i) can be considered as the case of an agriculture-dominated economy. In this case, the amount of land is the crucial factor. On the other hand, the assumption, $f(K) = aK$, in cases (ii) to (iv) implies that land is not a binding factor throughout the practical range of K. In case (iii), agglomeration economies prevail throughout the practical range of K. On the other hand, in case (iv), at first agglomeration economies and later diseconomies dominate.

Now let us compare, in each case, growth paths obtained under the three allocation rules. First, in the case of decreasing returns to scale, we obtain from (6), (92) and Lemma 6.7 that

$$p_r(t) \gtreqless p_s(t) \Leftrightarrow F'(K_r(t)) \gtreqless F'(K_s(t)) \Leftrightarrow g(K_r(t))f'(K_r(t))$$
$$\gtreqless g(K_r(t))f'(K_s(t)), \tag{96}$$

and hence we have the same growth path under any one of the above three allocation rules. Second, in the case of constant returns to scale, we see from (7), (93) and Lemma 6.9 that values of $p_l(t)$, $F'(K_l(t))$ $(= a)$ and $g(K_l(t))f'(K_l(t))$ $(= a)$ are independent of growth paths, and hence growth paths are independent of allocation rules. Third, in the case of increasing returns to scale, we have from (8), (94) and Lemma 6.8 that

$$\text{if } K_r(t) \neq K_s(t), \text{ then } p_r(t) \gtrless p_s(t) \Leftrightarrow F'(K_r(t)) \gtrless$$
$$F'(K_s(t)) \Leftrightarrow g(K_r(t))f'(K_r(t)) \gtrless g(K_s(t))f'(K_s(t)). \tag{97}$$

Hence, if $K_r^0 \neq K_s^0$, we have the same growth paths under any one of the three allocation rules. On the other hand, if $K_r^0 = K_s^0$, the PMP rule and the SMP rule give either one of the following solutions:

A. $\theta_r(t) = 1$ for all t, or $\theta_s(t) = 1$ for all t,
B. $\theta_r(t) = \theta_s(t) = \frac{1}{2}$ for all $t \in [0, \epsilon]$ with some $\epsilon > 0$, and
$\theta_r(t) = 1$ for $t \in (\epsilon, T]$ if $\theta_s(t) = 1$ for $t \in (\epsilon, T]$.

Solution B is possible, but it is a highly unstable solution. Hence if we limit the solution to stable ones, only A is possible. Therefore, in sum we have

Theorem 6.4. For both decreasing and increasing returns to scale, the three allocation rules (i.e., the Optimum, PMP and SMP rules) give the same growth path (note, in the case of increasing returns to scale, we have excluded unstable growth paths). In addition, for constant returns to scale, the three allocation rules give the same set of an infinite number of possible growth paths.

Finally we examine the case of variable returns to scale. The regional production function defined by (95) is depicted in Figure 6.11 (a). The relation between the social marginal productivity of capital, $F'(K)$, and the private marginal productivity of capital, $g(K)f'(K)$, is depicted in Figure 6.11(b).

Under the SMP rule described by (91), the allocation of investment follows the signal of the social marginal productivity of capital, $F'_l(K_l)$. Hence, switching the investment allocation from the more developed region to the less developed region takes place on the curve, $K_s = \hat{\psi}(K_r)$ or $K_r = \hat{\psi}(K_s)$, where function $\hat{\psi}$ is defined by (22) and depicted in Figure 6.12. We call this curve the *SMP switching line*.

Under the PMP rule described by (90), switching the investment allocation from the more developed region to the less developed region takes place on the curve, $K_s = \bar{\psi}(K_r)$ or $K_r = \bar{\psi}(K_s)$, where function $\bar{\psi}$ is defined by the equality of private marginal productivity of capital, namely, by

$$g(\bar{\psi}(K))f'(\bar{\psi}(K)) = g(K)f'(K).$$

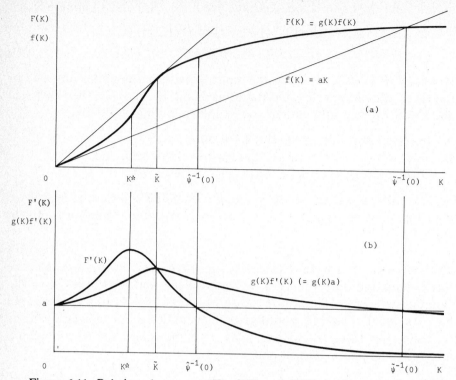

Figure 6.11. Relations between $g(K)$, $f(K)$ and $F(K)$ in the case of variable returns to scale.

But, since $f'(K) = a$ from (95), function $\tilde{\psi}$ is defined by

$$g(\tilde{\psi}(K)) = g(K) \quad \text{where } K \geq \tilde{K} \geq \tilde{\psi}(K), \tag{98}$$

where $\tilde{K}$ represents the amount of capital which gives the maximum value of the private marginal productivity of capital, $g(K)f'(K)$ $(= g(K)a)$. $\tilde{K}$ is shown in Figure 6.11. We call the curve represented by $K_s = \tilde{\psi}(K_r)$ and $K_r = \tilde{\psi}(K_s)$ the *PMP switching line*, and it is depicted in Figure 6.12.

In the case of the Optimal rule, the position of the switching depends on the length of the plan period as we have seen in Section 6.4. But let us limit ourselves to the case where the plan period is sufficiently long since it is our main concern in this study. Then the Optimum switching line is defined by (40), and depicted in Figure 6.12.

The three switching lines are summarized in Figure 6.12. Suppose for example, that the initial position of the economy is given at point *b* in Figure 6.12. Then the growth path corresponding to each allocation

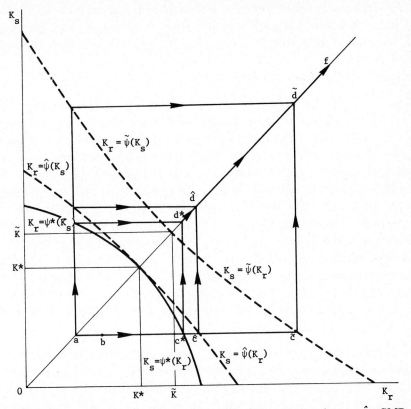

Figure 6.12. Comparison of three switching lines: ψ^*—Optimum, $\hat{\psi}$—SMP, and $\tilde{\psi}$—PMP.

rule is described by:

under the Optimum rule: $b \to c^* \to d^* \to f \to \cdots$,

under the SPM rule: $b \to \hat{c} \to \hat{d} \to f \to \cdots$,

under the PMP rule: $b \to \tilde{c} \to \tilde{d} \to f \to \cdots$.

Note that the path, $b \to \hat{c} \to \hat{d} \to f$, is less efficient than the path, $b \to c^* \to d^* \to f$, because of the following reason. As we see from the analysis in Section 6.4 and Figure 6.4(a), the optimal path is time-invariant and is determined uniquely by initial condition (K_r^0, K_s^0). Hence, any path which starts from b, reaches point c^*, but does not switch the allocation of investment from region r to region s at that time cannot be an optimal path, and hence must be less efficient than the optimal path, $b \to c^* \to d^* \to f$. Similarly, the path, $b \to \tilde{c} \to \tilde{d} \to f$, is less efficient than the path, $b \to \hat{c} \to \hat{d} \to f$. First, note that both paths reach $\hat{c}$ at the same time. Hence, considering this point, $\hat{c}$, as the (new) initial point, the optimal path starting from $\hat{c}$ is unique and given by $\hat{c} \rightarrow \hat{d} \rightarrow f$. Hence, the path, $b \rightarrow \bar{c} \rightarrow \hat{d} \rightarrow f$, is less efficient than the path $b \rightarrow \hat{c} \rightarrow \hat{d} \rightarrow f$.

Generalizing the above finding, we have

Theorem 6.5. In the case of variable returns to scale, suppose the plan period is sufficiently long. When the economy starts from a point below the SMP switching line ($K_s = \hat{\psi}(K_r)$ and $K_r = \hat{\psi}(K_s)$) in Figure 6.12, the growth path under the SMP rule is less efficient than the growth path under the Optimum rule, and the growth path under the PMP rule is the least efficient. When the economy starts from a point between the SMP switching line and the PMP switching line ($K_s = \tilde{\psi}(K_r)$ and $K_r = \bar{\psi}(K_s)$) in Figure 6.12, the Optimum rule and the SMP rule give the same growth path; and the growth path under the PMP rule is less efficient than that growth path. The three allocation rules give the same growth path only when the economy starts from a point above the PMP switching line. Therefore, under the rules of SMP and PMP, more capital tends to be concentrated in the more developed region than the optimal amount.

Comparing Theorems 6.4 and 6.5, we recognize the importance of investment control by the government for the case of variable returns to scale. Under conditions of variable returns to scale, the SMP rule is less efficient than the Optimum rule since the SMP rule cannot correctly take into account the effects of agglomeration economies and diseconomies in the future. The PMP rule is the least efficient since it cannot correctly account for the effects of future agglomeration economies and diseconomies, nor the effects of external economies and diseconomies at present. Hence, if we let investment allocation free in the competitive economy, the allocation of investment will be switched from the more developed region to the less developed region only after the more developed region becomes so congested that agglomeration economies accumulated in the initial stages are completely outweighed by agglomeration diseconomies accumulated in the later stages. To achieve the optimal growth path in the competitive economy the government should impose a sufficiently high congestion tax on investment in the more developed region so as to discourage the over-accumulation of capital. Such an investment tax is required when the economy is below the PMP switching line in Figure 6.12.

Chapter 7

CONCLUDING REMARKS

The objectives of this book have been to investigate the general character of optimal growth paths in spatial systems and to develop the related mathematical tools. But, as we have seen, these objectives were only partially attained, and the basic results achieved in this book constitute only a starting point in the study of growth planning in spatial systems. In this chapter, let us discuss what works remain for further study.

7.1. Within the framework of monotone space systems

In Part II, we mainly explored the duality and related characteristics of optimal growth paths, and the limiting behavior of optimal growth paths was investigated only in a very special model in Chapter 5. But, since the limiting behavior of optimal growth paths is a very important topic in growth planning, it should be more thoroughly studied in a general monotone space system framework. That is, we must develop general turnpike theorems in monotone space systems.

One possible approach to the development of general turnpike theorems in monotone spatial systems would be to utilize and extend the powerful limit theorem found in Rockafellar (1967) and Winter (1967).[1] In particular, a detailed analysis of eigen sets in monotone transformations – not only in the primal transformation but also in the inverse and the adjoint transformations may lead to a much simpler proof of general turnpike theorems (compared with, for example,

---

[1] See Theorem 4 in Part 8 of Rockafellar (1967) and Theorem on p. 73 of Winter (1967). Surprisingly, they developed quite similar theorems independently, and both were published in the same year.

McKenzie (1963) and (1968)). Another approach would be to apply Theorem 4.5 in Makarov and Rubinov (1970) to monotone space systems. This is one of the most robust turnpike theorems at present. Both approaches should be investigated in the future.

Not only are questions concerning limiting behavior important, but so are questions as to the transitionary behavior of optimal growth paths in space systems. Since the amount of each natural resource, especially land, is limited in each location, a location which has advantages in production and/or in transportation at earlier stages of development will decrease its advantage as the system grows. Thus it would be interesting to consider how the optimal spatial pattern of the system changes over time, according to the growth of the system.

The optimal network pattern of transportation arteries over time can also be investigated in an approximate way within the framework of monotone space systems (see Section 3.2.3). In connection with this, we must take into account the recent work by Andersson (1975) which proposes a way to treat time and space inseparably.

Another important topic is the investigation of the relationship between optimal growth paths, which are studied in this book, and competitive growth paths which would be realized in decentralized economies. It is clear from the general equilibrium theory in micro-economics that, when the convexity property prevails in production and transportation, both types of growth paths have a very close relationship. That is, it would be possible to show that each optimal growth path can be realized by a certain decentralized competitive economy, and vice versa. The usefulness of the study of optimal growth paths for actual economic planning will be enhanced by clarifying this interrelationship. In connection with this, we must refer to the recent work by Isard and Liossatos (1975) which has systematically investigated this interrelationship.

7.2. Extension of monotone space systems

In Section 2.5, we extended monotone programming to generalized monotone programming. Hence, it is not difficult to rewrite the results of Part II in terms of "generalized monotone space systems." Besides this extension, two other extensions are necessary to make monotone space systems more useful in actual spatial planning.

One is the introduction of "bads" into the monotone space systems. Namely, since conditions (v) and (vi) in the definition of m.t.cc. assume the free disposability of inputs and outputs, monotone space systems prohibit the consideration of those goods (bads) which are undesirable in production and/or consumption and are costly to be disposed. But, by changing the sign of the variables corresponding to these goods, it would be possible to extend monotone space systems to incorporate these goods (bads). This extension is necessary to handle environmental problems.

Another extension is to design the model to be able to incorporate restrictions on the ranges of commodity-variables and welfare-stock variables. For example, to introduce a political constraint which restricts the income disparities among regions to be within a certain range, we need to introduce a restriction on the allowable ranges of the welfare-stock variables in the system. One way to incorporate this type of restriction is suggested in Section 3.2.5. Another way would be to restrict the domain and range of each function in the model to certain convex cones. Both are not difficult to implement.

7.3. Beyond monotone space systems

As stated in Section 1.1, since monotone space systems are characterized by the convexity of functions and constraint-sets describing them, planning problems involving non-convex properties cannot be studied within the framework of monotone space systems. Sources of non-convex properties of space systems are, among others, indivisibility of commodities, increasing returns to scale, and external economies and diseconomies (in location theory, they are generally called agglomeration or scale economies and diseconomies). The study in Chapter 6 shows that monotone space systems are provided with exceptionally simple mathematical structures (compared with general non-convex spatial systems), and suggests that much of the interesting spatial phenomena can be observed only in non-convex spatial systems. Hence, it is vital to extend our study beyond monotone space systems.

Let us next examine what can be (should be) done within the framework of Chapter 6. Of course, Conjecture 6.1 must be proved. In addition, let us make the following interesting conjecture.

Conjecture 6.2 (spatial turnpike theorem).   In Problem A in Chapter 6, suppose that the number of regions is infinite (i.e., $n = \infty$) and that the initial condition is given such that

$K_l^0 > 0$ for $l = 1, 2, \ldots, n'$,

and $K_l^0 = 0$ for $l = n' + 1, \ldots, \infty$,

where n' is a finite integer, and we assume that $F(0) = 0$. Then, for any given $\epsilon > 0$, there exist positive numbers L_1 and L_2 with the following property: for every optimal growth path $\{(K_l(t))_1^\infty\}_0^T$ $(T > L_1 + L_2)$ beginning at $(K_l^0)_1^\infty$ we have

$$\left| \bar{\psi}^{-1}(0) - \frac{\sum\limits_{l=1}^{\infty} K_l(t)}{N(t)} \right| \leq \epsilon \quad \text{if } L_1 < t < T - L_2,$$

where $N(t)$ is the number of regions at time t which have a non-zero amount of capital at time t, and $\bar{\psi}$ is the function defined by (45) (i.e., (46)) in Chapter 6.

That is, when the number of regions is very large and the plan period is sufficiently long, then the average capital stock per region on the optimal path will most of the time be very close to capital stock level $\bar{\psi}^{-1}(0)$ at which each region has the maximum average productivity of capital.

There are two possible approaches for the proof of the above conjecture. One would be to apply the traditional method of the proof of turnpike theorems in non-spatial systems. The other approach would be, by using the methods in Chapter 6, to study in detail the properties of the optimal growth path when the plan period is relatively short compared to the number of regions in the system.

Though the above conjecture is theoretically interesting, the framework of Chapter 6 itself must be extended for the study of the case where the number of regions in the system is very large. That is, given a nation (or the world), the increase in the number of regions implies the decrease in the size of each region, which in turn implies that each region has many other regions within a relatively short distance from it. Then, important factors from the point of view of spatial analysis are not only the agglomeration economies and diseconomies within each region, but also these (dis)economies among regions.

There would be two different ways to express these agglomeration economies and diseconomies among regions in a model. One way is to follow Leontief (1965). We first divide the economy (a nation) into several regions (a priori), then divide each region into several local areas, and then divide each local area into several districts, and so on. Next, corresponding to this hierarchical spatial division of the economy, we introduce multiple orders of agglomeration economies and diseconomies: agglomeration economies and diseconomies among the regions (i.e., national level agglomeration economies and diseconomies), agglomeration economies and diseconomies among the local areas within a region (regional level agglomeration economies and diseconomies), and so on. Another way is to first divide the economies into a very large number of regions, and then express the agglomeration economies and diseconomies among regions by using potential type (and/or gravity type) measures. Whichever method we may choose, the optimal (competitive) growth path of the economy will assume much more complex and interesting spatial structures than the paths analyzed in Chapter 6.

In addition, we must introduce multiple economic sectors and factors into the model in Chapter 6.

The final step in the study of spatial systems would be to construct a general non-convex spatial system (which includes convex spatial systems as its special cases), and explore the general character of the optimal (competitive) growth path in that space system. But, this would not be easy since, for this purpose, we must first develop a general programming theory with non-convex structures. Hence, before that final step, we must perform many experimental works like the one in Chapter 6.

BASIC MATHEMATICS

In this appendix, some basic mathematical concepts necessary for the understanding of the analyses in this book are summarized in two sections. Topological properties of finite dimensional Euclidian spaces are explained in Section A.1, and some definitions and theorems related to convex sets and convex functions are summarized in Section A.2. Statements are given without proofs in this appendix. For a complete treatment of the topics, the reader is referred to, for example, Berge (1963) and Nikaido (1968).

A.1. Topological properties of Euclidian spaces

A.1.1. Euclidian spaces

The *n-dimensional Euclidian space* $\mathbf{R}^n$ is the set of all *n*-tuples of real numbers. Hence, each element x of $\mathbf{R}^n$ can be represented by

$$x = (x_1, x_2, \ldots, x_i, \ldots, x_n),$$

where each x_i is an appropriate real number.

By introducing the following conventional vectorial operations, we consider $\mathbf{R}^n$ to be the *n-dimensional vector space*, and each element of $\mathbf{R}^n$ is called a *vector*: Given two vectors

$$x = (x_1, x_2, \ldots, x_n), \quad y = (y_1, y_2, \ldots, y_n),$$

the *sum* $x + y$ is defined by

$$x + y = (x_1 + y_1, x_2 + y_2, \ldots, x_n + y_n).$$

Given a vector x and a real number α, the *scalar multiplication* αx is defined by

$$\alpha x = (\alpha x_1, \alpha x_2, \ldots, \alpha x_n).$$

Given two vectors x and y, the *innerproduct* $x \cdot y$ is defined by

$$x \cdot y = x_1 y_1 + x_2 y_2 + \cdots + x_n y_n.$$

The *(Euclidian) norm* $\|x\|$ of vector x is defined by

$$\|x\| = \sqrt{(x \cdot x)}.$$

Finally, the *(Euclidian) distance* between two vectors x and y is defined by the norm $\|x - y\|$ of vector $x - y$.

The Euclidian norm satisfies the following relations.

(i) $\|x\| \geqq 0$, with the equality holding if and only if $x = (0, 0, \ldots, 0)$,
(ii) $\|\alpha x\| = |\alpha| \|x\|$,
(iii) $\|x + y\| \leqq \|x\| + \|y\|$,
(iv) $|x \cdot y| \leqq \|x\| \|y\|$, with equality holding if and only if $x = \alpha y$ for some real number α,

where $x, y \in \mathbf{R}^n$, and $|\alpha|$ and $|x \cdot y|$ are, respectively, absolute values of α and $x \cdot y$.[1] Relation (iv) is called the Cauchy-Schwartz inequality.

In the following, an n-dimensional vector is often called a point of $\mathbf{R}^n$.

A.1.2. Open sets, closed sets and compact sets

Let

$$A_1, A_2, \ldots, A_k$$

be a collection of sets (in Euclidian space $\mathbf{R}^n$). The *union* of these sets is the set of elements which belong to at least one of these sets, and is denoted by

$$A_1 \cup A_2 \cup \cdots \cup A_k.$$

The *intersection* of these sets is the set of elements each of which

---

[1] As the norm of each vector, we can choose any real valued function on $\mathbf{R}^n$ which satisfies relations (i), (ii) and (iii). And, under this generalization, the concepts of open sets and closed sets presented in the next section also hold to be the same. For example, in applications in economics, it is often convenient to use the function, $\max_i |x_i|$, as the norm of vector x.

belongs simultaneously to all of these sets, and is denoted by

$$A_1 \cap A_2 \cap \cdots \cap A_k.$$

Given two sets A and B, if every element of A is also an element of B, then A is called a *subset* of B. This is expressed as $A \subset B$. Note that the expression $A \subset B$ does not exclude the possibility that A is equal to B. Of course, any set in $\mathbf{R}^n$ is a subset of $\mathbf{R}^n$. Finally, let A be a set in $\mathbf{R}^n$. Then the set of all points of $\mathbf{R}^n$ which do not belong to A is called the *complement* of A.

Next, let x be a point of $\mathbf{R}^n$, and r be a positive number. Then, the set of all points of $\mathbf{R}^n$ of which distances to x are less than r is called the *open ball* of radius r with center x, and it is denoted by $\mathring{B}_r(x)$. That is,

$$\mathring{B}_r(x) = \{y \mid \|y - x\| < r, y \in \mathbf{R}^n\}.$$

A set G in $\mathbf{R}^n$ is called an *open set* when, for any point x of G, there exists an open ball with center at x which is entirely contained in G. Each open set in $\mathbf{R}^n$ which contains point x is called a *neighborhood* of point x. Then, of course, every open ball with center at x is also a neighborhood of point x.

A point x in set A of $\mathbf{R}^n$ is called an *interior point* of set A if there exists a neighborhood of point x which is contained entirely in A; the set of interior points of A is denoted by $\mathring{A}$ and is called the *interior* of A. Clearly $\mathring{A} \subset A$. And, we easily see that a set G is an open set if and only if $\mathring{G} = G$, that is, if and only if every point of G is an interior point of G.

Example 1. An open interval, $(a, b) = \{x \mid a < x < b\}$, in $\mathbf{R}^1$ is an open set. Any open ball in $\mathbf{R}^n$ is an open set. The set G depicted in Figure A.1 is an open set in $\mathbf{R}^2$, and $U(a)$ is a neighborhood of a which is contained in G.

Next, let x be a point of $\mathbf{R}^n$, and A be a set in $\mathbf{R}^n$. Then, we call x a *point of accumulation* of set A if every neighborhood of x contains some point of A which is different from x. If x is a point of accumulation of A, every neighborhood of x contains an infinite number of points of A. A point of accumulation of set A is not necessarily contained in A. But, if a set A in $\mathbf{R}^n$ contains all of the points of accumulation of itself, then we call A a *closed set* in $\mathbf{R}^n$.

A point x in $\mathbf{R}^n$ is called a *point of closure* of set A if every neighborhood of x contains some point of A (which may be x itself

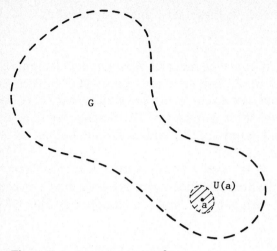

Figure A.1. An open set in $\mathbf{R}^2$.

when x is in A). The set of points of closure of set A is denoted by $\bar{A}$ and is called the *closure* of A. Clearly, $A \subset \bar{A}$. And, it is not difficult to see that closure $\bar{A}$ of any set A is a closed set, and that a set A is a closed set if and only if $\bar{A} = A$, that is, if and only if every point of closure of A is in A.

Example 2. A closed interval, $[a, b] = \{x \mid a \leqq x \leqq b\}$, in $\mathbf{R}^1$ is a closed set. The set of all points of $\mathbf{R}^n$ of which distance to x are less than or equal to r is called the *closed ball* of radius r with center x, and it is denoted by $B_r(x)$. That is,

$$B_r(x) = \{y \mid \|y - x\| \leqq r, y \in \mathbf{R}^n\},$$

and it is a closed set. Open ball $\mathring{B}_r(x)$ is the interior of closed ball $B_r(x)$, and the closure of $\mathring{B}_r(x)$ is $B_r(x)$.

Example 3. In Figure A.2, the set F which is the union of set A and point a is a closed set in $\mathbf{R}^2$. Point x is an interior point, a point of accumulation as well as a point of closure of F. Point y is not an interior point of F, but is a point of accumulation as well as a point of closure of F. Point a is neither an interior point nor a point of accumulation of F, but is a point of closure of F.

Important properties of open sets and closed sets are:

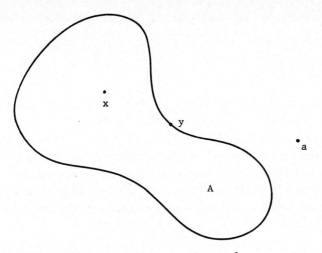

Figure A.2. A closed set $F = A \cup \{a\}$ in $\mathbf{R}^2$.

Theorem A.1. In the Euclidian vector space $\mathbf{R}^n$,

(i) the complement of any open set is a closed set, and the complement of any closed set is an open set,

(ii) the union of any number (possibly an infinite number) of open sets is an open set, and the intersection of any number (possibly an infinite number) of closed sets is a closed set,

(iii) the intersection of a finite number of open sets is an open set, and the union of a finite number of closed sets is a closed set,

(iv) $\mathbf{R}^n$ and the null set $\emptyset$ are open sets as well as closed sets.

Next, let

$$x^1, x^2, \ldots, x^j, \ldots, \tag{1}$$

be a *sequence* of elements of $\mathbf{R}^n$.[2] This sequence is often represented by $\{x^j\}_{j=1}^\infty$, or simply, by $\{x^j\}$. We say sequence (1) is *bounded* if there exists a number r such that $\|x^j\| < r$ for every x^j of (1). We say sequence (1) *converges* to a point $\bar{x}$ if

$$\lim_{j \to \infty} \|x^j - \bar{x}\| = 0,$$

[2]Note that, precisely, a sequence of elements of a set S is defined as a mapping from the set of all positive integers into set S. Two elements on a sequence assigned different numbers might be the same point of $\mathbf{R}^n$. Hence, a sequence of $\mathbf{R}^n$ is essentially different from the set of points in $\mathbf{R}^n$ consisting of points in that sequence. In particular, the latter set might contain only a finite number of points.

and $\bar{x}$ is called the *limit point* of sequence (1).[3] We often represent this convergence by

$$\lim_{j \to \infty} x^j = \bar{x}, \quad \text{or } x^j \to \bar{x}.$$

When sequence (1) converges to point $\bar{x}$ and contains an infinite number of different points of $\mathbf{R}^n$, then it can be shown that $\bar{x}$ is a point of accumulation of the set of all points contained in that sequence, and further that it is the unique point of accumulation of that set. A sequence in $\mathbf{R}^n$ cannot have more than one limit point. It can also be shown that if a sequence in $\mathbf{R}^n$ is bounded, we can choose a convergent sub-sequence from it. Using these properties, we can show that

Theorem A.2. In the Euclidian vector space $\mathbf{R}^n$,

(i) a bounded set which contains an infinite number of different points has a point of accumulation,

(ii) a point x is contained in set $\bar{A}$, if and only if there exists a sequence in A converging to x,

(iii) a set F is closed if and only if every convergent sequence in F has its limit point within F, that is, if and only if

$$x^j \in F, j = 1, 2, \ldots, x^j \to \bar{x} \text{ implies that } \bar{x} \in F.$$

Finally, one of the topological concepts that is frequently used in the text is compactness. A subset C of $\mathbf{R}^n$ is *compact* if any subset of C which contains an infinite number of different points has a point of accumulation within it. In terms of the concept of convergence, compactness can be equivalently defined as follows: a subset C of $\mathbf{R}^n$ is compact if any sequence of points in C contains a subsequence that converges to a point in C. It can be shown that

Theorem A.3. A subset C of $\mathbf{R}^n$ is compact if and only if C is bounded and closed in $\mathbf{R}^n$.

Note, in the above theorem, that a set in $\mathbf{R}^n$ is said to be *bounded* if it is contained in a certain closed ball with center at the origin of $\mathbf{R}^n$.

---

[3]That is, $\bar{x}$ is the limit point of sequence $\{x^j\}$ in $\mathbf{R}^n$ if for any $\epsilon > 0$, there exists an integer $n(\epsilon)$ such that $\|x^j - \bar{x}\| < \epsilon$ for every $j \geq n(\epsilon)$.

A.1.3. *Continuous mappings (single-valued)*

Let X and Y be any two sets. A *mapping f* of *X into Y* is a rule which assigns to each element x of X a uniquely defined element $f(x)$ of Y.[4] The element, $y = f(x)$, of Y is called the *image* of x under the mapping f. A shorthand notation for the statement that f is a mapping of X into Y is

$$f: X \rightarrow Y.$$

Mappings are often called *functions* or *transformations*.

Let A be a subset of X. Then, the *image $f(A)$* of A under a mapping $f: X \rightarrow Y$ is the set of all y in Y such that $y = f(x)$ for some x in A. Conversely, let B be a subset of Y. Then, the *inverse image $f^{-1}(B)$* of B under a mapping $f: X \rightarrow Y$ is the subset of X consisting of all the elements x of X such that $f(x) \in B$.

Next, let $\mathbf{R}^m$ and $\mathbf{R}^n$ be m-dimensional and n-dimensional Euclidian vector spaces, A be a subset of $\mathbf{R}^m$, and f be a mapping from A into $\mathbf{R}^n$. Mapping f is called *continuous at a point* $\bar{x}$ of A if for any positive real number ϵ, there exists a positive real number δ such that

$$\text{for any } x \in A, \quad \text{if } \|x - \bar{x}\| < \delta, \quad \text{then } \|f(x) - f(\bar{x})\| < \epsilon.$$

This definition of continuity is equivalent to either one of the following two statements.

(i) For each neighborhood $V(f(\bar{x}))$ of $f(\bar{x})$ in $\mathbf{R}^n$, there exists a neighborhood $U(\bar{x})$ of $\bar{x}$ in $\mathbf{R}^m$ such that

$$\text{if } x \in U(\bar{x}) \cap A, \quad \text{then } f(x) \in V(f(\bar{x})).$$

(ii) $x^j \rightarrow \bar{x}$ implies $f(x^j) \rightarrow f(\bar{x})$.

The function f is called *continuous on A* if it is continuous at every point of A.

The above definition of continuity is the mathematical formalization of the intuitive statement that points sufficiently close to $\bar{x}$ in set A are mapped by f into points sufficiently close to $f(\bar{x})$ in $\mathbf{R}^n$.

An important property of continuous functions is

---

[4]To differentiate from set-valued mappings discussed in the next section, we may call this type of mapping *single-valued mappings*. But, since we are only concerned in this section with single-valued mappings, we simply use the term, mapping.

Theorem A.4. Let $\mathbf{R}^m$ and $\mathbf{R}^n$ be Euclidian vector spaces, and C be a compact subset of $\mathbf{R}^m$. Then the image $f(C)$ under a continuous mapping $f: C \to \mathbf{R}^n$ is also compact.

If, in the above theorem, we take f to be a continuous numerical function (that is, $\mathbf{R}^n = \mathbf{R}^1$), then $f(C)$ is a bounded closed set (i.e., a compact set) of real numbers. Hence, we obtain the next theorem.

Theorem A.5. A real-valued continuous function defined on a compact subset of $\mathbf{R}^m$ takes a maximum value and a minimum value.

A.1.4. Closed mappings (multi-valued)

Let X and Y be two sets. If with each element x of X we associate a subset $F(x)$ of Y, we say that the correspondence $x \to F(x)$ is a *multi-valued mapping* of X into Y; the set $F(x)$ is called the image of x under the mapping F.

Multi-valued mappings are often called multi-valued functions or multi-valued transformations. In the rest of this section, they are simply called mappings.

If F is a mapping of X into Y and A is a non-empty subset of X, we write

$$F(A) = \bigcup_{x \in A} F(x).$$

If $A = \emptyset$, we unite $F(\emptyset) = \emptyset$. The set $F(A)$ is called the image of A under the mapping F. It is easy to see that $A \subset B$ implies that $F(A) \subset F(B)$. The *graph* of mapping F of X into Y is a subset of $X \times Y$ defined by[5]

$$\{(x, y) | x \in X, y \in Y, y \in F(x)\}.$$

The *lower inverse* F^{-1} of mapping F of X into Y is a mapping of Y into X defined by

$$F^{-1}(y) = \{x | x \in X, y \in F(x)\} \quad \text{for each } y \in Y.$$

---

[5] $X \times Y$ denotes the *Cartesian product* of sets X and Y, that is,

$\quad X \times Y = \{(x, y) | x \in X, y \in Y\}.$

In the text, a special type of multi-valued mapping, closed mappings, are frequently used, which are defined as follows.

Let A be a subset of $\mathbf{R}^m$. Then, a mapping F of A into $\mathbf{R}^n$ is said to be *closed* at $x \in A$, if

$$\left.\begin{array}{l} y^j \in F(x^j), j = 1, 2, \ldots, \\ y^j \to y, x^j \to x, \end{array}\right\} \quad \text{implies } y \in F(x).$$

Mapping F is called a *closed mapping* if it is closed at every $x \in A$. Then, it is obvious that a mapping F of A into $\mathbf{R}^n$ is closed if and only if the graph of F

$$\{(x, y) | x \in A, y \in F(x)\},$$

is a closed subset of $\mathbf{R}^m \times \mathbf{R}^n$. And, another immediate consequence of the definition is that if F is a closed mapping then the set $F(x)$ is closed in $\mathbf{R}^n$.

A.2. Convex sets and convex functions in Euclidian spaces

A.2.1. Convex sets

A subset C of $\mathbf{R}^n$ is said to be *convex* if

$$\left.\begin{array}{l} x, y \in C \\ 0 \leq \alpha \leq 1 \end{array}\right\} \quad \text{implies } \alpha x + (1 - \alpha)y \in C.$$

That is, a convex set is a set such that a segment joining any two points in the set is contained in the set.

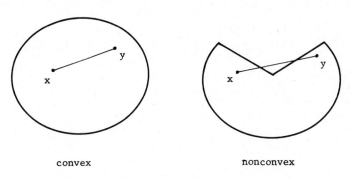

convex   nonconvex

Figure A.3. A convex set and a nonconvex set.

Example 4.   Examples of convex sets are

(a) a segment $[a, b]$ with endpoints a, b in $\mathbf{R}^n$ defined by $[a, b] = \{x \in \mathbf{R}^n | x = \alpha a + (1 - \alpha)b,\ 0 \leqq \alpha \leqq 1\}$,
(b) a ball (open or closed) in $\mathbf{R}^n$,
(c) the non-negative orthant, $\mathbf{R}^n_+ = \{x | x \in \mathbf{R}^n,\ x \geqq 0\}$, of $\mathbf{R}^n$,
(d) the half space of $\mathbf{R}^n$ given by the set, $\{x \in \mathbf{R}^n | a \cdot x \geqq \alpha$, where $a \in \mathbf{R}^n$,
(e) $\mathbf{R}^n$ itself.

It is not difficult to see that

Theorem A.6.   If A and B are two convex sets in $\mathbf{R}^n$, then

(i) their intersection $A \cap B$ is a convex set,
(ii) their *vectorial sum*, $A + B = \{x + y | x \in A,\ y \in B\}$, is a convex set,
(iii) the set, $\lambda A = \{\lambda x | x \in A\}$, is a convex set for any real number λ.

Next, we say that a set A in $\mathbf{R}^n$ is a *cone* if

$$x \in A \quad \text{and} \quad \lambda \geqq 0 \quad \text{implies } \lambda x \in A.$$

And, if a cone is convex, it is called a *convex cone*. That is, a set K in $\mathbf{R}^n$ is a convex cone if

$$\left. \begin{array}{l} x, y \in K \\ \alpha \geqq 0,\ \beta \geqq 0 \end{array} \right\} \text{ implies } \alpha x + \beta y \in K.$$

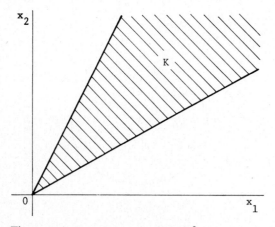

Figure A.4. A convex cone K in $\mathbf{R}^2$.

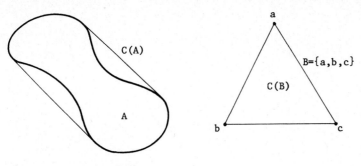

Figure A.5. Convex hulls.

An arbitrary set A in $\mathbf{R}^n$ need not be convex. But, one naturally expects the existence of a smallest convex superset of A that may be obtained by filling the hollows of X. That is, the smallest convex superset of a set A in $\mathbf{R}^n$, in the sense of set-theoretic inclusion, is termed the *convex hull* of A, and denoted by $C(A)$. The set $C(A)$ always exists, and is equal to the set of all convex linear combinations of points of A. Actually, $C(A)$ of a subset A of $\mathbf{R}^n$ is given by

$$C(A) = \left\{ \sum_{j=1}^{n+1} \alpha_j x^j \,\middle|\, \sum_{j=1}^{n+1} \alpha_j = 1, \alpha_j \geqq 0, x^j \in A, j = 1, 2, \ldots, n+1 \right\}.$$

That is, to form the convex hull of a subset A of $\mathbf{R}^n$, it is sufficient to choose $n+1$ points of A for the construction of a convex combination. In Figure A.5, $C(A)$ is the convex hull of set A, and $C(B)$ is the convex hull of three points, a, b and c.

A.2.2. Supporting hyperplanes and separation theorems

Let $p = (p_1, p_2, \ldots, p_i, \ldots, p_n) \in \mathbf{R}^n$ with $p \neq 0$. Then, the set of points x of $\mathbf{R}^n$ which satisfies the following equation

$$p \cdot x = \alpha, \quad \text{that is,} \ \sum_{i=1}^{n} p_i x_i = \alpha,$$

is called a *hyperplane* in $\mathbf{R}^n$ with *normal* p, where α is a real number. If $n = 2$, a hyperplane is a straight line, and if $n = 3$, a hyperplane is a plane. A hyperplane, $p \cdot x = \alpha$, in $\mathbf{R}^n$ divides $\mathbf{R}^n$ into the following two *closed half-spaces*:

$$\{x | p \cdot x \geqq \alpha, x \in \mathbf{R}^n\} \quad \text{and} \quad \{x | p \cdot x \leqq \alpha, x \in \mathbf{R}^n\}.$$

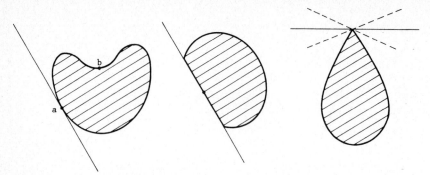

Figure A.6. Supporting hyperplanes.

Next, let A be a set in $\mathbf{R}^n$. A hyperplane $p \cdot x = \alpha$ is called a *supporting hyperplane* of A if (1) A lies in one of the two closed-half spaces (either $p \cdot x \geqq \alpha$ or $p \cdot x \leqq \alpha$) and (2) the hyperplane has a point in common with A. To be more specific, if a is one of the common points of A and the hyperplane, we use the expression, a "*supporting hyperplane at a*," or "the hyperplane supports A at a."

Examples of supporting hyperplanes are depicted in Figure A.6.

In the first example, set A is supported by a straight line at a, but there is no straight line which can support set A at b. Hence, a question of importance is the existence of a hyperplane which supports a set at a given point. The next theorem provides an answer to this question.

Theorem A.7. Let A be a convex set in $\mathbf{R}^n$. If a point a is not an interior point of A, there exists a hyperplane $p \cdot x = p \cdot a$ passing through point a such that A is contained in the half-space $p \cdot x \geqq p \cdot a$. Here $p \in \mathbf{R}^n$.

The following theorem is a slightly different version of the above theorem.

Theorem A.8. Let A be a non-empty closed convex set in $\mathbf{R}^n$. If a point a does not belong to A, then we have the following:

(i) There exists a hyperplane $p \cdot x = \alpha$ such that $p \cdot a < \alpha$, while the half-space $p \cdot x \geqq \alpha$ contains A.

(ii) For the same p above, the hyperplane $p \cdot x = p \cdot a$ produces the open half-space $p \cdot x > p \cdot a$ containing A.

The following theorem is particularly important for applications in economics.

Theorem A.9. Let A be a convex set in $\mathbf{R}^n$ containing no positive point. Then there is a separating hyperplane $p \cdot x = 0$ having a semipositive normal $p \geq 0$ such that the half space $p \cdot x \leq 0$ contains A. Here $p \in \mathbf{R}^n$.

A.2.3. Convex functions and concave functions

Let F be a real-valued function defined on a convex set C in $\mathbf{R}^n$. Then f is called a *concave function* if for all $x, y \in C$,

$$f(\alpha x + (1 - \alpha)y) \geq \alpha f(x) + (1 - \alpha)f(y),$$

for any $\alpha, 0 \leq \alpha \leq 1$. Then function f is called a *convex function* if for all $x, y \in C$,

$$f(\alpha x) + (1 - \alpha)y) \leq \alpha f(x) + (1 - \alpha)f(y),$$

for any $\alpha, 0 \leq \alpha \leq 1$. Thus f is convex if and only if $-f$ is concave. If the above inequalities are strict for $x \neq y$ and $\alpha \neq 0, 1$, we say that f is *strictly concave*, or *strictly convex*.

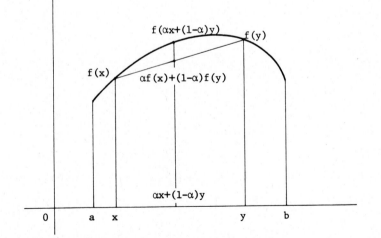

Figure A.7. A concave function on an interval $[a, b]$.

Example 5. A linear function $f(x) = \sum_{i=1}^{n} a_i x_i + b$ is both concave and convex on $\mathbf{R}^n$.

Some important properties of concave (convex) functions are:

Theorem A.10. Let f be a concave (convex) function on C in $\mathbf{R}^n$. Then for any fixed number α, the set

$$\{x | f(x) \geqq \alpha\} \quad (\text{resp. } \{x | f(x) \leqq \alpha\}),$$

is convex.

Theorem A.11. Let f be a concave (convex) function on C in $\mathbf{R}^n$. Then, f is continuous at any interior point of C.

Besides concave (convex) functions, homogeneous functions are often used in the text. Their definition is given as follows: a real-valued function $f(x)$ on a cone K in $\mathbf{R}^n$ (for example $\mathbf{R}^n_+$) is said to be *homogeneous of degree* $\alpha(\alpha \geqq 0)$ if

$$f(\lambda x) = \lambda^\alpha f(x) \quad \text{for all } x \in K \text{ and } \lambda > 0.$$

When $\alpha = 1$, f is said to be *linearly homogeneous.*

BASIC PROPERTIES OF MONOTONE TRANSFORMATIONS

The definitions of terms and the basic results in monotone programming are summarized here from Rockafellar (1967). The topics included are only those which are used in this book. See the original work by Rockafellar (1967) for the complete study of monotone programming and for results on such as sub-eigenvalues and growth rates, eigensets, and behavior in the limit. The parenthesized terms indicate the corresponding page in Rockafellar (1967).

B.1. Definition of monotone transformations

B-definition 1 (p. 9)[1]

A *monotone transformation T of concave type*[2] (hereafter abbreviated to m.t.cc. *T*) from $\mathbf{R}_+^m$ into $\mathbf{R}_+^n$ is a multivalued mapping such that

(i) $T(x^1 + x^2) \supset T(x^1) + T(x^2)$ for all $x^1 \in \mathbf{R}_+^m$, $x^2 \in \mathbf{R}_+^m$,

(ii) $T(\lambda x) = \lambda T(x)$ for all $x \in \mathbf{R}_+^m$ and $\lambda > 0$,

(iii) $T(0) = \{0\}$,

(iv) T is a closed mapping,

(v) $0 \leq x^1 \leq x^2$ implies $T(x^1) \subset T(x^2)$, and

(vi) $0 \leq y^1 \leq y^2 \in T(x)$ implies $y^1 \in T(x)$.

Dually, T is a *monotone transformation of convex type* (hereafter

---

[1](P. 9) means that this definition comes from page 9 in Rockafellar (1967).

[2]In Rockafellar (1967) this is called a *monotone process* of concave type. But the term, "process" has a special meaning in economics (see Koopmans, p. 76). Thus a slightly different name is used in this book.

abbreviated to m.t.cv.) if condition (iii) is omitted and conditions (v) and (vi) are replaced by

(v′) $x^1 \geqq x^2 \geqq 0$ implies $T(x^1) \subset T(x^2)$, and

(vi′) $y^1 \geqq y^2 \in T(x)$ implies $y^1 \in T(x)$.

This definition is slightly different from the one which is given in Rockafellar (1967), but clearly they are equivalent to each other.

The next property gives an alternative definition of these transformations.

B-property 1[3]

A multivalued mapping T from $\mathbf{R}_+^m$ into $\mathbf{R}_+^n$ is a m.t.cc. if and only if the following conditions are satisfied.

(a) graph T is a closed convex cone in $\mathbf{R}_+^m \times \mathbf{R}_+^n$,

(b) $(0, y) \in$ graph T implies $y = 0$,

(c) $(x, y) \in$ graph T, $x' \geqq x$, $y \geqq y' \geqq 0$ imply $(x', y') \in$ graph T.

For a m.t.cv., omit condition (b), and (c) is replaced by

(c′) $(x, y) \in$ graph T, $x \geqq x' \geqq 0$, $y' \geqq y$ imply $(x', y') \in$ graph T.

For the development of theories on monotone transformations, the following definitions are needed.

B-definition 2 (p. 10)

A *monotone concave gauge* on $\mathbf{R}_+^m$ is a continuous real-valued function f on $\mathbf{R}_+^m$ such that

(a) $f(\lambda x) = \lambda f(x)$ for all $x \geqq 0$ and $\lambda > 0$,

(b) $f(x^1 + x^2) \geqq f(x^1) + f(x^2)$ for all, $x^1 \geqq 0$, $x^2 \geqq 0$, and

(c) $f(x^1) \leqq f(x^2)$ for $0 \leqq x^1 \leqq x^2$.

A *monotone convex gauge* is subjected to the same requirements except that (b) is replaced by

(b′) $f(x^1 + x^2) \leqq f(x^1) + f(x^2)$ for all $x^1 \geqq 0$, $x^2 \geqq 0$.

---

[3]This property is not from Rockafellar (1967). But the equivalence of B-definition 1 and Property 1 is easily shown.

B-definition 3 (p. 10)

A *monotone set of concave type* in $\mathbf{R}_+^m$ is a non-empty closed bounded convex set C such that

$$0 \leq y^1 \leq y^2 \in 0 \quad \text{implies } y^1 \in C.$$

A *monotone set of convex type* is a non-empty closed convex set C such that

$$y^1 \geq y^2 \in C \quad \text{implies } y^1 \in C.$$

There exists the following relation between them.

B-property 2 (p. 12)

If T is a m.t.cc. (resp. m.t.cv.) from $\mathbf{R}_+^m$ into $\mathbf{R}_+^n$, then $T(x)$ is a monotone set of concave type (resp. convex type) in $\mathbf{R}_+^n$ for all $x \in \mathbf{R}_+^m$.

B.2. Adjoints and Kuhn–Tucker functions

Recall that each linear transformation A from $\mathbf{R}^m$ into $\mathbf{R}^n$ induces a bi-linear function

$$K(x, y^*) = \langle Ax, y^* \rangle \quad \text{on } \mathbf{R}^m \times \mathbf{R}^n,$$

and each bi-linear function K is of this form for a unique A. It follows from this fact, by reversing the roles of x and y^*, that for each A there exists a unique adjoint linear transformation A^* from $\mathbf{R}^n$ into $\mathbf{R}^m$ such that

$$\langle Ax, y^* \rangle = \langle x, A^*y^* \rangle \quad \text{for all } x, y^*.$$

In the following, B-properties 4 and 5 show the analogous properties for monotone transformations with the aid of B-definition 4 and B-property 3.

B-definition 4 (pp. 14, 17)

For a monotone set C of concave type in $\mathbf{R}_+^m$, $\langle C, x^* \rangle = \sup \{x \cdot x^* | x \in C\}$ for each $x^* \in \mathbf{R}_+^{*m}$, for that of convex type,

$$\langle C, x^* \rangle = \inf \{x \cdot x^* | x \in C\} \quad \text{for each } x^* \in \mathbf{R}_+^{*m}.$$

And given a monotone transformation T from $\mathbf{R}_+^m$ into $\mathbf{R}_+^n$ the function

$$\langle T(\cdot), \cdot \rangle,$$

on $\mathbf{R}_+^m \times \mathbf{R}_+^{*n}$ is called the *Kuhn–Tucker function* of T. Of course, by definition and B-property 2,

$\langle T(x), y^* \rangle = \sup \{ y \cdot y^* | y \in T(x) \}$ for T of concave type,

$\langle T(x), y^* \rangle = \inf \{ y \cdot y^* | y \in T(x) \}$ for T of convex type.

B-property 3 (p. 14)

If C is a monotone set of concave type in $\mathbf{R}_+^m$, then $\langle C, \cdot \rangle$ is a monotone convex gauge on $\mathbf{R}_+^{*m}$. Conversely, each monotone convex gauge f^* on $\mathbf{R}_+^{*m}$ is of this form for a unique monotone set C of concave type in $\mathbf{R}_+^m$, namely $f^* = \langle C, \cdot \rangle$

where $C = \{ x | \langle x, x^* \rangle \leq f^*(x^*), x \geq 0, \quad \text{for all } x^* \in \mathbf{R}_+^{*m} \}$.

Dually, the formulas

$f^*(x^*) = \langle C, x^* \rangle$ for all $x^* \in \mathbf{R}_+^{*m}$

$C = \{ x | \langle x, x^* \rangle \geq f^*(x^*), x \geq 0, \quad \text{for all } x^* \in \mathbf{R}_+^{*m} \}$.

define a one-to-one correspondence between the monotone sets of convex type in $\mathbf{R}_+^m$ and the monotone concave gauges on $\mathbf{R}_+^{*m}$.

With the aid of this property, the following properties are obtained which are analogous to linear transformations.

B-property 4 (p. 17)

If T is a m.t.cc. from $\mathbf{R}_+^m$ into $\mathbf{R}_+^n$, the $\langle T(x), y^* \rangle$ is a monotone convex gauge on $\mathbf{R}_+^{*n}$ for each fixed $x \in \mathbf{R}_+^m$ and is a monotone concave gauge on $\mathbf{R}_+^m$ for each fixed $y^* \in \mathbf{R}_+^{*n}$. Conversely, each function on $\mathbf{R}_+^m \times \mathbf{R}_+^{*n}$ with the latter properties is the Kuhn–Tucker function of a unique m.t.cc. T from $\mathbf{R}_+^m$ into $\mathbf{R}_+^n$.

Dually, the m.t.cv. from $\mathbf{R}_+^m$ into $\mathbf{R}_+^n$ corresponds one-to-one with the function K on $\mathbf{R}_+^m \times \mathbf{R}_+^{*n}$, such that $K(x, \cdot)$ is a monotone concave gauge on $\mathbf{R}_+^{*n}$ for each $x \in \mathbf{R}_+^m$ and $K(\cdot, y^*)$ is a monotone convex gauge on $\mathbf{R}_+^m$ for each $y^* \in \mathbf{R}_+^{*n}$.

From this property, the next one is directly obtained.

B-property 5 (p. 19)

Given any monotone transformation T from $\mathbf{R}_+^m$ into $\mathbf{R}_+^n$, there exists a unique monotone transformation T^* of opposite type from $\mathbf{R}_+^{*n}$ into $\mathbf{R}_+^{*m}$ such that

$$\langle T(x), y^* \rangle = \langle x, T^*(y^*) \rangle \quad \text{for all } x \in \mathbf{R}_+^m \text{ and } y^* \in \mathbf{R}_+^{*n}. \tag{1}$$

B-definition 5 (p. 19)

The monotone transformation T^* associated with the monotone transformation T by the above formula is called the *adjoint* of T. Of course,

$$(T^*)^* = T.$$

Thus the adjoint operation $T \to T^*$ sets up a one-to-one type-reversing order-preserving correspondence between the monotone transformations from $\mathbf{R}_+^m$ into $\mathbf{R}_+^n$ and the monotone transformation from $\mathbf{R}_+^{*n}$ into $\mathbf{R}_+^{*m}$.

An economic meaning of the adjoint transformation is obtained from the following property.

B-property 6 (p. 20)

For the adjoint transformation T^* of a m.t.cc. T, $x^* \in T^*(y^*)$ for each $y^* \in \mathbf{R}_+^{*n}$ if and only if

$$x^* \in \{x^* | x^* \geqq 0, x^* \cdot x \geqq y^* \cdot y \quad \text{for all } (x, y) \in \text{graph } T\}, \tag{2}$$

when T is a m.t.cv.,

$$x^* \in \{x^* | x^* \geqq 0, x^* \cdot x \leqq y^* \cdot y \quad \text{for all } (x, y) \in \text{graph } T\}. \tag{3}$$

In (2), if we consider that T transforms an input vector $x \in \mathbf{R}_+^m$ into a set of output vectors $T(x)$ which are feasible under the given technology, then $T^*(y^*)$ is a set of input prices each of which does not allow positive profit for any feasible activity $(x, y) \in \text{graph } T$, under the given output price vector y^*. Thus, in some sense, T^* performs the role of market under perfect competition.

B.3. Inverses and Polars

Given a monotone transformation T, there are two other important transformations associated with T besides its adjoint transformation. They are discussed in this section.

B-definition 6 (p. 22)

A m.t.cc. T is called *non-singular* if $T(x)$ has a non-empty interior for some x. If T is a m.t.cv., it is called non-singular if $0 \in T(x)$ only for $x = 0$.

B-property 7 (p. 22)

Let T be a m.t.cc. from $\mathbf{R}_+^m$ into $\mathbf{R}_+^n$. If T is non-singular, then $T(x)$ has a non-empty interior for every $x > 0$. On the other hand, if T is singular, the range of T is entirely contained in some proper suborthant of $\mathbf{R}_+^n$. In other words, there exists an index k such that, for every $x \in \mathbf{R}_+^m$ and every $y \in T(x)$, the kth coordinate of y vanishes.

Thus, singular transformations can be made non-singular by restricting their domains or ranges appropriately.

B-property 8 (p. 22)

Let T be a monotone transformation from $\mathbf{R}_+^m$ into $\mathbf{R}_+^n$ and for each $y \in \mathbf{R}_+^n$ let

$$T^{-1}(y) = \{x \mid y \in T(x)\}. \tag{4}$$

Then T^{-1} is a monotone transformation from $\mathbf{R}_+^n$ into $\mathbf{R}_+^m$, if and only if T is non-singular. In this event T^{-1} is of the opposite type and itself non-singular, with

$$(T^{-1})^{-1} = T.$$

B-definition 7 (p. 23)

If T is a non-singular monotone transformation from $\mathbf{R}_+^m$ into $\mathbf{R}_+^n$, the monotone transformation T^{-1} from $\mathbf{R}_+^n$ into $\mathbf{R}_+^m$ will be called the *inverse* of T.

From (4), $T^{-1}(y)$ is the set of input vectors from which y can be attained under transformation T.

B-property 9 (p. 24)

A monotone transformation T from $\mathbf{R}_+^m$ into $\mathbf{R}_+^n$ is non-singular if and only if, its adjoint is non-singular. In the non-singular case,

$$(T^{-1})^* = (T^*)^{-1}$$

is a monotone transformation $\mathbf{R}_+^{*m}$ into $\mathbf{R}_+^{*n}$ of the same type as T.

B-definition 8 (p. 24)

If T is a non-singular transformation from $\mathbf{R}_+^m$ into $\mathbf{R}_+^n$, the monotone transformation $(T^{-1})^* = (T^*)^{-1}$ from $\mathbf{R}_+^{*m}$ into $\mathbf{R}_+^{*n}$ will be called the *polar* of T.

From (3), when T is of concave type,

$$
\begin{aligned}
(T^{-1})^*(x^*) &= \{y^* \in \mathbf{R}_+^{*n} | y^* \cdot y \le x^* \cdot x \quad \text{for all } (y, x) \in \text{graph } T^{-1}\} \\
&= \{y^* \in \mathbf{R}_+^{*n} | x^* \cdot x \ge y^* \cdot y \quad \text{for all } (x, y) \in \text{graph } T\}. \quad (5)
\end{aligned}
$$

B-property 10 (p. 25)

If T is a non-singular m.t.cc. from $\mathbf{R}_+^m$ into $\mathbf{R}_+^n$ the Kuhn–Tucker function of its inverse is given for each $y \ge 0$ and $x^* \ge 0$ by minimax formula,

$$
\begin{aligned}
\langle T^{-1}(y), x^* \rangle &= \inf_{x \ge 0} \sup_{y^* \ge 0} \{x^* \cdot x + y^* \cdot y - \langle y^*, T(x) \rangle\} \\
&= \sup_{y^* \ge 0} \inf_{x \ge 0} \{x^* \cdot x + x^* \cdot y - \langle y^*, T(x) \rangle\}
\end{aligned}
$$

If T is a m.t.cv. instead, the formula is the same, except that one minimizes in y^* and maximizes in x.

B-definition 9 (p. 27)

A monotone set C of concave type will be called *non-singular* if it has a non-empty interior. If C is of convex type, it will be called *non-*

singular if it does not contain 0, i.e., if $C \neq \mathbf{R}_+^m$. A monotone convex gauge f will be called *non-singular* if $f(x) > 0$ for all non-zero $x \geq 0$. If f is concave, it will be called *non-singular* instead if it is not identically zero.

B.4. Monotone Convex Programs

In this section, the main results for monotone convex programs are summarized.

B-definition 11 (p. 30)

Let T be a non-singular m.t.cc. from $\mathbf{R}_+^m$ into $\mathbf{R}_+^n$. Let T^* be its adjoint, a m.t.cv. from $\mathbf{R}_+^{*n}$ into $\mathbf{R}_+^{*m}$. Fix any $\bar{x}^* \in \mathbf{R}_+^{*m}$ and $\bar{y} \in \mathbf{R}_+^n$. By *the dual monotone programs* associated with T, T^*, $\bar{y}$, $\bar{x}^*$, we shall mean the extremum problems

(I) minimize $\langle x, \bar{x}^* \rangle$ in x, subject to the constraints $x \geq 0$, $x \in T^{-1}(\bar{y})$,
(II) maximize $\langle \bar{y}, y^* \rangle$ in y^*, subject to the constraints $y^* \geq 0$, $y^* \in T^{*-1}(\bar{x}^*)$,

By the *minimax problem* associated with these same elements we shall mean

(III) minimaximize $\langle x, \bar{x}^* \rangle + \langle \bar{y}, y^* \rangle - \langle T(x), y^* \rangle$, minimizing subject to $x \geq 0$ and maximizing subject to $y^* \geq 0$.

(The obvious alterations must be made in the case where T is convex and T^* is concave.)

B-property 12 (p. 30)

For any non-singular monotone transformations T, T^*, and vectors $\bar{x}^* \geq 0$, $\bar{y} \geq 0$, the extrema in problems (I), (II) and (III) all exist and are equal to the same non-negative real number.[4]

---

[4]Note, the infimum in (I) exists always, but the minimum does not necessarily exist.

B-definition 12 (p. 31)

Let T, T^*, $\bar{x}^*$, $\bar{y}$ be as in B-definition 11. We say a vector $\bar{x}$ is a *solution* to (I) if the constrained infimum is attained at $x = \bar{x}$. Similarly, $\bar{y}^*$ is a solution to (II) if the constrained supremum is attained at $y^* = \bar{y}^*$. A solution to (III), on the other hand, is defined to be a saddle point of the function

$$K(x, y^*) = \langle x, \bar{x}^* \rangle + \langle \bar{y}, y^* \rangle - \langle T(x), y^* \rangle.$$

In other words, it is a pair $(\bar{x}, \bar{y}^*) \geq 0$ such that

$$K(\bar{x}, y^*) \leq K(\bar{x}, \bar{y}^*) \leq K(x, \bar{y}^*) \quad \text{for all } x \geq 0, y^* \geq 0.$$

(The obvious changes are necessary when T is of the opposite type.)

B-property 13 (p. 31)

Given any non-singular T, T^* and vectors $\bar{x}^* \geq 0$, $\bar{y} \geq 0$, the following conditions on a pair of vectors $\bar{x}$ and $\bar{y}^*$ are equivalent to each other:

(a) $\bar{x}$ is a solution to (I) and $\bar{y}^*$ is a solution to (II),
(b) $(\bar{x}, \bar{y}^*)$ is a solution to (III),
(c) $\bar{x} \in T^{-1}(\bar{y})$, $\bar{y}^* \in T^{*-1}(\bar{x}^*)$ and $\langle \bar{x}, \bar{x}^* \rangle = \langle \bar{y}, \bar{y}^* \rangle$.

B-property 14 (Reciprocity Principle) (p. 32)

A pair of vectors $\bar{x}$, $\bar{y}^*$ solves the problems (I), (II) and (III) associated with a given non-singular T, T^* and vectors $\bar{x}^*$, $\bar{y}$, if and only if, reciprocally, the pair $\bar{x}^*$, $\bar{y}$ solves the problems (I'), (II'), and (III') associated with T^{-1}, T^{*-1}, $\bar{x}$, $\bar{y}^*$. The common extremum value in (I), (II), and (III) then coincides with one in (I'), (II'), and (III').

B-property 15 (p. 33)

Let C be a monotone set of concave type in $\mathbf{R}_+^m$, and let D^* be a monotone set of convex type in $\mathbf{R}_+^{*m}$. Then

$$\mu = \sup_{x \in C} \inf_{x^* \in D^*} x^* \cdot x = \inf_{x^* \in D^*} \sup_{x \in C} x^* \cdot x$$

where the minimax μ is a non-negative real number.

B-definition 13 (p. 34)

If C and D^* are monotone sets of concave and convex type in $\mathbf{R}_+^m$ and $\mathbf{R}_+^{*m}$, respectively, we denote the common extremum μ in B-property 15 by $\langle C, D^* \rangle$. Thus, by definition,

$$\langle C, D^* \rangle = \sup_{x \in C} \langle x, D^* \rangle = \inf_{x^* \in D} \langle C, x^* \rangle.$$

When the types of C and D^* are interchanged, the "inf" and "sup" in this formula must be interchanged.

B-definition 14 (p. 35)

Let T be a monotone transformation from $\mathbf{R}_+^m$ to $\mathbf{R}_+^n$, and let C be a monotone set in $\mathbf{R}_+^m$ of the same type as T. Then we define,

$$T(C) = \cup\{T(x) | x \in C\}$$

with one exception: when T is a singular m.t.cv., we also need to apply the closure operation to the union on the right.

B-property 17 (p. 35)

In the context of B-definition 14, $T(C)$ is a monotone set in $\mathbf{R}_+^m$ of the same type as T and C. If T and C are non-singular, with their polars denoted by $(T^*)^{-1}$ and C^* as usual, then $T(C)$ is non-singular and its polar is $T^{*-1}(C^*)$.

B-property 18 (p. 37)

If C and D^* are monotone sets of opposite type and T is a monotone transformation of the same type as C, then $\langle T(C), D^* \rangle$ makes sense. In fact,

$$\langle T(C), D^* \rangle = \langle C, T^*(D^*) \rangle.$$

B-definition 15 (p. 37)

Let T be a non-singular m.t.cc. from $\mathbf{R}_+^m$ into $\mathbf{R}_+^n$, with adjoint T^*. Let C^* be a monotone set of concave type in $\mathbf{R}_+^{*m}$, and let D be a monotone set of convex type in $\mathbf{R}^n_+$. By the *dual monotone programs* associated with T, T^*, C^*, D, we shall mean the extremum problems:

(I) minimize $\langle x, C^* \rangle$ in x, subject to the constraints $x \geqq 0$ and $x \in T^{-1}(D)$,

(II) maximize $\langle D, y^* \rangle$ in y^*, subject to the constraints $y^* \geqq 0$ and $y^* \in T^{*-1}(C^*)$.

The associated *minimax problem* is

(III) minimaximize $\langle x, C^* \rangle + \langle D, y^* \rangle - \langle T(x), y^* \rangle$, minimizing subject to $x \geqq 0$ and maximizing subject to $y^* \geqq 0$.

The obvious changes must be made when types are reversed. *Solutions* to these problems are defined much as in B-definition 12.

B-property 19 (p. 37)

For any T, T^*, C^*, D as in B-definition 5, the extrema in (I), (II) and (III) all exist and are equal to the same non-negative real number.[5]

B-property 20 (p. 38)

Given any T, T^*, C^*, D as in B-definition 15, the following conditions on a pair of vectors $\bar{x} \in \mathbf{R}^m_+$ and $\bar{y}^* \in \mathbf{R}^{*n}_+$ are equivalent:

(a) $\bar{x}$ is a solution to (I) and $\bar{y}^*$ is a solution to (II),
(b) $(\bar{x}, \bar{y}^*)$ is a solution to (III),
(c) $T(\bar{x}) \cap D \neq \emptyset$ and $T^*(\bar{y}^*) \cap C^* \neq \emptyset$, and there exists a $\mu \geqq 0$ such that the hyperplane $\{y | \langle y, \bar{y}^* \rangle = \mu\}$ separates $T(\bar{x})$ and D, while the hyperplane $\{x^* | \langle \bar{x}, x^* \rangle = \mu\}$ separates $T^*(\bar{y}^*)$ and C^*.

(Then μ is the common extremum in the three problems.)

B.5. Combinatorial operations

Let $M^\wedge(\mathbf{R}^m_+, \mathbf{R}^n_+)$ denote the set of all monotone transformations of concave type from $\mathbf{R}^m_+$ to $\mathbf{R}^n_+$. Let $M^\vee(\mathbf{R}^m_+, \mathbf{R}^n_+)$ denote those of convex type.

---

[5] Again, the minimum in (I) does not always exist.

B-property 21 (p. 41)

If $T_1, T_2 \in M^\wedge(\mathbf{R}_+^m, \mathbf{R}_+^n)$, then

$$(T_1 \wedge T_2)(x) \equiv T_1(x) \cap T_2(x) \Rightarrow T_1 \wedge T_2 \in M^\wedge(\mathbf{R}_+^m, \mathbf{R}_+^n),$$
$$(T_1 \vee T_2)(x) \equiv \cup \{T_1(x-z) + T_2(z) | 0 \le z \le x\}$$
$$\Rightarrow T_1 \vee T_2 \in M^\wedge(\mathbf{R}_+^m, \mathbf{R}_+^n)$$

If $T_1, T_2 \in M^\vee(\mathbf{R}_+^m, \mathbf{R}_+^n)$, then

$$(T_1 \vee T_2)(x) \equiv T_1(x) \cap T_2(x) \Rightarrow T_1 \vee T_2 \in M^\vee(\mathbf{R}_+^m, \mathbf{R}_+^n),$$
$$(T_1 \wedge T_2)(x) \equiv \cup \{T_1(x-z) + T_2(z) | 0 \le z \le x\}$$
$$\Rightarrow M^\vee(\mathbf{R}_+^m, \mathbf{R}_+^n)$$

And for $T_1, T_2 \in M^\vee(\mathbf{R}_+^m, \mathbf{R}_+^n)$,

$$(T_1 \wedge T_2)^* = T_1^* \wedge T_2^* \in M^\wedge(\mathbf{R}_+^{*n}, \mathbf{R}_+^{*m}),$$
$$(T_1 \vee T_2)^* = T_1^* \vee T_2^* \in M^\wedge(\mathbf{R}_+^{*n}, \mathbf{R}_+^{*m}).$$

And for $T_1, T_2 \in M^\wedge(\mathbf{R}_+^m, \mathbf{R}_+^n)$,

$$(T_1 \wedge T_2)^* = T_1^* \wedge T_2^* \in M^\vee(\mathbf{R}_+^{*n}, \mathbf{R}_+^{*m}),$$
$$(T_1 \vee T_2)^* = T_1^* \vee T_2^* \in M^\vee(\mathbf{R}_+^{*n}, \mathbf{R}_+^{*m}).$$

B-definition 16 (p. 43)

For monotone transformations of the same type, *addition* and *inverse addition* are defined by

$$(T_1 + T_2)(x) = T_1(x) + T_2(x)$$
$$(T_1 \square T_2)(x) = \cup \{T_1(x-z) \cap T_2(z) | 0 \le z \le x\}.$$

B-property 22 (p. 42)

Both $M^\wedge(\mathbf{R}_+^m, \mathbf{R}_+^n)$ and $M^\vee(\mathbf{R}_+^m, \mathbf{R}_+^n)$ are closed under the binary operations of addition and inverse addition. These operations are commutative and associative. Both addition and inverse addition are preserved by the adjoint operation:

$$(T_1 + T_2)^* = T_1^* + T_2^* \quad \text{and} \quad (T_1 \square T_2)^* = T_1^* \square T_2^*.$$

B-definition 17 (p. 46)

The non-negative *scalar multiplication* of a monotone transformation T is defined by,

$$(\lambda T)(x) = \lambda \cdot T(x) \quad \text{for all } x \geqq 0 \text{ if } \lambda \geqq 0.$$

B-definition 18 (p. 47)

Let T and S be monotone transformations of the same type, from $\mathbf{R}_+^m$ to $\mathbf{R}_+^n$ and from $\mathbf{R}_+^n$ to $\mathbf{R}_+^r$, respectively. Then *binary multiplication ST* is defined by,

$$(ST)(x) = S(T(x)) \quad \text{(see B-definition 14).}$$

B-property 23 (p. 47)

In the context of B-definition 18, ST is another monotone transformation of the same type from $\mathbf{R}_+^m$ into $\mathbf{R}_+^r$. Binary multiplication is associative. It is reversed by the adjoint operation:

$$(ST)^* = T^*S^*.$$

B-property 24 (p. 49)

All six combinatorial operations for non-singular monotone transformations (except for the infinitary greatest lower bound and scalar multiplication by 0) preserve non-singularity. The inverse operation obeys the following laws for non-singular monotone transformations of both types:

(a) $(T_1 \wedge T_2)^{-1} = T_1^{-1} \vee T_2^{-1}$ and $(T_1 \vee T_2)^{-1} = T_1^{-1} \wedge T_2^{-1}$,

(b) $(T_1 + T_2)^{-1} = T_1^{-1} \square T_2^{-1}$ and $(T_1 \square T_2)^{-1} = T_1^{-1} + T_2^{-1}$,

(c) $(\lambda T)^{-1} = (1/\lambda)T^{-1}$ and $(ST)^{-1} = T^{-1}S^{-1}$.

PROOFS OF COROLLARIES AND LEMMAS IN CHAPTER 2

C.1. Proof of Corollary 2.2

(Sufficiency.) From the definition of the operation $\langle \cdot, \cdot \rangle$ by (6) and (7), $\langle D^*, \hat{x}_N \rangle \leq p \cdot \hat{x}_N$ for any $p \in D^*$, and $\langle \hat{p}_0, C \rangle \geq \hat{p}_0 \cdot x$ for any $x \in C$. And $\hat{p}_N \in D^*$ and $\hat{x}_0 \in C$ since $\{\hat{x}_t\}_0^N$ and $\{\hat{p}_t\}_0^N$ are feasible paths. Therefore,

$$\langle D^*, \hat{x}_N \rangle \leq \hat{p}_N \cdot \hat{x}_N, \quad \hat{p}_0 \cdot \hat{x}_0 \leq \langle \hat{p}_0, C \rangle.$$

On the other hand, the assumption says that

$$\langle D^*, \hat{x}_N \rangle \geq \hat{p}_N \cdot \hat{x}_N = \hat{p}_0 \cdot \hat{x}_0 \geq \langle \hat{p}_0, C \rangle.$$

Hence, it should be true that

$$\langle D^*, \hat{x}_n \rangle = \hat{p}_N \cdot \hat{x}_N = \hat{p}_0 \cdot \hat{x}_0 = \langle \hat{p}_0, C \rangle,$$

which implies $f(\hat{x}_N) = g(\hat{p}_0)$. Consequently, from the latter half of Theorem 2.1, $\{\hat{x}_t\}_0^N$ and $\{\hat{p}_t\}_0^N$ are optimal paths.

(Necessity.) This is true from (i) to (ii') of Theorem 2.1 by taking $t = 0$ for $A_t(C)$ and $t = N$ for $B_t^*(D^*)$ in (ii').

C.2. Proof of Lemma 2.2

In the context of Problem I, the Kuhn–Tucker Theorem in Gale (1967a,b) can be expressed as in (39). Hence, the equivalence of (a) and (b) is guaranteed by Theorem 2 in Gale (1967a) (see Appendix C.13 for the Kuhn–Tucker Theorem). Therefore, we here prove the equivalence of (b) and (c).

(c) $\Rightarrow$ (b): Suppose $\{\hat{p}_0\}_0^N$ is an optimal path for I(b). Then, for any optimal path $\{\hat{x}_t\}_0^N$ for I(a), we have from (37) that

$$\hat{p}_0 \cdot \hat{x}_0 = \langle B_0^*(D^*), \hat{x}_0 \rangle = \langle D^*, A_n(\hat{x}_0) \rangle.$$

And, since $\hat{p}_0 \in B_0^*(D^*)$ by definition of optimal path for I(b), it is true from definition (7) that

$$\hat{p}_0 \cdot x_0 \geq \langle B_0^*(D^*), x_0 \rangle = \langle D^*, A_N(x_0) \rangle \quad \text{for all } x_0 \in \mathbf{R}_+^m.$$

Consequently, from the above two relations, we obtain

$$\langle D^*, A_N(x_0) \rangle - \langle D^*, A_N(\hat{x}_0) \rangle \leq \hat{p}_0 \cdot (x_0 - \hat{x}_0) \quad \text{for all } x_0 \in \mathbf{R}_+^m.$$

Therefore, if $\{\hat{p}_0\}_0^N$ is an optimal path for I(b), the Kuhn–Tucker Theorem must hold for any optimal path for I(a) with price vector $\hat{p}_0$.

(b) $\Rightarrow$ (c): Suppose that the Kuhn–Tucker Theorem (39) holds for an optimal path $\{\hat{x}_t\}_0^N$ of I(a) with a non-negative vector $\hat{p}_0 \in \mathbf{R}_+^{*m}$.

Then, setting $x_0 = 0$ in (39), we obtain

$$\hat{p}_0 \cdot \hat{x}_0 \leq \langle D^*, A_N(\hat{x}_0) \rangle.$$

If we set $x_0 = 2\hat{x}_0$ in (39), we have

$$\begin{aligned}
\hat{p}_0 \cdot \hat{x}_0 &\geq \langle D^*, A_N(2\hat{x}_0) \rangle - \langle D^*, A_N(\hat{x}_0) \rangle \\
&= \langle D^*, A_N(\hat{x}_0) \rangle \quad \text{(from (ii) of m.t.cc.).}
\end{aligned}$$

Therefore, we have

$$\begin{aligned}
\hat{p}_0 \cdot \hat{x}_0 &= \langle D^*, A_N(\hat{x}_0) \rangle \\
&= \langle D^*, A_N(C) \rangle \quad \text{[since } \{\hat{x}_t\}_0^N \text{ is an optimal path of I(a)]} \\
&= \langle B_0^*(D^*), C \rangle \quad \text{(from Lemma 2.1).} \tag{1}
\end{aligned}$$

Hence, the first conclusion is that if (39) holds to be true, then

$$\hat{p}_0 \cdot \hat{x}_0 = \langle B_0^*(D^*), C \rangle, \quad \hat{x}_0 \in C.$$

Next, let us show by counterexample that $\hat{p}_0 \in B_0^*(D^*)$. Suppose $\hat{p}_0 \notin B_0^*(D^*)$. Then, since $B_0^*(D^*)$ is a closed convex set, there exists some $\bar{x} \in \mathbf{R}^m$ (see Theorem A.9) such that

$$\inf \{p_0 \cdot \bar{x} | p_0 \in B_0^*(D^*)\} > \hat{p}_0 \cdot \bar{x}$$

Then, since $B_0^*(D^*)$ is a monotone set of convex type which is unbounded above, it must be true that no components of $\bar{x}$ are negative, that is, it must be true that $\bar{x} \in \mathbf{R}_+^m$. Therefore, if $\hat{p}_0 \notin B_0^*(D^*)$, we have

$$\langle B_0^*(D^*), \bar{x} \rangle > \hat{p}_0 \cdot \bar{x}, \quad \bar{x} \in \mathbf{R}_+^m. \tag{2}$$

On the other hand, since $\hat{p}_0 \cdot \hat{x}_0 = \langle D^*, A_N(\hat{x}_0) \rangle$ from (1), the Kuhn–Tucker Theorem (39) becomes:

$$\langle D^*, A_N(x_0) \rangle \leq \hat{p}_0 \cdot x_0 \quad \text{for all } x_0 \in \mathbf{R}^m_+.$$

And, applying Lemma 2.1 to the case, $C = \{x | 0 \leq x \leq x_0\}$, we have $\langle D^*, A_N(x_0) \rangle = \langle B^*_0(D^*), x_0 \rangle$. Hence we have

$$\langle B^*_0(D^*), x_0 \rangle \leq \hat{p}_0 \cdot x_0 \quad \text{for all } x_0 \in \mathbf{R}^m_+.$$

This result and (2) contradict each other. Hence, it should be true that $\hat{p}_0 \in B^*_0(D^*)$.

Therefore, if $\hat{p}_0$ is a price vector satisfying the Kuhn–Tucker Theorem (39) at an optimal path $\{\hat{x}_t\}^N_0$, then it must be true that

$$\hat{p}_0 \cdot \hat{x}_0 = \langle B^*_0(D^*), C \rangle, \quad \hat{x}_0 \in C, \quad \hat{p}_0 \in B^*_0(D^*).$$

This implies $\hat{p}_0$ is an optimal solution for I(b) described in (32). Hence, a feasible path $\{p'_t\}^N_0$ with $p'_0 = \hat{p}_0$ for the original I(b) [which surely exists from (31)] is an optimal path for the original I(b).

C.3. Proof of Lemma 2.3

First let us prove that if a m.t.cc. T satisfies the Lipschitz condition with a constant k, then $\|T(C_1) - T(C_2)\| \leq k\|C_1 - C_2\|$ for any monotone set of concave type C_1 and C_2 in $\mathbf{R}^m_+$.

Consider any $x^1 \in C_1$. Then by definition there exists some $x^2 \in C_2$ such that $\|x^1 - x^2\| \leq \|C_1 - C_2\|$. Hence for any $y^1 \in T(x^1) \subset T(C_1)$ there must exist some $y^2 \in T(x^2) \subset T(C_2)$ such that $\|y^1 - y^2\| \leq \|T(x^1) - T(x^2)\| \leq k \cdot \|x^1 - x^2\| \leq k\|C_1 - C_2\|$. Then, since

$$T(C_1) = \cup\{T(x^1) | x^1 \in C_1\},$$

we have

$$\sup_{y^1 \in T(C_1)} \inf_{y^2 \in T(C_2)} \|y^1 - y^2\| \leq k\|C_1 - C_2\|.$$

Similarly, for any $x^2 \in C_2$ there is some $x^1 \in C_1$, with $\|x_1 - x_2\| \leq \|C_1 - C_2\|$, so that for any $y^2 \in T(x^2)$ there is some $y^1 \in T(x^1) \subset T(C_1)$ with $\|y^1 - y^2\| \leq \|T(x^1) - T(x^2)\| \leq k\|x^1 - x^2\| \leq k\|C_1 - C_2\|$. Thus, since

$$T(C_2) = \cup\{T(x^2) | x^2 \in C_2\},$$

we have

$$\sup_{y^2 \in T(C_2)} \inf_{y^1 \in T(C_1)} \|y^1 - y^2\| \leq k\|C_1 - C_2\|.$$

Hence we have

$$\|T(C_1) - T(C_2)\| \le k\|C_1 - C_2\|. \tag{3}$$

Next, suppose T_t satisfies the Lipschitz condition with $k = k_t$. Then, since $A_t(x) = (T_{t-1} \cdot \ldots \cdot T_1 \cdot T_0)(x) = T_{t-1}((T_{t-2} \cdot \ldots \cdot T_1 \cdot T_0)(x)) = T_{t-1}(A_{t-1}(x))$ and since $A_{t-1}(x)$ is a monotone set of concave type, from (3) we have

$$\|A_t(x^1) - A_t(x^2)\| = \|T_{t-1}(A_{t-1}(x^1)) - T_{t-1}(A_{t-1}(x^2))\|$$
$$\le k_{t-1}\|A_{t-1}(x^1) - A_{t-1}(x^2)\|.$$

Hence, finally we get

$$\|A_t(x^1) - A_t(x^2)\| \le \left(\prod_{\tau=0}^{t-1} k_\tau\right)\|x^1 - x^2\| \quad \text{for all } x^1, x^2 \in \mathbf{R}_+^m.$$

C.4. Proof of Corollary 2.3

Of course, from Theorem 2.1 (resp. Theorem 2.1') optimal paths satisfy (i), (ii), and (iii) in Theorem 2.1 (resp. Theorem 2.1'). And, quite similarly to Lemma 2.1, we can obtain the following relation:

$$\langle A_N^{*-1}(p_0), x_N\rangle = \cdots = \langle A_t^{*-1}(p_0), B_t^{-1}(x_N)\rangle = \cdots = \langle p_0, B_0^{-1}(x_N)\rangle$$
$$\text{for any } p_0 \in \mathbf{R}_+^{*m} \quad \text{and} \quad x_N \in \mathbf{R}_+^m.$$

Using this relation, the following equalities can be obtained similarly as the method to derive (37).

$$\langle A_t^{*-1}(\hat{p}_0), \hat{x}_t\rangle = \hat{p}_0 \cdot \hat{x}_0, \quad \langle \hat{p}_t, A_t^{-1}(\hat{x}_N)\rangle = \hat{p}_N \cdot \hat{x}_N.$$

And, from (i) in Theorem 2.1, $\hat{p}_0 \cdot \hat{x}_0 = \hat{p}_t \cdot \hat{x}_t = \hat{p}_N \cdot \hat{x}_N$. Hence, (iv) is true when $\tau' = N$ and $\tau = 0$. Then this is true for any $N \ge \tau' > t > \tau \ge 0$, since from relations (46), (47), (50), and (51) we have $\hat{x}_t \in B_{t\tau'}^{-1}(\hat{x}_{\tau'}) \subset B_t^{-1}(x_N)$ and $\hat{p}_t \in A_{\tau t}^{*-1}(\hat{p}_\tau) \subset A_t^{*-1}(\hat{p}_0)$.

C.5. Proof of Lemma 2.4

(b) $\Rightarrow$ (a): If (b) is true, there exists a vector $\hat{p}$ such that $\hat{p} \cdot \hat{x} = \max\{\hat{p} \cdot x | x \in X\} > 0$. Thus, $\hat{p} \cdot (\lambda\hat{x}) = \lambda\hat{p} \cdot \hat{x} > \hat{p} \cdot \hat{x}$ for any $\lambda > 1$. Hence, $\lambda\hat{x} \notin X$ for any $\lambda > 1$.

(a) $\Rightarrow$ (b): Let us denote by m' the dimension of X. Then, $m \geq m' \geq 1$ since $0 \in X$ and $X \neq \{0\}$. Since X is a monotone set of concave type, without loss of generality we can assume that X is a subset of $\mathbf{R}_+^{m'}$ and X has an interior point as a subset of $\mathbf{R}^{m'}$. We set $Y = X - \hat{x}$ where $\hat{x}$ is a frontier point of X in $\mathbf{R}_+^{m'}$.

Y does not contain any positive vector. We prove this by counterexample. Suppose there exists positive $y' \in Y$. Then there exists $x' \in X$ such that $x' > \hat{x}$. Then, since X is a monotone set of concave type, $C \subset X$ where $C = \{x | 0 \leq x \leq x'\}$. Hence $\lambda \hat{x} \in C \subset X$ for some $\lambda > 1$, and we get a contradiction.

Therefore, Y is a convex set in $\mathbf{R}^{m'}$ and it does not contain any positive vector. Hence, there is a separating hyperplane $\hat{p} \cdot y = 0$ such that [see Theorem A.9],

$$\hat{p} \cdot y \leq 0 \quad \text{for all } y \in Y \subset \mathbf{R}^{m'}, \text{ where } \hat{p} \in \mathbf{R}^{*m'} \text{ and } \hat{p} \geq 0,$$

and hence

$$\hat{p} \cdot (x - \hat{x}) \leq 0 \quad \text{for all } x \in X \subset \mathbf{R}^{m'}$$

so

$$\hat{p} \cdot \hat{x} = \max \{\hat{p} \cdot x | x \in X \subset \mathbf{R}^{m'}\}.$$

On the other hand, since X has an interior point as a subset of $\mathbf{R}^{m'}$ and $X \subset \mathbf{R}^{+m'}$, there is a positive vector in X. Hence $\hat{p} \cdot \hat{x} > 0$. Therefore now considering the facts that $X \subset \mathbf{R}^m$ and that $\hat{p}$ is a vector in $\mathbf{R}_+^{*m}$ (adding "zero" components if necessary), we obtain

$$\hat{p} \cdot \hat{x} = \max \{\hat{p} \cdot x | x \in X \subset \mathbf{R}_+^m\} > 0, \quad \hat{p} \geq 0.$$

C.6. Proof of Theorem 2.6

Let us abbreviate $K(\{x_t\}_1^N, \{p_t\}_0^{N-1})$ to $K(x, p)$ in this proof. Suppose $\{\hat{x}_t\}_0^N$ and $\{\hat{p}_t\}_0^N$ are optimal paths for I(a) and I(b), respectively. Then, from (iv) in Corollary 2.3 with $\tau = t - 1$, we have $\hat{p}_t \cdot \hat{x}_t = \langle T_{t-1}^{*-1}(\hat{p}_{t-1}), \hat{x}_t \rangle$ for $t = 1, 2, \ldots, N$. And from (i) in Theorem 2.1, $f(\hat{x}_N) = \hat{p}_N \cdot \hat{x}_N$. Hence we have $K(\hat{x}, \hat{p}) = g(\hat{p}_0)$. And, by definition, $f(x_N) = \langle D^*, x_N \rangle$. Therefore, for any $x \equiv \{x_t\}_1^N \geq 0$,

$$K(\hat{x}, \hat{p}) - K(x, \hat{p}) = \sum_{t=1}^{N-1} (\langle T_{t-1}^{*-1}(\hat{p}_{t-1}), x_t \rangle - \hat{p}_t \cdot x_t)$$
$$+ \langle T_{N-1}^{*-1}(\hat{p}_{N-1}), x_N \rangle - \langle D^*, x_N \rangle.$$

Next, since $T_{t-1}^{*-1}(\hat{p}_{t-1})$ is a monotone set of concave type and $\hat{p}_t \in T_{t-1}^{*-1}(\hat{p}_{t-1})$, we have

$$\langle T_{t-1}^{*-1}(\hat{p}_{t-1}), x_t \rangle \geqq \hat{p}_t \cdot x_t \quad \text{for any } x_t \geqq 0,$$

for $t = 1, 2, \ldots, N - 1$. Further, since $\hat{p}_{N-1} \in T_{N-1}^*(D^*)$, we see $T_{N-1}^{*-1}(\hat{p}_{N-1}) \cap D^* \neq \emptyset$. And $T_{N-1}^{*-1}(\hat{p}_{N-1})$ is a monotone set of concave type, and D^* is of convex type. Hence

$$\langle T_{N-1}^{*-1}(\hat{p}_{N-1}), x_N \rangle - \langle D^*, x_N \rangle \geqq 0.$$

From the above three relations, we have $K(\hat{x}, \hat{p}) \geqq K(x, \hat{p})$ for any $x \equiv \{x_t\}_1^N \geqq 0$.

Similarly, we can show that $K(\hat{x}, \hat{p}) \leqq K(\hat{x}, p)$ for any $p \equiv \{p_t\}_0^{N-1} \geqq 0$. Consequently, $(\{\hat{x}_t\}_1^N, \{\hat{p}_t\}_0^{N-1})$ is a saddle point of (52).

Conversely, suppose that $(\{\hat{x}_t\}_1^N, \{\hat{p}_t\}_0^{N-1})$ is a saddle point of (52). Then, from the left side of (53), we get

$$\langle D^*, x_N \rangle - \langle T_{N-1}^{*-1}(\hat{p}_{N-1}), x_N \rangle + \sum_{t=1}^{N-1} (\hat{p}_t \cdot x_t - \langle T_{t-1}^{*-1}(\hat{p}_{t-1}), x_t \rangle)$$

$$\leqq \langle D^*, \hat{x}_N \rangle - \langle T_{N-1}^{*-1}(\hat{p}_{N-1}), \hat{x}_N \rangle + \sum_{t=1}^{N-1} (\hat{p}_t \cdot \hat{x}_t - \langle T_{t-1}^{*-1}(\hat{p}_{t-1}), \hat{x}_t \rangle)$$

$$\text{for all } \{x_t\}_1^N \geqq 0.$$

Hence, by the positive linear-homogeneity of the left side in $\{x_t\}_1^N$, we get

$$\langle D^*, \hat{x}_N \rangle \geqq \langle T_{N-1}^{*-1}(\hat{p}_{N-1}), \hat{x}_N \rangle, \quad \hat{p}_t \cdot \hat{x}_t \geqq \langle T_{t-1}^{*-1}(\hat{p}_{t-1}), \hat{x}_t \rangle,$$
$$t = 1, 2, \ldots, N - 1,$$
$$\langle D^*, x_N \rangle \leqq \langle T_{N-1}^{*-1}(\hat{p}_{N-1}), x_N \rangle \quad \text{for all } x_N \geqq 0,$$

and

$$\hat{p}_t \cdot x_t \leqq \langle T_{t-1}^{*-1}(\hat{p}_{t-1}), x_t \rangle \quad \text{for all } x_t \geqq 0, t = 1, 2, \ldots, N - 1.$$

That is, $\hat{p}_{N-1} \in T_{N-1}^*(D^*)$, $\hat{p}_{t-1} \in T_{t-1}^*(\hat{p}_t)$ $(t = 1, 2, \ldots, N - 1)$, $\langle D^*, \hat{x}_N \rangle = \langle T_{N-1}^{*-1}(\hat{p}_{N-1}), \hat{x}_N \rangle$ and $\hat{p}_t \cdot \hat{x}_t = \langle T_{t-1}^{*-1}(\hat{p}_{t-1}), \hat{x}_t \rangle$ for $t = 1, 2, \ldots, N - 1$.

Similarly, from the right side of (53) we get $\hat{x}_1 \in T_0(C)$, $\hat{x}_t \in T_{t-1}(\hat{x}_{t-1})$, $t = 2, 3, \ldots, N$, and $\langle C, \hat{p}_0 \rangle = \langle \hat{p}_0, T_0^{-1}(\hat{x}_1) \rangle$, $\hat{p}_t \cdot \hat{x}_t = \langle \hat{p}_t, T_t^{-1}(\hat{x}_{t+1}) \rangle$ for $t = 1, 2, \ldots, N - 1$.

Therefore we get $\langle D^*, \hat{x}_N \rangle = \langle C, \hat{p}_0 \rangle$ and $\hat{p}_{N-1} \in T_{N-1}^*(D^*)$, $\hat{p}_{t-1} \in T_{t-1}^*(\hat{p}_t)$ $(t = 1, 2, \ldots, N - 1)$, $\hat{x}_t \in T_{t-1}(\hat{x}_{t-1})$ $(t = 2, 3, \ldots, N)$ and $\hat{x}_1 \in T_0(C)$. Thus, from the latter half of Theorem 2.1, for any $\hat{x}_0$ and $\hat{p}_N$

such that $\hat{x}_1 \in T_0(\hat{x}_0)$, $\hat{x}_0 \in C$, $\hat{p}_{N-1} \in T^*_{N-1}(\hat{p}_N)$ and $\hat{p}_N \in D^*$, a pair of sequences of vectors $\{\hat{x}_t\}_0^N$ and $\{\hat{p}_t\}_0^N$ are optimal paths for I(a) and I(b).

C.7. Proof of Lemma 2.5

Since $(\hat{p}_N, \hat{p}^N_{m+1}) = (0, 1)$, from (60) $(\hat{p}_{N-1}, \hat{p}^{N-1}_{m+1}) \in \mathscr{T}^*_{N-1}(\hat{p}_m, \hat{p}^N_{m+1})$ means $\hat{p}_{N-1} \cdot x_{N-1} + \hat{p}^{N-1}_{m+1} \cdot x^{N-1}_{m+1} \geq x^{N-1}_{m+1} + f_N(x_N, c_N)$ for all $(x_N, c_N) \in T_{n-1}(x_{N-1})$, $x_{N-1} \geq 0$ and $x^{N-1}_{m+1} \geq 0$. This is true if and only if

$$p^{N-1}_{m+1} \geq 1, \text{ and}$$

$$\hat{p}_{N-1} \cdot x_{N-1} \geq f_N(x_N, c_N) \quad \text{for all } (x_N, c_N) \in T_{N-1}(x_{N-1}), x_{N-1} \geq 0.$$

Hence, when $(\hat{p}_{N-1}, \hat{p}^{N-1}_{m+1})$ satisfies the above condition, $(\hat{p}_{N-1}, 1)$ also satisfies it. Thus,

$$(\hat{p}_{N-1}, 1) \in \mathscr{T}^*_{N-1}(\hat{p}_N, \hat{p}^N_{m+1}) \quad \text{and } \hat{p}^N_{m+1} \geq 1.$$

Next, $(\hat{p}_{N-2}, \hat{p}^{N-2}_{m+1}) \in \mathscr{T}^*_{N-2}(\hat{p}_{N-1}, \hat{p}^{N-1}_{m+1})$ means

$$\hat{p}_{N-2} \cdot x_{N-2} + \hat{p}^{N-2}_{m+1} x^{N-2}_{m+1} \geq \hat{p}_{N-1} \cdot x_{N-1} + \hat{p}^{N-1}_{m+1}(x^{N-2}_{m+1} + f_{N-1}(x_{N-1}, c_{N-1}))$$

for all $(x_{N-1}, c_{N-1}) \in T_{N-2}(x_{N-2})$, $x_{N-2} \geq 0$ and $x^{N-2}_{m+1} \geq 0$. Then, since $\hat{p}^{N-1}_{m+1} \geq 1$, it should be true that

$$\hat{p}_{N-2} \cdot x_{N-2} + \hat{p}^{N-2}_{m+1} x^{N-2}_{m+1} \geq \hat{p}_{N-1} \cdot x_{N-1} + (x^{N-2}_{m+1} + f_{N-1}(x_{N-1}, c_{N-1}))$$

for all $(x_{N-1}, c_{N-1}) \in T_{N-2}(x_{N-2})$, $x_{N-2} \geq 0$ and $x^{N-2}_{m+1} \geq 0$. This is true if and only if

$$\hat{p}^{N-2}_{m+1} \geq 1, \quad \text{and}$$

$$\hat{p}_{N-2} \cdot x_{N-2} \geq \hat{p}_{N-1} \cdot x_{N-1} + f_{N-1}(x_{N-1}, c_{N-1})$$
$$\text{for all } (x_{N-1}, c_{N-1}) \in T_{N-2}(x_{N-2}), x_{N-2} \geq 0.$$

Hence, we have

$$(\hat{p}_{N-2}, 1) \in \mathscr{T}^*_{N-1}(\hat{p}_{N-1}, 1) \quad \text{and } \hat{p}^{N-2}_{m+1} \geq 1.$$

Repeating this procedure, we see

$$(\hat{p}_{N-1}, 1) \in \mathscr{T}^*_{N-1}(\hat{p}_N, \hat{p}^N_{m+1}) \quad \text{and} \quad (\hat{p}_{t-1}, 1) \in \mathscr{T}^*_{t-1}(\hat{p}_t, 1),$$
$$t = 1, 2, \ldots, N-1. \quad \hat{p}^t_{m+1} \geq 1, t = 0, 1, \ldots, N-1.$$

Consequently, if $\{(\hat{p}_t, \hat{p}^N_{m+1})\}_0^N$ is an optimal path for II'(b), $\{(\hat{p}_t, 1)\}_0^N$ should also be an optimal path.

C.8. Proof of Lemma 2.6

The assumption says that there exist constants k_t and α_{t+1} such that

$$\|T_t(x) - T_t(x')\| \leq k_t \|x - x'\| \quad \text{for all } x, x' \in \mathbf{R}_+^m,$$

and

$$\|f_{t+1}(y, c) - f_{t+1}(y', c')\| \leq \alpha_{t+1} \|(y, c) - (y', c')\|$$
for all $(y, c), (y', c') \in \mathbf{R}_+^{m+n}$.

Next, the transformation $\mathcal{T}_t$ defined by (59) can be decomposed as follows.

$$\mathcal{T}_t(x_t, x_{m+1}^t) = \mathcal{T}_t^2(\mathcal{T}_t^1(x_t, x_{m+1}^t)) \quad \text{for each } (x_t, x_{m+1}^t) \in \mathbf{R}_+^{m+1},$$

where

$$\mathcal{T}_t^1(x_t, x_{m+1}^t) = \{(y_{t+1}, c_{t+1}, y_{m+1}^{t+1}) | (y_{t+1}, c_{t+1}) \in T_t(x_t), 0 \leq y_{m+1}^{t+1} \leq x_{m+1}^t\},$$
$$\mathcal{T}_t^2(y_{t+1}, c_{t+1}, y_{m+1}^{t+1}) = \{(x_{t+1}, x_{m+1}^{t+1}) | 0 \leq x_{t+1} \leq y_{t+1}, 0 \leq x_{m+1}^{t+1} \leq y_{m+1}^{t+1}$$
$$+ f_{t+1}(y_{t+1}, c_{t+1})\}.$$

Each of transformations $\mathcal{T}_t^1$ and $\mathcal{T}_t^2$ is a m.t.cc. And, when we define the norm of a vector $z \equiv (z_1, \ldots, z_i, \ldots, z_m) \in \mathbf{R}^m$ (m is any positive integer) by $\|z\| = \max_i |z_i|$, $\mathcal{T}_t^1$ satisfies the Lipschitz condition with the constant $\max\{k_t, 1\}$, and $\mathcal{T}_t^2$ satisfies that condition with the constant $(\alpha_{t+1} + 1)$. Hence, by the same approach as the proof of Lemma 2.3, we can prove that $\mathcal{T}_t$ satisfies the Lipschitz condition with the constant $\max\{k_t(\alpha_{t+1} + 1), (\alpha_{t+1} + 1)\}$.

C.9. Proof of Lemma 2.7

Note first that function $\eta f_t(x/\eta, c/\eta)$ is linearly homogeneous on $\mathbf{R}_+^{m+n} \times I$ where $I = (0, \infty)$. Next, take any two points (x^1, c^1, η^1), $(x^2, c^2, \eta^2) \in \mathbf{R}_+^{m+n} \times I$ (hence, $\eta^1, \eta^2 > 0$). Then, since f_t is concave on $\mathbf{R}_+^{m+n}$,

$$\frac{\eta^1}{\eta^1 + \eta^2} f_t\left(\frac{x_1}{\eta^1}, \frac{c^1}{\eta^1}\right) + \frac{\eta^2}{\eta^1 + \eta^2} f_t\left(\frac{x^2}{\eta^2}, \frac{c^2}{\eta^2}\right) \leq f_t\left(\frac{x^1 + x^2}{\eta^1 + \eta^2}, \frac{c^1 + c^2}{\eta^1 + \eta^2}\right),$$

and hence

$$\eta^1 f_t\left(\frac{x^1}{\eta^1}, \frac{c^1}{\eta^1}\right) + \eta^2 f_t\left(\frac{x^2}{\eta^2}, \frac{c^2}{\eta^2}\right) \leq (\eta^1 + \eta^2) f_t\left(\frac{x^1 + x^2}{\eta^1 + \eta^2}, \frac{c^1 + c^2}{\eta^1 + \eta^2}\right).$$

Therefore, taking into account the fact that function $\eta f_t(x/\eta, c/\eta)$ is linearly homogeneous on $\mathbf{R}_+^{m+n} \times I$, it is concave on $\mathbf{R}_+^{m+n} \times I$. On the other hand, since $f_t(x, c)$ is continuous on $\mathbf{R}_+^{m+n}$, $\eta f_t(x/\eta, c/\eta)$ is continuous on $\mathbf{R}_+^{m+n} \times I$. Finally, take any two points $0 \leq (x^1, c^1, \eta^1) \leq (x^2, c^2, \eta^2) \in \mathbf{R}_+^{m+n} \times I$ where $\eta^1, \eta^2 > 0$. Then, since f_t is concave and non-decreasing on $\mathbf{R}_+^{m+n}$,

$$\frac{\eta^2 - \eta^1}{\eta^2} f_t(0, 0) + \frac{\eta^1}{\eta_2} f_t\left(\frac{x^1}{\eta^1}, \frac{c^1}{\eta^1}\right) \leq f_t\left(\frac{x^1}{\eta^1}, \frac{c^1}{\eta^1}\right) \leq f_t\left(\frac{x^2}{\eta^2}, \frac{c^2}{\eta^2}\right).$$

Hence, since $f_t(0, 0) \geq 0$,

$$\eta^1 f_t\left(\frac{x^1}{\eta^1}, \frac{c^1}{\eta^1}\right) \leq \eta^2 f_t\left(\frac{x^2}{\eta^2}, \frac{c^2}{\eta^2}\right).$$

Therefore, function $\eta f_t(x/\eta, c/\eta)$ is non-decreasing on $\mathbf{R}_+^{m+n} \times I$. In sum, function $\eta f_t(x/\eta, c/\eta)$ is continuous, concave, linearly homogeneous and non-decreasing on $\mathbf{R}_+^{m+n} \times I$. Therefore, the function $h_t(x, c, \eta)$ of which the graph is given by the closure of set (65) is real-valued, continuous, concave, linearly homogeneous and non-decreasing on $\mathbf{R}_+^{m+n} \times \mathbf{R}_+$, and $h_t(x, c, \eta) = \eta f_t(x/\eta, c/\eta)$ for all $(x, c) \in \mathbf{R}_+^{m+n}$ and $\eta > 0$. Hence, by definition, it is a monotone concave gauge on $\mathbf{R}_+^{m+n+1}$, and $h(x, c, 1) = f_t(x, c)$ for all $(x, c) \in \mathbf{R}_+^{m+n}$.

The later half of Lemma 2.7 can be proved similarly, and its proof is omitted.

C.10. Proof of Lemma 2.9

Let us first prove the initial half of Lemma 2.9. If f_t is a monotone concave gauge, h_t clearly satisfies the Lipschitz condition. Hence, let us assume that there exists a real number K_t such that $f_t(x, c) \leq K_t$ on $\mathbf{R}_+^{m+n}$.

Take any two points $(x^1, c^1, \eta^1), (x^2, c^2, \eta^2) \in \mathbf{R}_+^{m+n+1}$. Without loss of generality, we can assume that

$$\eta^1 f_t\left(\frac{x^1}{\eta^1}, \frac{c^1}{\eta^1}\right) \geq \eta^2 f_t\left(\frac{x^2}{\eta^2}, \frac{c^2}{\eta^2}\right)$$

where, since $h(x, c, \eta) = 0$ when $\eta = 0$ because $f_t(x, c) \leq K_t$ on $\mathbf{R}_+^{m+n}$, we define $\eta f_t(x/\eta, c/\eta) = 0$ when $\eta = 0$ and $(x, c) \in \mathbf{R}_+^m$. Suppose $\eta^2 \geq$

η^1. Then since $\eta f_t(x/\eta, c/\eta)$ is non-decreasing function of η,

$$\eta^1 f_t\left(\frac{x^1}{\eta^1}, \frac{c^1}{\eta^1}\right) - \eta^2 f_t\left(\frac{x^2}{\eta^2}, \frac{c^2}{\eta^2}\right) \leq \eta^2 f_t\left(\frac{x^1}{\eta^2}, \frac{c^1}{\eta^2}\right) - \eta^2 f_t\left(\frac{x^2}{\eta^2}, \frac{c^2}{\eta^2}\right).$$

Next, since f_t satisfies the Lipschitz condition, there exists a number σ_t such that

$$|f_t(x, c) - f_t(x', c')| \leq \sigma_t \|(x,c) - (x', c')\|$$
$$\text{for any } (x, c), (x', c') \in \mathbf{R}_+^{m+n}.$$

Hence

$$\eta^1 f_t\left(\frac{x^1}{\eta^1}, \frac{c^1}{\eta^1}\right) - \eta^2 f_t\left(\frac{x^2}{\eta^2}, \frac{c^2}{\eta^2}\right) \leq \eta^2 \sigma_t \left\|\left(\frac{x^1}{\eta^2}, \frac{c^1}{\eta^2}\right) - \left(\frac{x^2}{\eta^2}, \frac{c^2}{\eta^2}\right)\right\|$$
$$= \sigma_t \|(x^1, c^1) - (x^2, c^2)\| \leq \sigma_t \|(x^1, c^1, \eta^1) - (x^2, c^2, \eta^2)\|.$$

That is,

$$|h_t(x^1, c^1, \eta^1) - h_t(x^2, c^2, \eta^2)| \leq \sigma_t \|(x^1, c^1, \eta^1) - (x^2, c^2, \eta^2)\|.$$

Next, suppose $\eta^1 > \eta^2$. Then,

$$\eta^1 f_t\left(\frac{x^1}{\eta^1}, \frac{c^1}{\eta^1}\right) - \eta^2 f_t\left(\frac{x^2}{\eta^2}, \frac{c^2}{\eta^2}\right)$$
$$= (\eta^1 - \eta^2) f_t\left(\frac{x^1}{\eta^1}, \frac{c^1}{\eta^1}\right) + \eta^2 f_t\left(\frac{x^1}{\eta^1}, \frac{c^1}{\eta^1}\right) - \eta^2 f_t\left(\frac{x^2}{\eta^2}, \frac{c^2}{\eta^2}\right)$$
$$\leq (\eta^1 - \eta^2) f_t\left(\frac{x^1}{\eta^1}, \frac{c^1}{\eta^1}\right) + \eta^2 f_t\left(\frac{x^1}{\eta^2}, \frac{c^1}{\eta^2}\right) - \eta^2 f_t\left(\frac{x^2}{\eta^2}, \frac{c^2}{\eta^2}\right)$$
$$\leq (\eta^1 - \eta^2) K_t + \eta^2 \sigma_t \left\|\left(\frac{x^1}{\eta^2}, \frac{c^1}{\eta^2}\right) - \left(\frac{x^2}{\eta^2}, \frac{c^2}{\eta^2}\right)\right\|$$
$$\leq (\eta^1 - \eta^2) K_t + \sigma_t \|(x^1, c^1) - (x^2, c^2)\|$$
$$\leq \max(K_t, \sigma_t)((\eta^1 - \eta^2) + \|(x^1 - c^1) - (x^2 - c^2)\|)$$
$$\leq 2 \max(K_t, \sigma_t) \|(x^1, c^1, \eta^1) - (x^2, c^2, \eta^2)\|.$$

Hence,

$$|h_t(x^1, c^1, \eta^1) - h_t(x^2, c^2, \eta^2)|$$
$$\leq 2 \max(K_t, \sigma_t) \|(x^1, c^1, \eta^1) - (x^2, c^2, \eta^2)\|$$

for any $(x^1, c^1, \eta^1), (x^2, c^2, \eta^2) \in \mathbf{R}_+^{m+n+1}$, which means that h_t satisfies the Lipschitz condition.

The latter half of Lemma 2.9 can be proved similarly, and its proof is omitted.

C.11. Proof of Lemma 2.10

(i) The set $A_t(\bar{x}_0)$ is a monotone set of concave type in $\mathbf{R}_+^m$, and from assumption (72) $A_t(\bar{x}_0) \neq \{0\}$. Thus, the maximization problem

$$\max \{p \cdot x | x \in A_t(\bar{x}_0)\} \quad \text{where } p > 0,$$

has a solution $\hat{x}$ with $p \cdot \hat{x} > 0$. Then, from Lemma 2.4, $\hat{x}$ is a frontier point of $A_t(\bar{x}_0)$. Hence, $F_t \neq \emptyset$. F_t is bounded since it is a subset of $A_t(\bar{x}_0)$. The fact that F_t is closed can be proved as follows.

Conversely, suppose that F_t is not closed. Then there exists a sequence $\{x^j\}$ such that

$$x^j \in F_t, \quad j = 1, 2, \ldots, \lim_{j \to \infty} x^j = \bar{x} \quad \text{and} \quad \bar{x} \notin F_t. \tag{4}$$

Since $x^j \in F_t \subset A_t(\bar{x}_0)$ and $A_t(\bar{x}_0)$ is closed, it is true that $\bar{x} \in A_t(\bar{x}_0)$. Then from Lemma 2.4, the fact, $\bar{x} \notin F_t$, implies there exists a scalar λ such that

$$\lambda \bar{x} \in A_t(\bar{x}_0), \quad \lambda > 1.$$

By choosing λ appropriately, we can assume that

$$\lambda \bar{x} \notin F_t, \lambda \bar{x} \in A_t(\bar{x}_0), \quad \lambda > 1. \tag{5}$$

Then, by using the fact that $A_t(\bar{x}_0)$ is a monotone set of concave type, it is not difficult to show that (4) and (5) are inconsistent.

Consequently, F_t is a nonempty closed subset of $A_t(\bar{x}_0)$, and hence is a nonempty compact subset of $\mathbf{R}_+^m$.

(ii) Conversely, suppose that there exists x_{t-1} such that

$$x_{t-1} \in A_{t-1}(\bar{x}_0), \quad x_{t-1} \notin F_{t-1} \quad \text{and} \quad T_{t-1}(x_{t-1}) \cap F_t \neq \emptyset.$$

This implies there exists x_t such that

$$x_t \in T_{t-1}(x_{t-1}) \subset A_t(\bar{x}_0) \quad \text{and} \quad x_t \in F_t.$$

On the other hand, $x_{t-1} \in A_{t-1}(\bar{x}_0)$ and $x_{t-1} \notin F_{t-1}$ implies there exists a scalar λ such that $\lambda x_{t-1} \in A_{t-1}(\bar{x}_0)$ and $\lambda > 1$. Then,

$$\lambda x_t \in \lambda T_{t-1}(x_{t-1}) = T_t(\lambda x_{t-1}) \subset A_t(\bar{x}_0), \quad \lambda > 1.$$

Thus $\hat{x}_t \notin F_t$, and we get a contradiction.

Consequently, if $x_{t-1} \in A_{t-1}(\bar{x}_0)$ and $x_{t-1} \notin F_{t-1}$, then $T_{t-1}(x_{t-1}) \cap F_t = \emptyset$.

(iii) $X(\bar{x}_0)$ is non-empty since the path $(x_0 = \bar{x}_0, x_t = 0$ for all $t \geq 1)$ is included in $X(\bar{x}_0)$. To prove the compactness of $X(\bar{x}_0)$, observe first that

$$X(\bar{x}_0) \subset \bar{x}_0 \times A_1(\bar{x}_0) \times \cdots \times A_t(\bar{x}_0) \times \cdots.$$

Here, the set in the right hand side is the Cartesian product of $\bar{x}_0, A_1(\bar{x}_0), \ldots$, each of which is regarded as a subset of $\mathbf{R}^m$. Note that $A_t(\bar{x}_0)$ is compact $(t = 1, 2, \ldots)$. Hence, to prove the compactness of $X(\bar{x}_0)$, from Tychonoff's Theorem it is sufficient to show that $X(\bar{x}_0)$ is closed.

Consider a sequence of continuous mappings

$$\phi_0(\{x_t\}_0^\infty) = x_0 : \Omega \to \mathbf{R}^m$$
$$\phi_t(\{x_t\}_0^\infty) = (x_{t-1}, x_t) : \Omega \to \mathbf{R}^m \times \mathbf{R}^m, \quad t = 1, 2, \ldots.$$

Then,

$$X(\bar{x}_0) = \phi_0^{-1}(\bar{x}_0) \cap \left(\bigcap_{t=1}^{\infty} \phi_t^{-1}(\text{graph } T_{t-1}) \right).$$

Since $\bar{x}_0$ is a closed set and ϕ_0 is continuous, $\phi_0^{-1}(\bar{x}_0)$ is closed. Likewise, since graph T_{t-1} is closed in $\mathbf{R}_+^m \times \mathbf{R}_+^m$ and hence in $\mathbf{R}^m \times \mathbf{R}^m$ and ϕ_t is continuous, $\phi_t^{-1}(\text{graph } T_{t-1})$ is closed in Ω $(t = 1, 2, \ldots)$. Therefore $X(\bar{x}_0)$ is closed.

Consequently, $X(\bar{x}_0)$ is a nonempty compact subset of Ω.

(iv) Consider a sequence of continuous mappings

$$\pi_t(\{x_t\}_0^\infty) = x_t : \Omega \to \mathbf{R}^m, \quad t = 1, 2, \ldots$$

Then,

$$\hat{X}_t(\bar{x}_0) = X(\bar{x}_0) \cap \pi_t^{-1}(F_t).$$

From (iii), $X(\bar{x}_0)$ is compact. F_t is compact in $\mathbf{R}_+^m$ and hence in $\mathbf{R}^m$, and π_t is continuous. Thus, $\pi_t^{-1}(F_t)$ is closed in Ω. Therefore, $\hat{X}_t(\bar{x}_0)$ which is the intersection of a compact set $X(\bar{x}_0)$ and a closed set $\pi_t^{-1}(F_t)$ is compact. $\hat{X}_t(\bar{x}_0)$ is nonempty since both of $X(\bar{x}_0)$ and $\pi_t^{-1}(F_t)$ include a feasible path $\{x_t\}_0^\infty$ such that

$$x_t \in F_t, x_\tau = 0 \quad \text{for } \tau > t,$$

where F_t is nonempty from Lemma 2.10(i).

Consequently, $\hat{X}_t(\bar{x}_0)$ is a nonempty compact subset of Ω.

C.12. Proof of Lemma 2.11

Since $\hat{x}_t \in F_t$ and $A_t(\bar{x}_0) \neq \{0\}$, from Lemma 2.4 there exists $\hat{p}_t$ such that

$$\hat{p}_t \cdot \hat{x}_t = \max \{\hat{p}_t \cdot x_t | x_t \in A_t(\bar{x}_0)\} > 0.$$

Then, from (i) in Theorem 2.1′ and (b) in Theorem 2.3, there exists a vector $\hat{p}_0 \in A_t^*(\hat{p}_t)$ such that

$$0 < \hat{p}_t \cdot \hat{x}_t = \hat{p}_0 \cdot \bar{x}_0 = \max \{\hat{p}_t \cdot x_t | x_t \in A_t(\bar{x}_0)\}$$
$$= \min \{p_0 \cdot \bar{x}_0 | p_0 \in A_t^*(\hat{p}_t)\}$$

where A_t^* is defined by (28b). Therefore, from Theorem 2.4 (reciprocity principle) we have

$$0 < \hat{p}_t \cdot \hat{x}_t = \hat{p}_0 \cdot \bar{x}_0 = \min \{\hat{p}_0 \cdot x_0 | x_0 \in A_t^{-1}(\hat{x}_t)\}$$
$$= \max \{p_t \cdot \hat{x}_t | p_t \in A_t^{*^{-1}}(\hat{p}_t)\}.$$

Consequently, $N_t(\bar{x}_0)$ is nonempty.

Next, by the definition of $N_t(\bar{x}_0)$ in (73) and since $\bar{x}_0 > 0$ from assumption,

$$N_t(\bar{x}_0) \cup \{0\} = \{\hat{p}_0 | \hat{p}_0 \cdot (x_0 - \bar{x}_0) \geqq 0 \quad \text{for all } x_0 \in A_t^{-1}(\hat{x}_t)\}.$$

That is, the set $N_t(\bar{x}_0) \cup \{0\}$ constitutes the positive polar cone of the set, $A_t^{-1}(\hat{x}_t) - \bar{x}_0$. Hence, $N_t(\bar{x}_0) \cup \{0\}$ is a closed convex cone since the positive polar cone of any set is a closed convex cone (see Luenberger [1969], Proposition 6.6).

Now consider the set

$$N_t'(\bar{x}_0) \equiv (N_t(\bar{x}_0) \cup \{0\}) \cap \left\{ p \in \mathbf{R}_+^{*m} \middle| \sum_{i=1}^m p_i = 1 \right\}$$
$$= N_t(\bar{x}_0) \cap \left\{ p \in \mathbf{R}_0^{*m} \middle| \sum_{i=1}^m p_i = 1 \right\}.$$

Since $N_t(\bar{x}_0) \cup \{0\}$ is closed, $N_t'(\bar{x}_0)$ is a compact subset of $\mathbf{R}^{*m}$. It is nonempty since $N_t(\bar{x}_0) \neq \{0\}$.

Finally let us show that the interaction

$$\bigcap_{t=1}^\infty N_t'(\bar{x}_0)$$

is nonempty. Observe that since $\{\hat{x}_t\}_0^\infty$ is a feasible path

$$A_1^{-1}(\hat{x}_1) \subset A_2^{-1}(\hat{x}_2) \subset \cdots \subset A_t^{-1}(\hat{x}_t) \subset \cdots,$$

thus

$$N_1(\bar{x}_0) \supset N_2(\bar{x}_0) \supset \cdots \supset N_t(\bar{x}_0) \supset \cdots,$$

and hence

$$N_1'(\bar{x}_0) \supset N_2'(\bar{x}_0) \supset \cdots \supset N_t'(\bar{x}_0) \supset \cdots.$$

Consequently, $\cap_{t=1}^{\infty} N_t'(\bar{x}_0)$ is a compact subset of $\mathbf{R}^{*m}$ since each $N_t'(\bar{x}_0)$ is compact, and it is nonempty from the finite intersection axiom. Therefore, since $N_t'(\bar{x}_0) \subset N_t(\bar{x}_0)$, the set $N(\bar{x}_0) = \cap_{t=1}^{\infty} N_t(\bar{x}_0)$ is nonempty.

C.13. Regular technologies and Kuhn–Tucker theorem (from Section 5 in Gale [1967a], also see Gale [1967b])

Let f be a concave function on $\mathbf{R}^m$, then the *steepness* σ of this function at a point x is defined by

$$\sigma = \sup_{x'} [(f(x') - f(x))/\|x - x'\|]$$

where $\|x - x'\|$ may be any convenient norm on $\mathbf{R}^m$. And, given a technology set $\mathscr{T}$ which is a convex subset in $\mathbf{R}_+^m \times \mathbf{R}_+^m$ and given a concave function $u(x, y)$ on $\mathscr{T}$, we consider the next problem: Find $(\bar{x}, \bar{y})$ in $\mathscr{T}$ such that

$$u(\bar{x}, \bar{y}) = \max_{x \leq y_0} u(x, y) \quad \text{for } (x, y) \text{ in } \mathscr{T}. \tag{6}$$

Let us assume the set $\{(x, y) | (x, y) \in \mathscr{T} \text{ and } x \leq y_0\}$ is compact and convex. Then this maximum always exists and we say that the *Kuhn–Tucker Theorem* holds for the above problem if there exists a non-negative vector $\bar{p}$ such that

$$u(x, y) + \bar{p} \cdot (y_0 - x) \leq u(\bar{x}, \bar{y}) \quad \text{for all } (x, y) \text{ in } \mathscr{T}.$$

And, for any m-vector $z \geq 0$ define the function $\mu_1(z)$ by the rule

$$\mu_1(z) = \max_{x \leq z} u(x, y) \quad \text{for } (x, y) \text{ in } \mathscr{T}.$$

Then we have:

Theorem 2 (p. 8 in Gale [1967a])

The Kuhn–Tucker Theorem holds for problem (6) if and only if the function μ_1 has bounded steepness σ at $z = y_0$. In this case the vector $\bar{p} = (\bar{p}_1, \ldots, \bar{p}_m)$ can be chosen so that $\Sigma \, \bar{p}_i = \sigma$.

Next, the set valued function T is defined by the rule

$$T(z) = \{(x, y) | (x, y) \in \mathcal{T} \text{ and } x \leqq z\}.$$

And the technology $\mathcal{T}$ is called *regular* if the function T satisfies a Lipschitz condition, that is, if there is a constant k such that

$$\|T(z) - T(z')\| \leqq k \|z - z'\|$$

where $\|T(z) - T(z')\|$ is the Hausdorf metric. Then we get,

Lemma 1 (p. 8, Gale [1967a])

If $\mathcal{T}$ is regular and u is any concave function with finite steepness, the function μ_1 has finite steepness for all $z \geqq 0$.

PROOFS OF STATEMENTS IN CHAPTER 3 TO CHAPTER 5

D.1. Proof of Theorem 3.1

(i) This is true from B-property 23.

(ii) T_1 is equivalent to

$$X \to \left\{ Y \mid Y \in \sum_{l=1}^{n} F'_l(X) \right\}$$

where

$F'_l(X) = \{Y \mid Y = (0, 0, \dots, 0, y_l, 0, \dots, 0) \text{ for some } y_l \in F_l(x_l),$
$0 \in \mathbf{R}^{m'}, \text{ where } (x_l)_1^n = X\}, \quad l = 1, \dots, n.$

Each F'_l $(l = 1, \dots, n)$ is a m.t.cc. from $\mathbf{R}_+^{mn}$ into $\mathbf{R}_+^{m'n}$ since each F_l $(l = 1, \dots, n)$ is a m.t.cc. from $\mathbf{R}_+^m$ into $\mathbf{R}_+^{m'}$. Thus, from B-property 22, T_1 is a m.t.cc. from $\mathbf{R}_+^{mn}$ into $\mathbf{R}_+^{m'n}$.

(iii) T_2 is equivalent to

$$Y \to \left\{ X \mid X \in \sum_{l=1}^{n} G'_l(Y) \right\}$$

where

$G'_l(Y) = \{X_l \mid X_l \in G_l(y_l), \quad \text{where } (y_l)_1^n = Y\}, \quad l = 1, \dots, n.$

Each G'_l $(l = 1, \dots, n)$ is a m.t.cc. from $\mathbf{R}_+^{m'n}$ into $\mathbf{R}_+^{mn}$ since each G_l $(l = 1, \dots, n)$ is a m.t.cc. from $\mathbf{R}_+^{m'}$ into $\mathbf{R}_+^{mn}$. Thus from B-property 22, T_2 is a m.t.cc. from $\mathbf{R}_+^{m'n}$ into $\mathbf{R}_+^{mn}$.

(iv) G_l is equivalent to

$$y_l \to \{X_l \mid X_l \in (G'_{l1} \vee G'_{l2} \vee \cdots \vee G'_{ls} \vee \cdots \vee G'_{ln}) y_l\}$$

where

$$(G'_{l1} \vee \cdots \vee G'_{ls} \vee \cdots \vee G'_{ln})y_l = \cup \left\{ \sum_{s=1}^{n} G'_{ls}(y_{ls}) \right|$$

$$\left. \sum_{s=1}^{n} y_{ls} = y_l, \, y_{ls} \geqq 0, \quad s = 1, \ldots, n \right\}$$

in which

$$G'_{ls}(y_{ls}) = \{X_{l.} | X_{l.} = (0, 0, \ldots, 0, x_{ls}, 0, \ldots, 0)$$
$$\text{for some } x_{ls} \in G_{ls}(y_{ls}), 0 \in \mathbf{R}^m \}.$$

Each G'_{ls} $(s = 1, \ldots, n)$ is a m.t.cc. from $\mathbf{R}_+^{m'}$ into $\mathbf{R}_+^{mn}$ since each G_{ls} $(s = 1, \ldots, n)$ is a m.t.cc. from $\mathbf{R}_+^{m'}$ into $\mathbf{R}_+^{m}$. Thus, from B-property 21, G_l is a m.t.cc. from $\mathbf{R}_+^{m'}$ into $\mathbf{R}_+^{mn}$.

D.2. Proof of Lemma 4.1

Since the time index t is irrelevant in the proof, let us omit it here.

(i) The assumption says that there exists number σ_l such that

$$|f_l(x_l, c_l) - f_l(x'_l, c'_l)| \leq \sigma_l \|(x_l, c_l) - (x'_l, c'_l)\|$$

for any $(x_l, c_l), (x'_l, c'_l) \in \mathbf{R}_+^{m+m_c}$, $l = 1, 2, \ldots, n$. Thus

$$\sum_{l=1}^{n} |f_l(x_l, c_l) - f_l(x'_l, c'_l)| \leq \sum_{l=1}^{n} \sigma_l \|(x_l, c_l) - (x'_l, c'_l)\|$$

$$\leq \sigma \sum_{l=1}^{n} \|(x_l, c_l) - (x'_l, c'_l)\|,$$

where $\sigma = \max_l \sigma_l$. Next, we define the norm $\|z\|$ of any vector $z = (z_i)$ by $\max_i |z_i|$. Then

$$\sum_{l=1}^{n} \|(x_l, c_l) - (x'_l, c'_l)\| \leq n \|((x_l)_1^n, (c_l)_1^n) - ((x'_l)_1^n, (c'_l)_1^n)\|.$$

Finally,

$$\left| \sum_{l=1}^{n} f_l(x_l, c_l) - \sum_{l=1}^{n} f_l(x'_l, c'_l) \right| \leq \sum_{l=1}^{n} |f_l(x_l, c_l) - f_l(x'_l, c'_l)|.$$

Consequently, we have

$$\left| \sum_{l=1}^{n} f_l(x_l, c_l) - \sum_{l=1}^{n} f_l(x'_l, c'_l) \right| \leq n\sigma \|((x_l)_1^n, (c_l)_1^n) - ((x'_l)_1^n, (c'_l)_1^n)\|.$$

(ii) The assumption says that there exist k_l, $l = 1, 2, \ldots, n$, such that

$$\|F_l(x_l) - F_l(x_l')\| \le k_l \|x_l - x_l'\|,$$

for any $x_l, x_l' \in \mathbf{R}_+^m$, where $\|F_l(x_l) - F_l(x_l')\|$ is the distance between two sets $F_l(x_l)$ and $F_l(x_l')$ under the Hausdolf metric. Hence, when we use the same norm as in (i),

$$\sum_l \|F_l(x_l) - F_l(x_l')\| \le \sum_l k_l \|x_l - x_l'\|$$
$$\le nk \|(x_l)_1^n - (x_l')_1^n\|,$$

where $k = \max_l k_l$. Next, by definition,

$$\|T_1(X) - T_1(X')\| = \|T_1((x_l)_1^n) - T_1((x_l')_1^n)\|$$

$$= \max \left\{ \sup_{(y_l)_1^n \in T_1((x_l)_1^n)} \inf_{(y_l')\in T_1((x_l')_1^n)} \|(y_l)_1^n - (y_l')_1^n\|, \right.$$
$$\left. \sup_{(y_l')_1^n \in T_1((x_l')_1^n)} \inf_{(y_l)_1^n \in T_1((x_l)_1^n)} \|(y_l)_1^n - (y_l')_1^n\| \right\}$$

$$\le \max \left\{ \sup_{(y_l)_1^n \in T_1((x_l)_1^n)} \inf_{(y_l')_1^n \in T_1((x_l')_1^n)} \sum_l \|y_l - y_l'\|, \right.$$
$$\left. \sup_{(y_l')_1^n \in T_1((x_l')_1^n)} \inf_{(y_l)_1^n \in T_1((x_l)_1^n)} \sum_l \|y_l - y_l'\| \right\}$$

$$\le \sum_{l=1}^n \max \left\{ \sup_{y_l \in F_l(x_l)} \inf_{y_l' \in F_l(x_l')} \|y_l - y_l'\|, \right.$$
$$\left. \sup_{y_l' \in F_l(x_l')} \inf_{y_l \in F_l(x_l)} \|y_l - y_l'\| \right\}$$

$$= \sum_{l=1}^n \|F_l(x_l) - F_l(x_l')\|.$$

Therefore, we have

$$\|T_1(X) - T_1(X')\| \le nk \|X - X'\|$$

for any $X, X' \in \mathbf{R}_+^{mn}$.

(iii) The proof of (iii) is quite similar to that of (ii), and hence it is omitted.

(iv) Each G_l can be decomposed as follows:

$$G_l(y_l) = G_l^2 \cdot G_l^1(y_l),$$

where

$$G_i^1(y_i) = \left\{ (y_{l1}, \ldots, y_{ls}, \ldots, y_{ln}) \middle| \quad \text{for some} \right.$$

$$\sum_{s=1}^{n} y_{ls} = y_l, \, y_{ls} \geq 0, \quad s = 1, 2, \ldots, n \right\},$$

$$G_i^2(y_{l1}, \ldots, y_{ls}, \ldots, y_{ln}) = \{(z_{l1}, \ldots, z_{ls}, \ldots, z_{ln}) \middle| \text{ for some}$$

$$z_{ls} \in G_{ls}(y_{ls}), \quad s = 1, 2, \ldots, n \}.$$

$G_i^1(y_i)$ is clearly a m.t.cc. and it satisfies the Lipschitz condition. G_i^2 is also a m.t.cc. when each $G_{ls}(s = 1, \ldots, n)$ is a m.t.cc. And the fact that G_i^2 satisfies the Lipschitz condition is shown quite similarly to the proof of (ii). Hence, using exactly the same method as the proof of Lemma 2.3, it can be shown that each G_l satisfies the Lipschitz condition.

(v) This is obvious and the proof is omitted.

D.3. Proof of (41) in Chapter 5

Since the plane (40) passes through $\bar{z}_r$ and $\bar{z}_s$, we get

$$p_1^*(a_1^r + b_1^r) + p_2^*b_2^r - \bar{\lambda}(p_1^*a_1^r + p_2^*a_2^r) = 0,$$
$$p_1^*(a_1^s + b_1^s) + p_2^*b_2^s - \bar{\lambda}(p_1^*a_1^s + p_2^*a_2^s) = 0.$$

That is,

$$p_1^*(a_1^r + b_1^r - \bar{\lambda}a_1^r) = p^*(\bar{\lambda}a_2^r - b_2^r),$$
$$p_1^*(a_1^s + b_1^s - \bar{\lambda}a_1^s) = p_2^*(\bar{\lambda}a_2^s - b_2^s).$$

We will show that if $p_1^* = 0$, then we would obtain a contradiction. First, if $p_1^* = 0$, then $p_2^* \neq 0$, otherwise (40) is not the equation of a plane. Hence, we obtain from the above equations that $a_1^r + b_1^r - \bar{\lambda}a_1^r = 0$ and $a_1^s + b_1^s - \bar{\lambda}a_1^s = 0$, and thus

$$\bar{\lambda} = \frac{a_1^r + b_1^r}{a_1^r} = \frac{a_1^s + b_1^s}{a_1^s} \tag{1}$$

On the other hand, from (28) and (30)

$$\bar{\lambda} = \frac{\bar{\alpha}b_2^r + (1 - \bar{\alpha})b_2^s}{\bar{\alpha}a_2^r + (1 - \bar{\alpha})a_2^s} \tag{2}$$

Next, we see $b_2^r/a_2^r \neq b_2^s/a_2^s$ because of the following reason. If

$b_2^r/a_2^r = b_2^s/a_2^s$, we get from (1) and (2) that

$$\bar{\lambda} = \frac{a_1^r + b_1^r}{a_1^r} = \frac{a_1^s + b_1^s}{a_1^s} = \frac{b_2^r}{a_2^r} = \frac{b_2^s}{a_2^s}.$$

Hence $(a_1^r + b_1^r, b_2^r) = \bar{\lambda}(a_1^r, a_2^r)$ and $(a_1^s + b_1^s, b_2^s) = \bar{\lambda}(a_1^s, b_2^s)$, which is a contradiction to A5 plus A6.

Therefore, suppose for example, $b_2^r/a_2^r > b_2^s/a_2^s$. Then, from (2), $b_2^r/a_2^r > \bar{\lambda}$ since $1 > \bar{\alpha} > 0$ from A6. This implies, taking into account (1), that

$$[(a_1^r, a_2^r), \bar{\lambda}(a_1^r, a_2^r)] \leqq [(a_1^r, a_2^r), (a_1^r + b_1^r, b_2^r)] \in S_r.$$

Hence, since the transformation F_r is a m.t.cc. and thus has the property of the free disposability of outputs, we have $[(a_1^r, a_2^r), \bar{\lambda}(a_1^r, a_2^r)] \in S_r$. This also is a contradiction to A5 plus A6. If we assume $b_2^r/a_2^r < b_2^s/a_2^s$, we get $[(a_1^s, a_2^s), \bar{\lambda}(a_1^s, a_2^s)] \in S_s$, and hence we also have a contradiction.

Consequently, it must be true that $p_2^* > 0$. Similarly, we can prove that $p_1^* > 0$.

D.4. Proof of the fact that only the set of activities defined by (42) in Chapter 5 break even under (p_1^*, p_2^*)

Since the technology has the property of constant returns to scale, it is enough to examine activities in the set S. Suppose there exists an activity $\tilde{z}$ such that $\pi(\tilde{z}) = 0$, $\tilde{z} \in S$ and $\tilde{z} \neq \alpha \tilde{z}_r + (1 - \alpha)\tilde{z}_s$ for any $0 \leqq \alpha \leqq 1$. This says that point $\tilde{z}$ is on the plane (40) but is not on the line generated by $\tilde{z}_r$ and $\tilde{z}_s$ because of A5. Hence, the set which consists of the convex hull of the three points, $\tilde{z}_r, \tilde{z}_s, \tilde{z}$, is not a line segment but a triangle set and it is on the plane (40). But $L(\bar{\lambda})$ is on the plane (40) and it passes through a point on the relative interior of the line segment $\overline{\tilde{z}_r \tilde{z}_s}$ because of A6. Hence $L(\bar{\lambda}) \cap S$ is not a point and this is a contradiction to A5. Hence we conclude that only the set of activities generated by convex combination of $\tilde{z}^r$ and $\tilde{z}^s$ break even under (p_1^*, p_2^*) among activities in S. Thus, generally, only the set of activities defined by (42) break even under (p_1^*, p_2^*).

PROOFS OF STATEMENTS IN CHAPTER 6

The proof of Lemma 6.1 is omitted since it is relatively simple.

Lemma 6.2. Define $\Delta p_{rs} = p_r(t) - p_s(t)$. Then, from (21), $\Delta\dot{p}_{rs}(t) = \dot{p}_r(t) - \dot{p}_s(t) = -s \max(p_r(t), p_s(t))(F'(K_r(t)) - F'(K_s(t)))$. Hence, from (9) and Lemma 6.1, we have

$$\text{sign } \Delta\dot{p}_{rs}(t) = \text{sign } (K_r(t) - K_s(t)) \quad \text{when}$$
$$K_r(t) \geq K^* \text{ and } K_s(t) \geq K^*. \tag{E.1}$$

In addition, from (20),

$$\text{if } \Delta p_{rs}(t) > 0, \quad \text{then } \dot{K}_r(t) > \dot{K}_s(t) = 0. \tag{E.2}$$

Now we examine each case.

Case A. Suppose we have that $K_r(t') > K_s(t') \geq K^*$ and $\Delta p_{rs}(t') > 0$ for some $t' \geq 0$. Then from (E.1) and (E.2), we have $\Delta p_{rs}(t) > 0$ and $K_r(t) > K_s(t) \geq K^*$ for all $t \in [t', T]$, and hence $p_r(T) > p_s(T)$ and $K_r(T) > K_s(T) \geq K^*$, which contradicts the terminal condition (14) under (9).

Case B. Suppose we have that $K_r(t') > K_s(t') \geq K^*$ and $\Delta p_{rs}(t') = 0$ for some $t' \geq 0$. Then, since $K_r(t) > K_s(t)$ for $t \in [t', t' + \epsilon]$ with some $\epsilon > 0$, from (E.1) we have that $\Delta\dot{p}_{rs}(t) > 0$ for $t \in [t', t' + \epsilon]$, and hence $p_r(t' + \epsilon) > p_s(t' + \epsilon)$ and $K_r(t' + \epsilon) > K_s(t' + \epsilon) \geq K^*$. But, this is the same case as A, and hence we have a contradiction.

Case C. Suppose we have that $K_r(t') = K_s(t') \geq K^*$ and $p_r(t') > p_s(t')$. Then, since $p_r(t) > p_s(t)$ for $t \in [t', t' + \epsilon]$ for some $\epsilon' > 0$, from (E.1) and (E.2) we have that $\Delta p_{rs}(t' + \epsilon) > \Delta p_{rs}(t') > 0$ and $K_r(t' + \epsilon) > K_s(t' + \epsilon)$, which reduces to Case A, and we have a contradiction.

Therefore, if $K_r(t) > K_s(t) \geq K^*$, then $p_r(t) < p_s(t)$. And if $K_r(t) = K_s(t) \geq K^*$, then $p_r(t) = p_s(t)$.

Lemma 6.3. The first half is clear from Lemma 6.2(a). Therefore, consider the case where $K_r(t') = K_s(t') \geq K^*$. In this case, suppose we have, for example, $\theta_r(t) > \theta_s(t)$ for $t \in [t', t' + \epsilon)$ for some $\epsilon > 0$. Then $K_r(t) > K_s(t)$ for $t \in (t', t' + \epsilon)$. This implies, from Lemma 6.2(a), that $p_r(t) < p_s(t)$ for $t \in (t', t' + \epsilon)$. Hence we have a contradiction of (20). Therefore, when $K_r(t') = K_s(t') \geq K^*$, $\theta_r(t) = \theta_s(t)$ at least for $t \in (t', t' + \epsilon)$ for some $\epsilon > 0$. Then, repeating the same process, we have that if $K_r(t') = K_s(t') \geq K^*$, then $\theta_r(t) = \theta_s(t)$ for $t \in [t', T]$ except for time intervals which measure zero.

Lemma 6.4. Define $\Delta p_{sr}(t)$ by $\Delta p_{sr}(t) = p_s(t) - p_r(t)$ where $t \in [0, T]$. Then, from (21), $\Delta \dot{p}_{sr}(t) = - \max(p_r(t), p_s(t)) s(F'(K_s(t)) - F'(K_r(t)))$. Hence, from (9) and Lemma 6.1, we have

$$\text{sign } \Delta \dot{p}_{sr}(t) = \text{sign } (K_r(t) - K_s(t)) \quad \text{when } K^* > K_r(t)$$
$$\text{and } K^* > K_s(t). \tag{E.3}$$

We now prove the lemma with the following three steps.

Step 1. Suppose we have

$$K^* > K_r(t') > K_s(t') \quad \text{and} \quad p_r(t') < p_s(t') \quad \text{for some } t' \in [0, T] \tag{E.4}$$

Then, from (E.3) and the allocation rule (20), we have $p_r(t) < p_s(t)$ and $\theta_r(t) = 0$, $\theta_s(t) = 1$ when $K_s(t) \leq K_r(t')$ and $t \geq t'$. Hence there are two possibilities:

$$K_s(T) \leq K_r(t') = K_r(T) \quad \text{and} \quad p_s(T) > p_r(T), \quad \text{or} \tag{E.5}$$
$$K_s(t'') = K_r(t') = K_r(t'') \quad \text{and} \quad p_s(t'') > p_r(t'') \tag{E.6}$$
for some t'' where $t' < t'' < T$.

(E.5) is a contradiction of the terminal condition (14). Hence, if (E.4) were true, we would have (E.6). Then, from the allocation rule (20) and from the continuity of $p_i(t)$ with respect to t, we have

$$\theta_r(t) = 0, \theta_s(t) = 1 \quad \text{for } t \in [t', t'' + \epsilon] \text{ for some } \epsilon > 0, \tag{E.7}$$

where $t'' + \epsilon < T$. Hence, if we take ϵ sufficiently small, we have

$$K_r(t'' + \epsilon) = K_r(t') \quad \text{and} \quad K_r(t') < K_s(t'' + \epsilon) < K^*. \tag{E.8}$$

On the other hand, instead of (E.7), consider the following allocation rule.

$$\theta_r(t) = 1, \theta_s(t) = 0 \quad \text{for } t \in [t', t' + \alpha), \quad \text{and}$$

$$\theta_r(t) = 0, \theta_s(t) = 1 \quad \text{for } t \in [t' + \alpha, t'' + \epsilon] \tag{E.9}$$

where α is defined by

$$K_r(t' + \alpha)|_{\text{under (E.9)}} = K_s(t'' + \epsilon)|_{\text{under (E.7)}} \tag{E.10}$$

We will confirm later that

$$t' + \alpha < t'' + \epsilon. \tag{E.11}$$

Let us denote the capital accumulation path under (E.7) by $K_r(t|\theta)$ and $K_s(t|\theta)$, and the capital accumulation path under (E.9) by $K_r(t|\theta^*)$ and $K_s(t|\theta^*)$, where $t' \leq t \leq t'' + \epsilon$. We will show in step 2 that

$$K_s(t'' + \epsilon|\theta^*) > K_r(t'). \tag{E.12}$$

Then, since the naming of the regions is arbitrary in our problem, we *exchange the subscripts r and s for each other on the path under (E.9) at time $t'' + \epsilon$*, and we have

$$K_r(t'' + \epsilon|\theta^*) > K_r(t') = K_r(t'' + \epsilon|\theta), \quad \text{and}$$

$$K_s(t'' + \epsilon|\theta^*) = K_r(t' + \alpha|\theta^*) = K_s(t'' + \epsilon|\theta).$$

In the rest of the plan period, $(t'' + \epsilon, T]$, we apply the same allocation ratios, $\theta_r(t)$ and $\theta_s(t)$ (which are supposed to be optimal) for both paths. Then, by using the same notation as before for the distinction of the two paths, we have $F(K_r(t|\theta^*)) + F(K_s(t|\theta^*)) > F(K_r(t|\theta)) + F(K_s(t|\theta))$ for $t'' + \epsilon \leq t \leq T$, and $K_r(t|\theta^*) > K_r(t|\theta)$ and $K_s(t|\theta^*) > K_s(t|\theta)$ for $t'' + \epsilon < t \leq T$. Hence, $F(K_r(T|\theta^*)) + F(K_s(T|\theta^*)) > F(K_r(T|\theta)) + F(K_s(T|\theta))$, which is a contradiction of the definition of the optimal path for Problem A.

Therefore, if $K^* > K_r(t') > K_s(t')$, it should be true that $p_r(t') \geq p_s(t')$. Now suppose $K_r(t') > K_s(t')$ and $p_r(t') = p_s(t')$. Then, from (E.3), we have, $K^* > K_r(t' + \mu) > K_s(t' + \mu)$ and $p_r(t' + \mu) < p_s(t' + \mu)$ for a sufficiently small $\mu > 0$. But this is the same case as (E.4) and we get a contradiction in this situation. Therefore, we conclude that if $K^* > K_r(t') > K_s(t')$, it should be true that $p_r(t') > p_s(t')$.

Step 2. Let us prove relation (E.12). First, in general, we have

$$F'(K_l(t' + \beta)) = F'(K_l(t')) + F''(K_l(t'))(K_l(t' + \beta) - K_l(t'))$$
$$+ 0(K_l(t' + \beta) - K_l(t')),$$

where $\quad 0(K_l(t'+\beta)-K_l(t'))/|K_l(t'+\beta)-K_l(t')| \to 0 \quad$ as $\quad K_l(t'+\beta) \to K_l(t')$. Therefore, under allocation rule (E.7), we have

$$\sum_{l=r,s} F'(K_l(t'+\beta|\theta)) = \sum_{l=r,s} F'(K_l(t')) + F''(K_s(t'))(K_s(t'+\beta|\theta)$$
$$- K_s(t')) + 0(K_s(t'+\beta|\theta) - K_s(t'))$$

$$= \sum_{l=r,s} F'(K_l(t')) + F''(K_s(t'))$$

$$\times \int_{t'}^{t'+\beta} s\left[\sum_{l=r,s} F'(K_l(t'))\right.$$

$$+ F''(K_s(t'))(K_s(\tau|\theta) - K_s(t')) + 0(K_s(\tau|\theta)$$

$$\left. - K_s(t'))\right]\mathrm{d}\tau + 0(K_s(t'+\beta|\theta) - K_s(t'))$$

$$= \sum_{l=r,s} F'(K_l(t')) + F''(K_s(t'))s$$

$$\times \sum_{l=r,s} F'(K_l(t'))\beta + 0_s(\beta),$$

where $0_s(\beta)/\beta \to 0$ as $\beta \to 0$.

Similarly, under allocation rule (E.9), we have

$$\sum_{l=r,s} F'(K_l(t'+\beta)|\theta^*) = \sum_{l=r,s} F'(K_l(t')) + F''(K_r(t'))s$$

$$\times \sum_{l=r,s} F'(K_l(t'))\beta + 0_r(\beta),$$

where $0_r(\beta)/\beta \to 0$ as $\beta \to 0$, and $\beta \le \alpha$.

Then, since $K^* > K_r(t') > K_s(t')$ by assumption (E.4) and hence $F''(K_r(t')) > F''(K_s(t'))$ from (9), at least for a sufficiently small $\beta > 0$ we have

$$\sum_{l=r,s} F'(K_l(t|\theta^*)) > \sum_{l=r,s} F'(K_l(t|\theta)) \quad \text{for } t \in (t', t'+\beta].$$

Suppose $\beta < \alpha$. Then, since in general $\int_{t'}^{t'+\beta} (\dot{K}_r(t) + \dot{K}_s(t))\,\mathrm{d}t = \int_{t'}^{t'+\beta} s(F(K_r(t)) + F(K_s(t)))\,\mathrm{d}t$, under allocation rules (E.7) and (E.9) we have

$$K_r(t'+\beta|\theta) = K_r(t'), \quad K_s(t'+\beta|\theta^*) = K_s(t')$$
$$K_r(t'+\beta|\theta^*) - K_r(t') > K_s(t'+\beta|\theta) - K_s(t').$$

Therefore, from the convexity of function F in the interval where $K \leqq K^*$, we have

$$\sum_{l=r,s} F'(K_l(t'+\beta|\theta^*)) > \sum_{l=r,s} F'(K_l(t'+\beta|\theta)),$$

$$F''(K_r(t'+\beta|\theta^*)) > F''(K_s(t'+\beta|\theta)),$$

$$\sum_{l=r,s} F'(K_l(t'+\beta|\theta^*)) - \sum_{l=r,s} F'(K_l(t'+\beta|\theta))$$

$$> \sum_{l=r,s} F'(K_l(t'|\theta^*)) - \sum_{l=r,s} F'(K_l(t'|\theta)) = 0.$$

Hence, by the same method as before, we can prove that

$$\sum_{l=r,s} F'(K_l(t|\theta^*)) > \sum_{l=r,s} F'(K_l(t|\theta)) \quad \text{for } t \in [t'+\beta, t'+\beta+\gamma],$$

for a sufficiently small $\gamma > 0$. In addition,

$$\sum_{l=r,s} F'(K_l(t'+\beta+\gamma)|\theta^*) - \sum_{l=r,s} F'(K_l(t'+\beta+\gamma|\theta)$$

$$> \sum_{l=r,s} F'(K_l(t'+\beta|\theta^*)) - \sum_{l=r,s} F'(K_l(t'+\beta|\theta)).$$

Hence if $\beta + \gamma < \alpha$, we can generate a sequence, $\beta, \beta + \gamma, \ldots.$, which does not converge before α and we get

$$\sum_{l=r,s} F'(K_l(t|\theta^*)) > \sum_{l=r,s} F'(K_l(t|\theta)) \quad \text{for } t \in (t', t+\alpha]. \tag{E.13}$$

Therefore, we have

$$\int_{t'}^{t'+\alpha} (\dot{K}_r(t|\theta^*) + \dot{K}_s(t|\theta^*)) \, dt > \int_{t'}^{t'+\alpha} (\dot{K}_r(t|\theta) + \dot{K}_s(t|\theta)) \, dt$$

and hence, under allocation rules (E.7) and (E.9) we have

$$K_r(t'+\alpha|\theta^*) - K_r(t') > K_s(t'+\alpha|\theta) - K_s(t'). \tag{E.14}$$

This fact, remembering Definition (E.10), confirms relation (E.11). Further, by repeating the same procedures for $t \in [t'+\alpha, t''+\epsilon]$ under (E.7) and (E.9), we get

$$\sum_{l=r,s} F'(K_l(t|\theta^*)) > \sum_{l=r,s} F'(K_l(t|\theta)) \quad \text{for } t \in [t'+\alpha, t''+\epsilon].$$

Therefore, by combining the above with (E.13), we have

$$\int_{t'}^{t'+\epsilon} (\dot{K}_r(t|\theta^*) + \dot{K}_s(t|\theta^*))\,dt > \int_{t'}^{t''+\epsilon} (\dot{K}_r(t|\theta) + \dot{K}_s(t|\theta))\,dt. \qquad \text{(E.15)}$$

In addition, under (E.7) and (E.9) we have $K_r(t'' + \epsilon|\theta) = K_r(t'|\theta) = K_r(t')$, $K_s(t'|\theta^*) = K_s(t'|\theta) = K_s(t')$, and $K_r(t'' + \epsilon|\theta^*) = K_r(t' + \alpha|\theta^*) = K_s(t'' + \epsilon|\theta)$ which follows from Definition (E.10). Hence, from (E.15), we get

$$K_s(t'' + \epsilon|\theta^*) > K_r(t'|\theta^*) = K_r(t'),$$

which is the required relation.

Step 3. We now prove (b) in Lemma 6.4. Suppose we have $K^* > K_r(t') = K_s(t')$ for some $t' \in [0, T]$. In this case, we have the following three possibilities.

$$p_r(t') = p_s(t'), \quad p_r(t') > p_s(t'), \quad \text{or} \quad p_r(t') < p_s(t').$$

First take the case of $p_r(t') = p_s(t')$. In this case, suppose, for example, that $\theta_r(t) > \frac{1}{2} > \theta_s(t)$ for $t \in [t', t' + \epsilon)$ for some $\epsilon > 0$. Then, $\dot{K}_r(t) > \dot{K}_s(t)$ for $t \in [t', t' + \epsilon)$, and hence $K_r(t) > K_s(t)$ for $t \in (t', t' + \epsilon)$. Hence, from (A.3) we have $p_r(t) < p_s(t)$ for $t \in (t', t' + \epsilon)$. But, the fact, $\theta_r(t) > \frac{1}{2} > \theta_s(t)$ and $p_r(t) < p_s(t)$ for $t \in (t', t' + \epsilon)$, contradicts with (20). Hence, we know that if $p_r(t') = p_s(t')$, it cannot be true that $\theta_r(t) \neq \theta_s(t)$ for $t \in [t', t' + \epsilon)$ for any $\epsilon > 0$. Therefore we conclude that if $p_r(t') = p_s(t')$, $\theta_r(t) = \theta_s(t) = \frac{1}{2}$ as long as $K_r(t) = K_s(t) < K^*$. But, by using a similar method as in Step 2, we can easily show that such an allocation cannot be the optimal one. Hence we conclude that if $K_r(t') = K_s(t')$, $p_r(t') \neq p_s(t')$.

Therefore, one of the last two relations should be true. But, since the naming of regions is arbitrary in Problem A, we must choose arbitrarily one of the two relations.

Lemma 6.5. The proof of this lemma immediately follows from Lemma 6.4.

Lemma 6.6. Taking (23) into account, we find that the proof of (a) is quite similar to the proof of (a) in Lemma 6.2. Hence we give here only the proof of (b). Suppose we have $K_r(t') > K^* > K_s(t') = \hat{\psi}(K_r(t'))$. If $t' = T$, from terminal condition (14) we have $p_r(t') = p_s(t')$. Hence, suppose $0 \leqq t' < T$. Then, since $K_s(t') = \hat{\psi}(K_r(t'))$ and since $\hat{\psi}(K)$ is a decreasing function of K, under any allocation of investment we have

$$K_r(t) > K_s(t) \geqq K_s(t') = \psi(K_r(t')) \geqq \psi(K_r(t)) \quad \text{for } t \in (t', t' + \epsilon)$$

for some $\epsilon > 0$. In addition, since $\dot{K}_r(t) + \dot{K}_s(t) > 0$ and $\psi(K)$ is a decreasing function, $\psi(K_r(t)) \neq K_s(t)$ for any $t \in (t', t' + \epsilon)$. Therefore we have

$$K_r(t) > K_s(t) > \psi(K_r(t)) \quad \text{for } t \in (t', t' + \epsilon) \quad \text{for some } \epsilon > 0.$$

Then, by definition of function ψ, $F'(K_r(t)) < F'(K_s(t))$ and hence $\dot{p}_r(t) > \dot{p}_s(t)$ for $t \in (t', t' + \epsilon)$. Hence, if $p_r(t') \geqq p_s(t')$, we have $K_r(t) > K_s(t) > \psi(K_r(t))$ and $p_r(t) > p_s(t)$, which is a contradiction of Lemma 6.6(a). Therefore, if $K_r(t') > K^* > K_s(t') = \psi(K_r(t'))$ and $t' \neq T$, we should have $p_r(t') < p_s(t')$.

Theorem 6.1, Corollary 6.1. The validity of these two properties is clear from the text.

Theorem 6.2(a), (b). For notational convenience, we rewrite (42) as follows:

$$\int_{K'_s}^{K'_r} \frac{F'(K_s) - F'(K'_r)}{(F(K'_r) + F(K_s) + M)^2} \, dK_s = 0 \quad \text{and} \quad K'_r \geqq K^* \geqq K'_s. \tag{E.16}$$

And, let us first prove that $\partial \psi^*(K, M)/\partial K < 0$, namely, $dK'_s / dK'_r |_{M = \text{constant}} < 0$. Let us set

$$f(K_s, K'_r, M) = (F'(K_s) - F'(K'_r))/(F(K'_r) + F(K_s) + M)^2. \tag{E.17}$$

Then, by taking the total derivative of Equation (E.16) with respect to K'_r and K'_s, and by using relations $f(K'_r, K'_r, M) = 0$ and $\partial f(K_s, K'_r, M)/\partial K'_s = 0$, we have

$$\frac{dK'_s}{dK'_r}\bigg|_{M = \text{constant}} = \int_{K'_s}^{K'_r} \frac{\partial f(K_s, K'_r, M)}{\partial K'_r} \, dK_s / f(K'_s, K'_r, M). \tag{E.18}$$

From (9) and Definition (E.16),

$$f(K'_s, K'_r, M) < 0. \tag{E.19}$$

And, from (E.16)

$$\int_{K_s'}^{K_r'} \frac{\partial f(K_s, K_r', M)}{\partial K_r'}\, dK_s = \int_{K_s'}^{K_r'} \frac{-F''(K_r')}{(F(K_r') + F(K_s) + M)^2}\, dK_s$$

$$-2\int_{K_s'}^{K_r'} \frac{(F'(K_s) - F'(K_r'))F'(K_r')}{(F(K_r') + F(K_s) + M)^3}\, dK_s. \qquad \text{(E.20)}$$

From assumption (9), $F''(K_r') < 0$, and hence

$$\int_{K_s'}^{K_r'} \frac{-F''(K_r')}{(F(K_r') + F(K_s) + M)^2}\, dK_s > 0. \qquad \text{(E.21)}$$

Next, let us set

$$\Delta p(K_s, K_r', M) = \int_{K_s}^{K_r'} \frac{F'(K) - F'(K_r')}{(F(K_r') + F(K) + M)^2}\, dK,$$

$$g(K_s, K_r', M) = \frac{F'(K_r')}{F(K_r') + F(K) + M}. \qquad \text{(E.22)}$$

Then, we have

$$\int_{K_s'}^{K_r'} \frac{(F'(K_s) - F'(K_r'))F'(K_r')}{(F(K_r') + F(K_s) + M)^3}\, dK_s$$

$$= \int_{K_s'}^{K_r'} -\frac{\partial \Delta p(K_s, K_r', M)}{\partial K_s}\, g(K_s, K_r', M)\, dK_s$$

$$= -[\Delta p(K_s, K_r', M) g(K_s, K_r', M)]_{K_s'}^{K_r'}$$

$$+ \int_{K_s'}^{K_r'} \Delta p(K_s, K_r', M) \frac{\partial g(K_s, K_r', M)}{\partial K_s}\, dK_s$$

$$= \int_{K_s'}^{K_r'} \Delta p(K_s', K_r', M) \frac{\partial g(K_s, K_r', M)}{\partial K_s}\, dK_s \quad \text{(from (37))}$$

$$= \int_{K_s'}^{K_r'} \Delta p(K_s', K_r', M) \frac{-F'(K_r')F'(K_s)}{(F(K_r') + F(K_s) + M)^2}\, dK_s,$$

which is negative since $F'(K_r') > 0$ and $F'(K_s) > 0$ from (9), and since from (9), (E.16) and (E.22), $\Delta p(K_s, K'_r, M) > 0$ for $K_s \in (K'_s, K'_r)$. Hence, taking (E.19) and (E.21) into account, we conclude that $dK'_s/dK'_r|_{M=\text{constant}} < 0$ along the switching curve, $K'_s = \psi^*(K'_r, M)$, that is, $\partial \psi^*(K, M)/\partial K < 0$.

Next we show that $\partial \psi^*(K, M)/\partial M < 0$. From the total derivative of (E.16) with respect to K'_s and M, we have

$$\frac{dK'_s}{dM}\bigg|_{K'_r=\text{constant}} = \frac{\displaystyle\int_{K'_s}^{K'_r} \frac{\partial f(K_s, K'_r, M)}{\partial M}\, dK_s}{f(K'_s, K'_r, M)}.$$

From (E.16) and (E.17), $f(K'_s, K'_r, M) < 0$. And, we have

$$\int_{K'_s}^{K'_r} \frac{\partial f(K_s, K'_r, M)}{\partial M}\, dK_s = \int_{K'_s}^{K'_r} - \phi(K_s) f(K_s, K'_r, M)\, dK_s,$$

where $\phi(K_s) = 2/(F(K'_r) + F(K_s) + M)$. From (E.16), we have $\int_{K'_s}^{K'_r} f(K_s, K'_r, M)\, dK_s = 0$. And, by the definition of function $\hat{\psi}$, $f(K_s, K'_r, M) < 0$ for $K_s \in [K'_s, \hat{\psi}(K'_r))$, $f(K_s, K'_r, M) > 0$ for $K_s \in (\hat{\psi}(K'_r), K'_r)$. Moreover, $\phi(K_s) > 0$ and $d\phi(K_s)/dK_s < 0$ for $K_s \in [K'_s, K'_r]$. Therefore, we see that $\int_{K'_s}^{K'_r} \phi(K_s) f(K_s, K'_r, M)\, dK_s < 0$. Hence,

$$\int_{K'_s}^{K'_r} \frac{\partial f(K_s, K'_r, M)}{\partial M}\, dK_s > 0. \tag{E.23}$$

Consequently, we have

$$\frac{dK'_s}{dM}\bigg|_{K'_r=\text{constant}} < 0,$$

that is, $\partial \psi^*(K, M)/\partial M < 0$.

Next, from the total derivative of (E.16) with respect to K'_r and M, we have

$$\frac{dK'_r}{dM}\bigg|_{K'_s=\text{constant}} = - \frac{\displaystyle\int_{K'_s}^{K'_r} \frac{\partial f(K_s, K'_r, M)}{\partial M}\, dK_s}{\displaystyle\int_{K'_s}^{K'_r} \frac{\partial f(K_s, K'_r, M)}{\partial K'_r}\, dK_s},$$

which is negative from (E.20) and (E.23) (recall that (E.20) was shown to be positive). Hence,

$$\frac{dK_r'}{dM}\bigg|_{K_s'=\text{constant}} < 0.$$

The rest of the relations in (a) and (c) immediately follow from those which have been proved in the above. And, (d) is clear from the definition of each function.

Lemmas 6.7 to 6.9.   They are clear in the text.

Theorem 6.3.   We consider here that n and T are finite numbers. We first show that, under assumption (9), any feasible path for Problem A lies in a compact subset of $\mathbf{R}^n$ which has an interior point. For this purpose, given $F(K)$, choose constants a and b such that

$$F(K) \leqq a + bK \quad \text{for all } K \in [0, \infty).$$

Such constants surely exist under assumption (9). And we define function $\hat{F}(K)$ on $\mathbf{R}_+$ by

$$\hat{F}(K) = a + bK.$$

Consider a new problem, Problem A′, which is obtained by substituting $F(K)$ with $\hat{F}(K)$ in Problem A. Then, in this new problem, we have

$$\sum_{l=1}^{n} \dot{K}_l(t) = s\left(na + b\sum_{l=1}^{n} K_l(t)\right).$$

Hence, if we set $K(t) = \sum_{l=1}^{n} K_l(t)$, then

$$\dot{K}(t) = sna + sbK(t), \quad \text{and} \quad K(0) = \sum_{l=1}^{n} K_l^0.$$

Solving this differential equation, we set $K(T) \equiv K_f$. Then, of course, K_f is finite. Now, since $F(K) \leqq \hat{F}(K)$ for all $K \in [0, \infty)$, it is easy to see that any feasible path for Problem A under assumption (9) lies within the set

$$D \equiv \overbrace{K_f \times K_f \times \cdots \times K_f}^{n}.$$

By definition, D is a compact subset of $\mathbf{R}^n$, and it has an interior point.

Next, define the set $G(\{K_l\}_1^n)$ by

$$G(\{K_l\}_1^n) = \left\{ \left[\sum_{l=1}^{n} F'(K_l)\dot{K}_l, \{\dot{K}_l\}_1^n \right] \Big| \dot{K}_l \right.$$

$$= \theta_l s \sum_{j=1}^{n} F(K_j), \sum_{l=1}^{n} \theta_l = 1, \theta_l \geqq 0, l = 1, 2, \ldots, n \left. \right\}.$$

Then $G(\{K_l\}_1^n)$ is a convex set for each $\{K_l\}_1^n$.

Therefore, we see that, under assumption (9), Problem A satisfies all the assumptions required for Lemma 2 of Sugimoto (1967, p. 119). Hence, from the lemma, we see that the set of all points $\sum_{l=1}^{n} F(K_l(T))$ which are feasible in Problem A constitutes a compact subset of **R**. Hence, there exists an optimal path for Problem A under assumption (9).

Lemma 6.10. This can be proved by using basically the same method as in the proofs of Lemmas 6.2, 6.4(a) and 6.6. Hence its proof is omitted.

Lemmas 6.11 and 6.12. They can be proved, respectively, quite similarly as Lemma 6.3(b) and Lemma 6.4(b) were proved, and hence these proofs are omitted.

Lemmas 6.13 to 6.20. They are clear, respectively, in the text.

Lemma 6.21. We adopt Rule 6.1. Suppose that, before region n catches up to region $n - 1$, region $\alpha + 1$ catches up to region α, where $1 \leqq \alpha \leqq n - 2$. And suppose that, before region n catches up to region $n - 1$, no two regions between $\alpha + 1$ and $n - 1$ come together. That is, we assume that regions α and $\alpha + 1$ are the closest two regions to region $n - 1$ (in terms of the numbering of the regions) among the regions which come together before region n catches up to region $n - 1$ (refer to Figure E.1).

And, we denote by $K'_l, \alpha \leqq l \leqq n - 1$, the point (on the K-axis) at which the allocation of investment is switched from region l to region $l + 1$ (refer to Figure E.1). And, denote by $t_l, \alpha \leqq l \leqq n - 1$, the time at which the allocation of investment is switched from region l to region $l + 1$; and by $t'_l, \alpha + 1 \leqq l \leqq n$, the time at which region l reaches region $l - 1$ (i.e., the time at which regions l and $l - 1$ come together). Then, recalling Lemma 6.10(b) and Lemma 6.19, from the initial set of

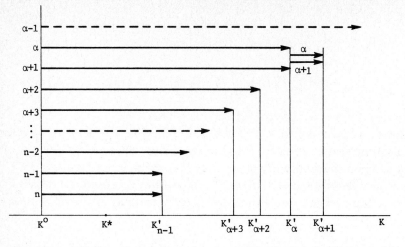

Figure E.1. Relation of capital stocks assumed in the proof of Lemma 6.21.

assumptions we have

$$t_\alpha < t_{\alpha+1} < \cdots < t_{n-1} < t'_n < t'_{n-1} < t'_{n-2} < \cdots < t'_{\alpha+2} < t'_{\alpha+1}, \qquad \text{(E.24)}$$

$$K'_\alpha \leqq K'_{\alpha+1}, \qquad \text{(E.25)}$$

$$K_l(t_{\alpha+1}) = K_l(t'_{\alpha+1}) \geqq K_\alpha(t_{\alpha+1}) = K_{\alpha+1}(t_{\alpha+1}) \equiv K'_{\alpha+1}$$
$$= K_{\alpha+1}(t'_{\alpha+2}) = K_\alpha(t'_{\alpha+2})$$
$$> K_{\alpha+2}(t_{\alpha+2}) \equiv K'_{\alpha+2} = K_{\alpha+2}(t'_{\alpha+3}) = K_{\alpha+3}(t'_{\alpha+3}) = \cdots$$
$$= K_n(t'_{\alpha+3})$$
$$> K_{\alpha+3}(t_{\alpha+3}) \equiv K'_{\alpha+3} = K_{\alpha+3}(t'_{\alpha+4}) = K_{\alpha+4}(t'_{\alpha+4}) = \cdots$$
$$= K_n(t'_{\alpha+4})$$
$$\vdots$$
$$> K_{n-1}(t_{n-1}) \equiv K'_{n-1} = K_n(t'_n) > K^*, \qquad \text{(E.26)}$$

where $l = 1, 2, \ldots, \alpha + 2$.

Next, according to the initial set of assumptions, $p_{\alpha+1}(t)$-path and $p_{\alpha+2}(t)$-path intersect at times $t_{\alpha+1}$ and $t'_{\alpha+2}$ (recall the allocation rule (16) and the continuity of $p_l(t)$-path with respect to t). Hence, it should be true that

$$0 = \int\limits_{t_{\alpha+1}}^{t'_{\alpha+2}} (\dot{p}_{\alpha+1}(t) - \dot{p}_{\alpha+2}(t)) \, dt$$

$$= \int\limits_{t_{\alpha+1}}^{t_{\alpha+2}} (\dot{p}_{\alpha+1}(t) - \dot{p}_{\alpha+2}(t))\, dt + \int\limits_{t_{\alpha+2}}^{t'_{\alpha+3}} (\dot{p}_{\alpha+1}(t) - \dot{p}_{\alpha+2}(t))\, dt$$

$$+ \int\limits_{t'_{\alpha+3}}^{t'_{\alpha+2}} (\dot{p}_{\alpha+1}(t) - \dot{p}_{\alpha+2}(t))\, dt.$$

Hence, if we set

$$A = \int\limits_{t_{\alpha+1}}^{t_{\alpha+2}} (\dot{p}_{\alpha+1}(t) - \dot{p}_{\alpha+2}(t))\, dt, \qquad B = \int\limits_{t'_{\alpha+2}}^{t'_{\alpha+3}} (\dot{p}_{\alpha+1}(t) - \dot{p}_{\alpha+2}(t))\, dt,$$

$$C = \int\limits_{t'_{\alpha+3}}^{t'_{\alpha+2}} (\dot{p}_{\alpha+1}(t) - \dot{p}_{\alpha+2}(t))\, dt,$$

then,

$$A + B + C = 0. \tag{E.27}$$

From (18) and (E.26), we get

$$B = \int\limits_{t_{\alpha+2}}^{t'_{\alpha+3}} p(t)s(F'(K_{\alpha+2}(t)) - F'(K_{\alpha+1}(t)))\, dt,$$

which is positive from (9), (E.26) and Lemma 6.1 (recalling Definition (15)). Hence, from (E.27), we have

$$A + C < 0. \tag{E.28}$$

Next, by the same method as in Section 6.5.2, we obtain

$$A = \int\limits_{K^0}^{K'_{\alpha+2}} \left(\frac{F'(K_{\alpha+2}) - F'(K'_{\alpha+1})}{F(K'_{\alpha+1}) + F(K_{\alpha+2}) + M_{\alpha+2}} p_{\alpha+2}(t_{\alpha+2}) \right.$$

$$\left. \times \exp\left(\int\limits_{K_{\alpha+2}}^{K'_{\alpha+2}} \frac{F'(K)}{F(K'_{\alpha+1}) + F(K) + M_{\alpha+2}}\, dK \right) \right) dK_{\alpha+2}$$

$$= p_{\alpha+2}(t_{\alpha+2})(F(K'_{\alpha+1}) + F(K'_{\alpha+2}) + M_{\alpha+2})$$

$$\times \int\limits_{K^0}^{K'_{\alpha+2}} \frac{F'(K_{\alpha+2}) - F'(K'_{\alpha+1})}{(F(K'_{\alpha+1}) + F(K_{\alpha+2}) + M_{\alpha+2})^2}\, dK_{\alpha+2},$$

$$C = \int_{K'_{\alpha+2}}^{K'_{\alpha+1}} \left(\frac{F'(K_{\alpha+2}) - F'(K'_{\alpha+1})}{[Y'/(n-\alpha-1)] + F(K_{\alpha+2})} p_{\alpha+2}(t_{\alpha+2}) \right.$$

$$\times \exp\left(-\int_{K^0}^{K'_{\alpha+3}} \frac{F'(K)}{Y' + F(K'_{\alpha+2}) + F(K) + (n-\alpha-3)F(K^0)} dK\right)$$

$$\times \exp\left(-\int_{K^0}^{K'_{\alpha+4}} \frac{F'(K)}{Y' + F(K'_{\alpha+2}) + F(K'_{\alpha+3}) + F(K) + (n-\alpha-4)F(K^0)} dK\right)$$

$$\vdots$$

$$\times \exp\left(-\int_{K^0}^{K'_{n-1}} \frac{F'(K)}{Y' + F(K'_{\alpha+2}) + \cdots + F(K'_{n-2}) + F(K) + F(K^0)} dK\right)$$

$$\times \exp\left(-\int_{K^0}^{K'_{n-1}} \frac{F'(K)}{Y' + F(K'_{\alpha+2}) + \cdots + F(K'_{n-1}) + F(K)} dK\right)$$

$$\times \exp\left(-\int_{K'_{n-1}}^{K'_{n-2}} \frac{F'(K)}{[(Y' + F(K'_{\alpha+2}) + \cdots + F(K'_{n-2}))/2] + F(K)} dK\right)$$

$$\vdots$$

$$\times \exp\left(-\int_{K'_{\alpha+3}}^{K'_{\alpha+2}} \frac{F'(K)}{[(Y' + F(K'_{\alpha+2}))/(n-\alpha-2)] + F(K)} dK\right)$$

$$\times \exp\left(-\int_{K'_{\alpha+2}}^{K_{\alpha+2}} \frac{F'(K)}{[Y'/(n-\alpha-1)] + F(K)} dK\right) \Bigg) dK_{\alpha+2}$$

$$= p_{\alpha+2}(t_{\alpha+2})(F(K'_{\alpha+1}) + F(K'_{\alpha+2}) + M_{\alpha+2})$$

$$\times \int_{K'_{\alpha+2}}^{K'_{\alpha+1}} \frac{F'(K_{\alpha+2}) - F'(K'_{\alpha+1})}{[Y' + (n-\alpha-1)F(K_{\alpha+2})]^2/(n-\alpha-1)} dK_{\alpha+2}$$

where

$$M_{\alpha+2} = \sum_{l=1}^{\alpha} F(K_l(t_{\alpha+1})) + (n-\alpha-2)F(K^0), \quad Y' = \sum_{l=1}^{\alpha+1} F(K_l(t_{\alpha+1})).$$

Therefore, since $p_{\alpha+1}(t_{\alpha+2}) > 0$ from Lemma 6.1, condition (E.28) can be rewritten as follows:

$$\int_{K^0}^{K'_{\alpha+2}} a(K_{\alpha+2})\,dK_{\alpha+2} + \int_{K'_{\alpha+2}}^{K'_{\alpha+1}} c(K_{\alpha+2})\,dK_{\alpha+2} < 0, \qquad (E.29)$$

where

$$a(K_{\alpha+2}) = \frac{F'(K_{\alpha+2}) - F'(K'_{\alpha+1})}{(F(K'_{\alpha+1}) + F(K_{\alpha+2}) + M_{\alpha+2})^2},$$

$$c(K_{\alpha+2}) = \frac{F'(K_{\alpha+2}) - F'(K'_{\alpha+1})}{[F(K'_{\alpha+1}) + (n-\alpha-1)F(K_{\alpha+2})]^2/(n-\alpha-1)}.$$

Next, since region $\alpha + 1$ directly catches up to region α at K'_α, we have

$$\int_{K^0}^{K'_\alpha} \frac{F'(K_{\alpha+1}) - F'(K'_\alpha)}{(F(K'_\alpha) + F(K_{\alpha+1}) + M_{\alpha+1})^2}\,dK_{\alpha+1} = 0,$$

where $M_{\alpha+1} = \Sigma_{l=1}^{\alpha-1} F(K_l(t_\alpha)) + (n-\alpha-1)F(K^0)$. Therefore, since $K'_{\alpha+1} = K'_\alpha$ and $M_{\alpha+2} > M_{\alpha+1}$, from (b) of Theorem 6.2, we get

$$0 < \int_{K^0}^{K'_{\alpha+1}} a(K_{\alpha+2})\,dK_{\alpha+2} = \int_{K^0}^{K'_{\alpha+2}} a(K_{\alpha+2})\,dK_{\alpha+2} + \int_{K'_{\alpha+2}}^{K'_{\alpha+1}} a(K_{\alpha+2})\,dK_{\alpha+2}.$$

Hence,

$$\int_{K^0}^{K'_{\alpha+2}} a(K_{\alpha+2})\,dK_{\alpha+2} + \int_{K'_{\alpha+2}}^{K'_{\alpha+1}} c(K_{\alpha+2})\,dK_{\alpha+2}$$

$$> \int_{K'_{\alpha+2}}^{K'_{\alpha+1}} (c(K_{\alpha+2}) - a(K_{\alpha+2}))\,dK_{\alpha+2}.$$

Thus, from (E.29), it should be true that

$$\int_{K'_{\alpha+2}}^{K'_{\alpha+1}} (c(K_{\alpha+2}) - a(K_{\alpha+2}))\,dK_{\alpha+2} < 0. \qquad (E.30)$$

Next, we have $a(K_{\alpha+2}) > 0$ and $c(K_{\alpha+2}) > 0$ for $K_{\alpha+2} \in [K'_{\alpha+2}, K'_{\alpha+1})$, and we have for $K_{\alpha+2} \in [K'_{\alpha+2}, K'_{\alpha+1}]$ that

$$
\begin{aligned}
\frac{a(K_{\alpha+2})}{c(K_{\alpha+2})} &= \frac{(Y' + (n - \alpha - 1)F(K_{\alpha+2}))^2}{(n - \alpha - 1)(F(K'_{\alpha+1}) + F(K_{\alpha+2}) + M_{\alpha+2})^2} \\
&= \frac{(Y' + (n - \alpha - 1)F(K_{\alpha+2}))^2}{(n - \alpha - 1)(Y' + F(K_{\alpha+2}) + (n - \alpha - 2)F(K^0))^2} \\
&< \frac{(Y' + (n - \alpha - 1)F(K_{\alpha+2}))^2}{(n - \alpha - 1)(Y' + F(K_{\alpha+2}))^2} \\
&\leqq \frac{(Y' + (n - \alpha - 1)F(K'_{\alpha+1}))^2}{(n - \alpha - 1)(Y' + F(K'_{\alpha+1}))^2} \\
&= \frac{(1 + (n - \alpha - 1)[F(K'_{\alpha+1})/Y'])^2}{(n - \alpha - 1)(1 + [F(K'_{\alpha+1})/Y'])^2} \\
&\leqq \frac{(1 + (n - \alpha - 1)[1/(\alpha + 1)])^2}{(n - \alpha - 1)(1 + [1/(\alpha + 1)])^2} \quad \left(\text{since } \frac{F(K'_{\alpha+1})}{Y'} \leqq \frac{1}{\alpha + 1} \right) \\
&= \frac{n^2}{(n - \alpha - 1)(\alpha + 2)^2}.
\end{aligned}
$$

Therefore, for (E.30) to be true, it must (at least) be true that

$$
\frac{n^2}{(n - \alpha - 1)(\alpha + 2)^2} > 1. \tag{E.31}
$$

In sum, for the processes described at the beginning of this proof to be able to occur on the optimal path, it is necessary that condition (E.31) is satisfied. And,

$$
\begin{aligned}
n^2 &- (n - \alpha - 1)^2(\alpha + 2)^2 \\
&= (\alpha - n + 2) \left(\alpha - \frac{-3 + \sqrt{(1 + 4n)}}{2} \right) \left(\alpha - \frac{-3 - \sqrt{(1 + 4n)}}{2} \right).
\end{aligned}
$$

Hence, recalling that $n \geqq 3$ by assumption, we see that condition (E.31) is satisfied only when

$$
\alpha < \frac{-3 + \sqrt{(1 + 4n)}}{2} \quad \text{or} \quad n - 2 < \alpha < n - 1
$$

Thus, since $\alpha \geqq n - 2$ by assumption, we conclude that the processes which are described at the beginning of this proof may happen only when

$$
\alpha < \frac{-3 + \sqrt{(1 + 4n)}}{2}.
$$

Consequently, we obtain Lemma 6.21.

Lemma 6.23. (a) This can be easily proved by applying the method for the proof of Lemma 6.4(a) (i.e., by neglecting the discussion and the information about the decreasing phase).

(b) If $n' = 1$, then this is implied by (a). If $n' = 2$, this is essentially the combination of (a) and Lemma 6.4(b). When $n' \geq 3$, we can prove this by using the following method of counterexample: Suppose the investment is allocated among more than two regions during a finite interval of time. Then, pick any two regions, say $l = 1$ and 2, among them, and show, by using the method for the case of the two-region economy in Section 6.4, that the final total income of these two regions can be increased by concentrating that investment, which is allocated to these two regions, to one of them. Hence, on the optimal path, if $K_1(t) = K_2(t) \equiv \max_l K_l(t)$, then one of the two regions must stay at $K_1(t)$ $(= K_2(t))$ for the rest of the plan period. Then, if $p_1(t) = p_2(t)$, the solution of (18) can not satisfy terminal condition (14) for both of regions 1 and 2. Hence we have an inconsistency. Consequently, if $K_1(t) = K_2(t) = \max_l K_l(t)$, then $p_1(t) > \max_{l \neq 1} p_l(t)$ or $p_2(t) > \max_{l \neq 2} p_l(t)$.

REGIONAL ALLOCATION OF INVESTMENT UNDER ALTERNATIVE OBJECTIVE FUNCTIONS

The shape of the optimal growth path, of course, depends on the form of the objective function. Then, it is essential from a planning point of view to examine how the shape of the optimal growth path is sensitive to the form of the objective function. In this appendix, we briefly examine this question by using the two-region economy, and show that when the plan period is sufficiently long, the shape of the optimal growth path is relatively stable regardless of the form of the objective function employed in the problem and the main conclusions obtained in Chapter 6 still remain to be true.

Let us consider a problem which is obtained from Problem A in Chapter 6 by replacing (27) with the following objective function, where we assume $n = 2$.

$$\int_0^T U(C(t), t)\, dt + w(F(K_r(T)) + F(K_s(T))), \tag{F.1}$$

where

$$C(t) = (1 - s(t))(F(K_r(t)) + F(K_s(t))). \tag{F.2}$$

In Equation (F.1), w is a positive constant, and U is the temporary-welfare-function of the economy which is a function of the total consumption, $C(t)$, at each time. In this new problem we first consider that saving ratio $s(t)$ is a given function of time as in Chapter 6, and it is assumed to be positive at each time. Let us call this new problem *Problem B*. Here we assume that

U is concave with respect to C, and $\partial U(C, t)/\partial C > 0$

$$\text{for all } t \ (\geqq 0) \text{ and for all } C \ (\geqq 0). \tag{F.3}$$

The Hamiltonian function, H, corresponding to Problem B is given by

$$H = U(C(t), t) + (\theta_r(t)p_r(t)$$
$$+ \theta_s(t)p_s(t))s(t)(F(K_r(t)) + F(K_s(t))). \tag{F.4}$$

Hence we see that the optimal allocation ratios, $\theta_r(t)$ and $\theta_s(t)$, which maximize function H are decided by the same rule, (20) in Chapter 6.

Next, from condition $\dot{p}_l = -\partial H/\partial K_l$, we get

$$\dot{p}_r(t) = -\left[\frac{\partial U(C(t), t)}{\partial C}(1 - s(t))\right.$$
$$\left. + (\theta_r(t)p_r(t) + \theta_s(t)p_s(t))s(t)\right] F'(K_r(t)),$$

$$\dot{p}_s(t) = -\left[\frac{\partial U(C(t), t)}{\partial C}(1 - s(t))\right. \tag{F.5}$$
$$\left. + (\theta_r(t)p_r(t) + \theta_s(t)p_s(t))s(t)\right] F'(K_s(t)).$$

Finally, the terminal condition is given by

$$p_r(T) = wF'(K_r(T)), \qquad p_s(T) = wF'(K_s(T)). \tag{F.6}$$

Now, by examining the proofs of Lemmas 6.1 to 6.6, it is not difficult to see that all these properties are also true under Problem B. Therefore, Lemmas 6.7 to 6.9 are also true under Problem B, and hence there is no essential change in the properties of the optimal growth paths for the cases of decreasing returns to scale, increasing returns to scale, or constant returns to scale.

Next, for variable returns to scale, we define switching function ψ^* by the condition:

$$p_r(t) = p_s(t) \quad \text{when } K_s(t) = \psi^*(K_r(t), t), \text{where } K_r(t) \geq K^* \geq K_s(t). \tag{F.7}$$

Since Equation (F.7) includes the time variable t explicitly as a component of $U(C, t)$, the switching function ψ^* in the case of Problem B is generally a function of K_r and t (or, K_s and t). Hence, the switching line for Problem B is a curve in (K_r, K_s, t)-space. Suppose that point $(K_r(t), K_s(t), t) \equiv (K_r', K_s', t)$ is a point on the switching curve. Then, from Lemma 6.6(b) we get

$$K_s' < \hat{\psi}(K_r') \quad \text{and hence } F'(K_r') > F'(K_s'). \tag{F.8}$$

Namely, the switching curve in Problem B lies under the curve,

$K_s = \hat{\psi}(K_r)$ (and $K_r = \hat{\psi}(K_s)$), in (K_r, K_s, t)-space. Therefore, the most important conclusion in Chapter 6, Theorem 6.5, is still valid under Problem B.

Finally, let us consider a problem where saving ratio $s(t)$ in the objective function defined by (F.1) and (F.2) is another control variable in addition to $\theta_r(t)$ and $\theta_s(t)$. Call this problem *Problem C*. In this new problem, the optimal value of $s(t)$ should be chosen at each time so as to maximize Hamiltonian function (F.4) subject to the restriction $0 \leqq s(t) \leqq 1$. Therefore, if we assume that the optimal value of $s(t)$ is internal at each time, that is,

$$0 < s(t) < 1 \quad \text{for all } t, \tag{F.9}$$

then we get, from $\partial H / \partial s = 0$, the following condition:

$$\frac{\partial U(C(t), t)}{\partial C} = \max\left(p_r(t), p_s(t)\right). \tag{F.10}$$

Hence, (F.5) is simplified as follows:

$$\begin{aligned}
\dot{p}_r(t) &= -\max\left(p_r(t), p_s(t)\right) F'(K_r(t)), \\
\dot{p}_s(t) &= -\max\left(p_r(t), p_s(t)\right) F'(K_s(t)).
\end{aligned} \tag{F.11}$$

Therefore, it is not difficult to show by using (19) in Chapter 6 and (F.11) that Theorem 6.5 is still valid under Problem C. In particular, suppose that on the optimal path

$$s(t) = s \text{ (a positive constant)} \quad \text{for all } t \in [0, T], \tag{F.12}$$

that is, the optimal saving ratio is a constant throughout the planning horizon.[1] Then, by using (19) in Chapter 6 and (F.11), we see that the switching function $\psi^*(K, t) \equiv \psi^*(K)$ is obtained by solving

$$\int_{\psi^*(K)}^{K} \frac{F'(K_s) - F'(K)}{s(F(K) + F(K_s))^2} \, dK_s = 0 \quad \text{where } K \geqq K^* \geqq \psi^*(K). \tag{F.13}$$

Comparing (F.13) with (40) in Chapter 6, we see that there are no essential differences between the characteristics of the switching function for Problem A and those for Problem C under assumption (F.12).

---

[1] This would not be a bad assumption in practice. Namely, in actual planning problems, a given utility function $U(C, t)$ in (F.1) would be accepted only when it brings on an appropriate value of the saving ratio. Hence, in practice, we might first choose the optimal saving ratio s and just assume that condition (F.10) is always satisfied.

REFERENCES

Andersson, Å.E., 1975, "A closed nonlinear growth model for international inter-regional trade and location", *Regional Science & Urban Economics*, vol. 5(4), pp. 427–444.

Arrow, K.J. and M. Kurz, 1970, *Public investment, the rate of return and fiscal policy*, The Johns Hopkins University Press.

Athans, M. and P.L. Falb, 1966, *Optimal control, an introduction to the theory and its applications*, McGraw-Hill.

Atsumi, H., 1965, "Neoclassical growth and the efficient program of capital ac-cumulation", *Review of Economics Studies*, vol. 32(2), pp. 127–136.

Bellman, R., 1957, *Dynamic programming*, Princeton University Press.

Berge, C., 1963, *Topological spaces* (English translation), Oliver and Boyd.

Bhatia, V.G., 1961, *Interregional allocation of investment in development programming*, Unpublished Ph.D. dissertation, Harvard University.

Burmeister, E. and A.R. Dobell, 1970, *Mathematical theories of economic growth*, The Macmillan Company.

Canon, M.D., C.D. Cullum, Jr., and E. Polak, 1970, *Theory of optimal and mathema-tical programming*, McGraw-Hill.

Cass, D. and K. Shell (eds.), 1976, *The Hamiltonian approach to dynamic economics*, Academic Press.

Datta-Chaudhuri, M., 1967, "Optimum allocation of investment and transportation in a two-region economy", in *Essays on the theory of optimal economic growth* (K. Shell ed.), The M.I.T. Press, pp. 129–140.

Dorfman, R., P.A. Samuelson, and R.M. Solow, 1958, *Linear programming and economic analysis*, McGraw-Hill.

Fan, L.T. and C.S. Wang, 1964, *The discrete maximum principle*, John Wiley & Sons, Inc.

Fujita, M., 1972, *Optimum growth in monotone space systems*, unpublished Ph.D. dissertation, University of Pennsylvania.

Fujita, M., 1973, "Optimum growth in two-region, two-good space systems: the final state problem", *Journal of Regional Science*, vol. 13(3), pp. 385–407.

Fujita, M., 1974, "Duality and maximum principle in multi-period convex program-ming", *Journal of Mathematical Economics*, vol. 1(3), pp. 295–326.

Fujita, M., 1975, "On optimal development in a multi-commodity space system", *Regional Science & Urban Economics*, vol. 5(1), 59–89.

Gale, D., 1956, "The closed linear model of production", in *Linear inequalities and related systems* (H.W. Kuhn and A.W. Tucker, eds.), Princeton University Press, pp. 285–303.

Gale, D., 1967a, "On optimal development in a multi-sector economy", *Review of Economic Studies*, vol. 34(1), pp. 1–18.

Gale, D., 1967b, "A geometric duality theorem with economic applications", *Review of Economic Studies*, vol. 34(97), pp. 19–24.

Halkin, H., 1974, "Necessary condition for optimal control problems with infinite horizon", *Econometrica*, vol. 42(2), pp. 267–273.

Haurie, A., 1976, "Optimal control on an infinite time horizon: the turn-pike approach", *Journal of Mathematical Economics*, vol. 3(1), pp. 81–92.

Hicks, J.R., 1965, *Capital and growth*, Clarendon Press.

Intriligator, M.D., 1964, "Regional allocation of investment: Comment", *Quarterly Journal of Economics*, vol. 78, pp. 659–662.

Isard, W., 1958, "Interregional linear programming: an elementary presentation and a general model", *Journal of Regional Science*, vol. 1(1), pp. 1–59.

Isard, W. and P. Liossatos, 1975, "General micro behavior and optimal macro space-time planning", *Regional Science & Urban Economics*, vol. 5(3), pp. 285–323.

Karlin, S., 1959, *Mathematical methods and theory in games, programming and economics*, Addison-Wesley.

Koopmans, T.C., 1951, "Analysis of production as an efficient combination of activities", in *Activity analysis of production and allocation* (Koopmans ed.), John Wiley & Sons, Inc., pp. 33–97.

Koopmans, T.C., 1957, *Three essays on the state of economic science*, McGraw-Hill.

Koopmans, T.C., 1964, "Economic growth at a maximal rate", *Quarterly Journal of Economics*, vol. 78, pp. 355–394.

Koopmans, T.C. and S. Reiter, 1951, "A model of transportation", in *Activity analysis of production and allocation* (Koopmans ed.), John Wiley & Sons, Inc., pp. 222–259.

Koopmans, T.C. and M. Beckmann, 1957, "Assignment problems and the location of economic activities", *Econometrica*, vol. 25(1), pp. 53–76.

Lefever, L., 1958, *Allocation in space*, North-Holland.

Leontief, W., A. Morgan, K. Polenske, D. Simpson, and E. Tower, 1965, "The economic impact – industrial and regional – of an arms cut", *Review of Economics and Statistics*, vol. 47(3), pp. 217–241.

Łos, J. and H. Łos (eds.), 1974, *Mathematical models of economics*, North-Holland.

Luenberger, D.G., 1969, *Optimization by vector space methods*, John Wiley & Sons, Inc.

Malinvaud, E., 1953, "Efficient capital accumulation and efficient allocation of resources", *Econometrica*, vol. 21(2), pp. 233–268.

Makarov, V.L. and A.M. Rubinov, 1970, "Superlinear point-set maps and models of economic dynamics", *Russian Mathematical Surveys*, 25.

McKenzie, L.W., 1963, "Turnpike theorems for a generalized Leontief model", *Econometrica*, vol. 31(1–2), pp. 165–180.

McKenzie, L.W., 1968, "Accumulation programs of maximum utility and von Neumann facet", in *Value, capital, and growth* (J.N. Wolfe, ed.), Aldine, pp. 353–383.

Morishima, M., 1961, "Proof of a turnpike theorem: the no-joint production case", *Review of Economic Studies*, vol. 28(2), pp. 89–97.

Morishima, M., 1964, *Equilibrium, stability and growth*, Oxford University Press.

Morishima, M., 1969, *Theory of economic growth*, Clarendon Press.

Moses, L.N., 1960, "A general equilibrium model of production, interregional trade and location of industry", *Review of Economic Studies*, vol. 27, pp. 373–399.

Negishi, T., 1972, *General equilibrium theory and international trade*, North-Holland.

Nikaido, H., 1968, *Convex structures and economic theory*, Academic Press.

Ohtsuki, Y., 1971, "Regional allocation of public investment in a *n*-region economy", *Journal of Regional Science*, vol. 11(2), pp. 225–233.

Pontryagin, L.S., V.G. Boltyanskii, R.V. Gamkrelidze, and E.F. Mishenko, 1962, *Mathematical theory of optimal process* (English translation), Interscience.

Radner, R., 1961, "Paths of economic growth that are optimal with regard only to final states: a turnpike theorem", *Review of Economic Studies*, vol. 28(1), pp. 98–104.

Radner, R., 1966, "Optimal growth in a linear-logarithmic economy", *International Economic Review*, vol. 7(1), pp. 1–33.

Radner, R., 1967, "Efficiency prices for infinite horizon production programmes", *Review of Economic Studies*, vol. 34(1), pp. 51–66.

Rahman, A., 1963, "Regional allocation of investment". *Quarterly Journal of Economics*, vol. 77, pp. 36–39.

Rahman, A., 1966, "Regional allocation of investment: the continuous version", *Quarterly Journal of Economics*, vol. 80, pp. 159–160.

Reiter, S. and G.R. Sherman, 1962, "Allocating indivisible resources affording external economies or diseconomies", *International Economic Review*, vol. 3(1), pp. 108–135.

Rockafellar, R.T., 1967, *Monotone processes of convex and concave type*, American Mathematical Society Memoir, No. 77, 1967.

Rockafellar, R.T., 1970, *Convex analysis*, Princeton University Press.

Rockafellar, R.T., 1974, "Convex algebra and duality in dynamic models of production", in *Mathematical models of economics* (J. Łos and M. Los eds.), North-Holland.

Sakashita, N., 1967, "Regional allocation of public investment", *Papers and Proceedings of the Regional Science Association*, vol. 19, pp. 161–182.

Samuelson, P.A., 1952, "Spatial price equilibrium and linear programming", *American Economic Review*, vol. 42, pp. 283–303.

Stevens, B.H., 1958, "An interregional linear programming model", *Journal of Regional Science*, vol. 1(1), pp. 60–98.

Stevens, B.H., 1961, "Linear programming and location rent", *Journal of Regional Science*, vol. 3(1), pp. 15–21.

Stevens, B.H., 1968, "Location theory and programming models: the von Thünen Case", *Papers and Proceedings of the Regional Science Association*, vol. 21.

Sugimoto, M., 1967, *Optimization problems* (in Japanese), Kyoritsu Publishing Co.

Takayama, A., 1967, "Regional allocation of investment: a further analysis", *Quarterly Journal of Economics*, vol. 81, pp. 330–337.

Takayama, A., 1974, *Mathematical Economics*, The Dryden Press.

Takayama, T. and G.G. Judge, 1970, "Alternative spatial equilibrium models", *Journal of Regional Science*, vol. 10(1), pp. 1–12.

Takayama, T. and G.G. Judge, 1971, *Spatial and temporal price and allocation models*, North-Holland.

Tsukui, J., 1966, "Turnpike theorem in a generalized dynamic input–output system", *Econometrica*, vol. 34(2), pp. 396–407.

Vietorisz, T., 1963, "Industrial development planning model with economies of scale and indivisibilities", *Papers and Proceedings of the Regional Science Association*, vol. 12, pp. 157–192.

von Neumann, J., 1945, "A model of general economic equilibrium" (English translation), *Review of Economic Studies*, vol. 13, pp. 1–9.

Winter, S.G., 1967, "The norm of a closed technology and the straight-down-the-turnpike theorem", *Review of Economic Studies*, vol. 34, pp. 67–84.

AUTHOR INDEX

SUBJECT INDEX